CAMBRIDGE LIBRARY COLLECTION

Books of enduring scholarly value

Life Sciences

Until the nineteenth century, the various subjects now known as the life sciences were regarded either as arcane studies which had little impact on ordinary daily life, or as a genteel hobby for the leisured classes. The increasing academic rigour and systematisation brought to the study of botany, zoology and other disciplines, and their adoption in university curricula, are reflected in the books reissued in this series.

Hortus Veitchii

The Veitch dynasty, originally from Scotland, owned plant nurseries in Devon and London throughout the nineteenth century. By commissioning several expeditions to search for new and exotic flora for British gardens, they were instrumental in bringing many previously unknown plants into cultivation in Britain. James Herbert Veitch (1868–1907), who became managing director of the firm, spent time in Germany and France studying the techniques of horticulture, and later travelled the world himself collecting plants for the nursery in Chelsea. This work, published in 1906, gives a detailed account of the family business and of the men that the firm sent to South America, Japan, China and India during the period 1840–1906, including distinguished plant finders such as William Lobb, his brother Thomas, who first introduced various types of orchids from India to Britain for cultivation, and Richard Pearce, who brought back tuberous begonias from South America.

Hortus Veitchii

A History of the Rise and Progress of the Nurseries of Messrs. James Veitch and Sons

James Herbert Veitch

CAMBRIDGE UNIVERSITY PRESS

Cambridge, New York, Melbourne, Madrid, Cape Town,
Singapore, São Paolo, Delhi, Tokyo, Mexico City

Published in the United States of America by Cambridge University Press, New York

www.cambridge.org
Information on this title: www.cambridge.org/9781108037365

This edition first published 1906
This digitally printed version 2011

ISBN 978-1-108-03736-5 Paperback

HORTUS VEITCHII

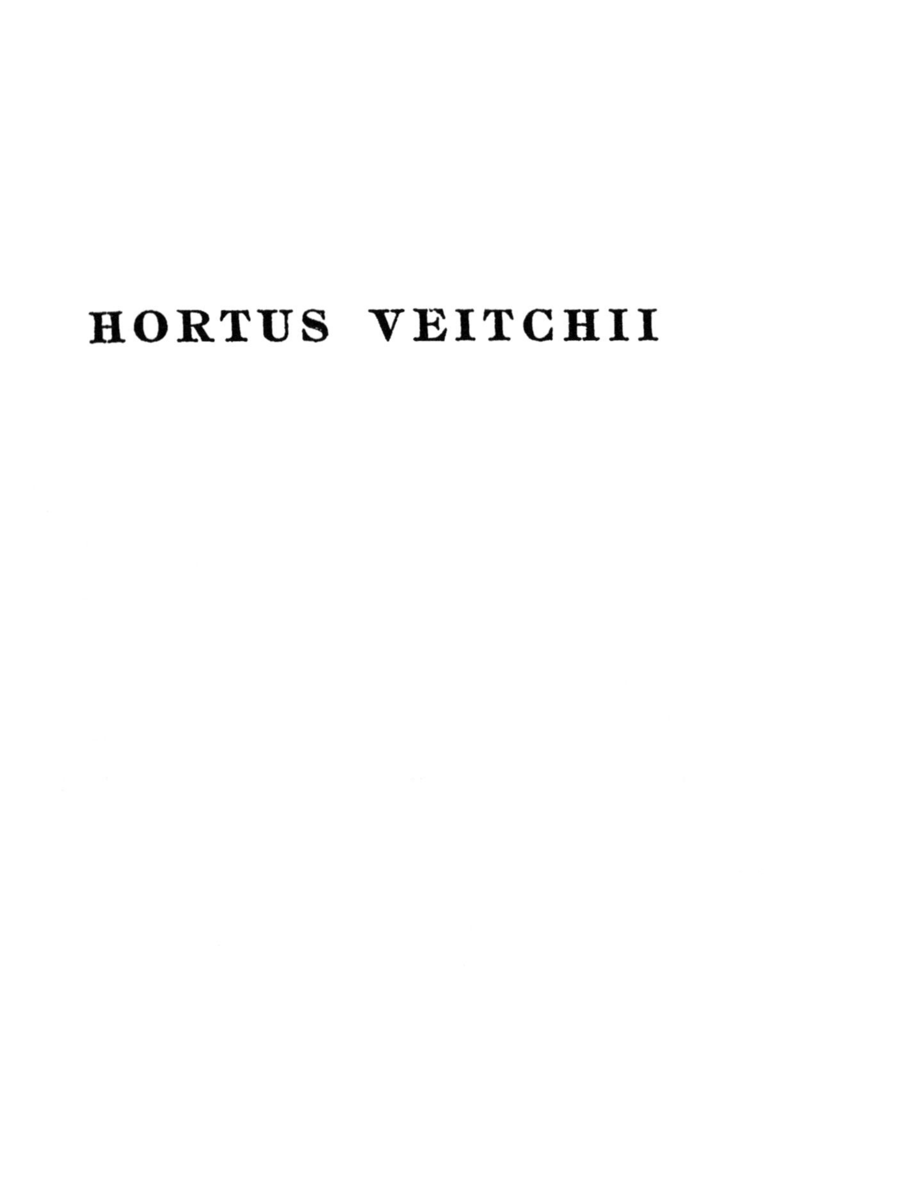

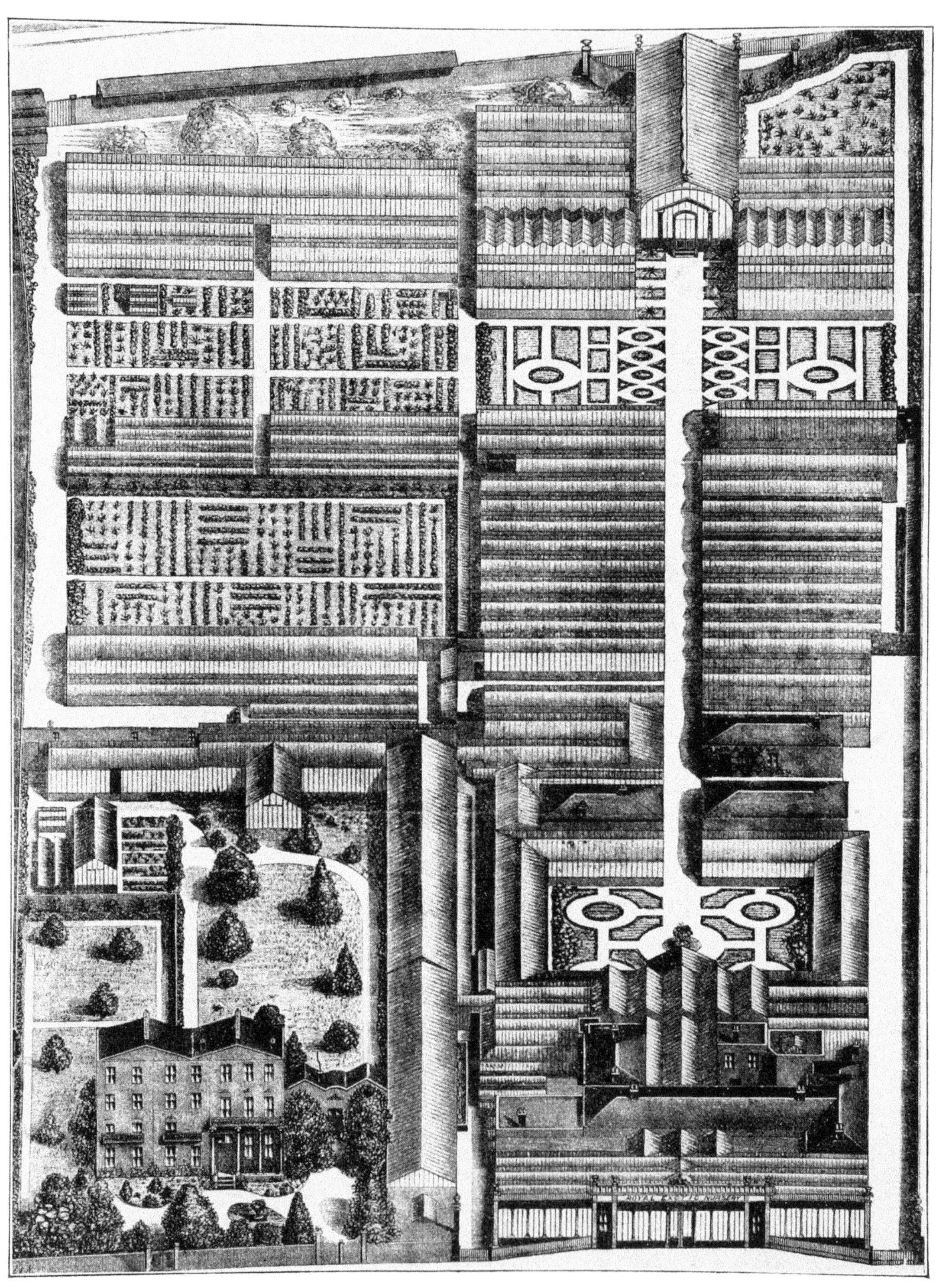

THE ORIGINAL NURSERY AT CHELSEA

HORTUS VEITCHII

A HISTORY

OF THE

RISE AND PROGRESS OF THE NURSERIES OF MESSRS. JAMES VEITCH AND SONS, TOGETHER WITH AN ACCOUNT OF THE BOTANICAL COLLECTORS AND HYBRIDISTS EMPLOYED BY THEM AND A LIST OF THE MOST REMARKABLE OF THEIR INTRODUCTIONS

BY

JAMES H. VEITCH

With Fifty Illustrations

London
JAMES VEITCH & SONS LIMITED, CHELSEA
1906

PREFACE

I AM indebted in this work for help to Mr. Harry J. Veitch, Mr. P. C. M. Veitch, Mr. J. G. Veitch, Dr. Maxwell T. Masters, Editor of the Gardeners' Chronicle, Mr. George Nicholson, the late Curator of the Royal Gardens, Kew, who revised the nomenclature; to Sir William T. Thiselton-Dyer, the late Director of Kew, for permission to photograph various plants in the Royal Gardens, and to Baron Sir Henry Schröder and other owners of large estates for similar kindness; to various members of the staff, past and present, and to my assistant, Herman Spooner.

INTRODUCTION

The following pages contain a record of continuous work for over three-quarters of a century in the field of Horticulture by one family,—work which justly may be claimed to have permanently benefited every garden.

The good fortune to usually find the right assistants for both home and foreign service, and the signal loyalty and capability of those selected, largely contributed to success, and the names of several are well known to all with any knowledge of plants.

To the representatives seeking unknown plants at one period or another in almost every clime, fortune has not invariably been kind, but the work of such men as Thomas Lobb, William Lobb, the late John Gould Veitch, Charles Maries and E. H. Wilson, has been a gain in every way; whilst the efforts in hybridizing and selecting of John Dominy, John Seden, V.M.H., and John Heal, V.M.H., have given a wider interest to all cultivators.

It would be strange if results were not forthcoming when such practically virgin lands as California, certain parts of South America, Japan and Central China were offered to men of the calibre of the Lobbs in the early forties of the last century, the late John Gould Veitch in the early sixties, and Wilson but recently; or when such hybridists as those named devoted their life efforts to the improvement of many now well-known garden plants.

It is difficult to realize the gardens of this and other

countries without many of their now ordinary inhabitants, yet it is within the memory of living men that Lilium auratum and Ampelopsis Veitchii (Vitis inconstans) were unknown, to name but two of many hundreds of plants, and but sixty years since Conifers with certain few exceptions were little more than rarities.

Gardening, as understood to-day, in its broadest aspects, was not possible: the material was not available.

The commencement of the nursery business of the firm of Veitch was on a limited scale, but records exist that Mr. John Veitch, who came from Jedburgh, Scotland, to enter the employ of Sir Thomas Acland at Killerton, Devon, held land, presumably for nursery purposes, in 1808.

His work, as well as that of his son Mr. James Veitch senior, seems to have been entirely in the neighbourhood of Exeter.

There may be read in the *Cottage Gardener* of January 9th 1855 the following notice of Mr. James Veitch, the son of the founder:—

"The history of botany furnishes us with several instances of enterprising men, who devoted a large measure of their means, or personal enterprise, to the enrichment of the botanical collections of this country with the vegetable products of foreign regions. To such men the present generation is greatly indebted; and thus it is that the names of the Tradescants, Peter Collinson, Dr. Anderson, John Frazer, James Lee and the Loddiges are so familiar to the minds and the memories of all true lovers of botanical science.

"For nearly half a century, however, that spirit of private enterprise has, except in a few instances, given way to the united efforts of corporate bodies and government officials; and it was not till the bold and energetic course which has been pursued by a provincial nurseryman of England was adopted, that a new era in botanical discovery was begun which has placed the name of 'Veitch of Exeter' among the worthies of science in our own times.

"The father of Mr. James Veitch was a native of Jedburgh, in Scotland, and towards the close of the last century he came to England, where he acted for many years as land-steward on the property of Sir Thomas Acland, at Killerton, and there the subject of the present notice was born on the 25th of January 1792.

"The success which attended the formation of the Killerton nursery was so great, that, in course of time, Mr. James Veitch found the distance of eight miles from Exeter disadvantageous to the interests of the establishment, as it prevented him from competing with those which were nearer the city; and accordingly, in 1832, he purchased that large extent of ground formerly called Mount Radford, but now converted into what is better known by the name of the 'Exeter Nursery,' an establishment which, by the industry and energy of Mr. Veitch and his son, has attained such a position as to be justly regarded as the finest of the kind ever known in England.

"In the year 1837 there were, in the Killerton nursery, two young men named William and Thomas Lobb, who were gardeners, and who were remaining there with a view to inprovement in their profession.

"In this same year, Mr. William Lobb was sent by Messrs. Veitch as gardener to Stephen Davey, Esq., of Redruth, in Cornwall, and after remaining there for three years, he was appointed by Mr. Veitch to proceed to the Brazils as a botanical collector, and he accordingly left England in 1840.

"The singular success which rewarded his researches is, perhaps, unparalleled in the history of botanical discovery;

the labours of David Douglas not even forming an exception. In the first parcel sent home were those two justly popular plants, Dipladenia splendens and Hindsia violacea; and from these, down to the later arrivals, including the wonderful Wellingtonia gigantea, what a mass of interest and beauty has been added to the gardens of Great Britain!

"About three years after Mr. William Lobb left, his brother, Thomas Lobb, who was then in the Exeter nursery, was sent by Mr. Veitch to Java, and the success which attended his efforts were not short of that of his brother.

"In the first parcel he sent home was that magnificent orchid Phalænopsis grandiflora, not before known in England, Vanda suavis, and numerous others.

"To enumerate all the plants that these gentlemen have discovered, or which Mr. Veitch has been the means of introducing, would occupy more space than we can afford to devote; but we shall furnish a list of some of the most remarkable of these introductions, showing to what an extent the country is now indebted to the enterprise of Mr. Veitch.

"It may be worth recording that all these new introductions, whether in the shape of seeds or living plants, are on their arrival taken under Mr. Veitch's care. He sows all seeds with his own hands, watches and tends them, and it is not until they are beyond all danger that they are committed to the management of others.

"In April 1853 the old establishment of Messrs. Knight & Perry, of the King's Road, Chelsea, being about to be relinquished, was offered to Messrs. Veitch & Son of Exeter, who shortly afterwards became its possessors; and now in this wonderful establishment may be seen one of the most extensive and valuable stocks of exotic plants which is to be met with in any private establishment in this country.

SEQUOIA GIGANTEA

POLTIMORE, DEVON

INTRODUCTION

"The following is a list of a few of the most remarkable plants introduced to this country by Messrs. Veitch & Son :—

Abutilon vitifolium.
Ærides Lobbii.
Æschynanthus Lobbiana.
„ pulcher.
Anectochilus Lobbii.
Befaria æstuans.
„ coarctata.
Begonia coccinea.
Berberis Darwinii.
Bulbophyllum Lobbii.
Calanthe vestita.
„ curculigoides.
Cantua dependens.
Ceratostemma longiflora.
Collinsia heterophylla.
Cryptomeria Lobbii.
Cypripedium barbatum superbum.
„ caudatum.
Dendrobium albo-sanguineum.
„ chrysotoxum.
„ Farmerii.
„ Kuhlii.
„ tortile.
„ transparens.
„ Veitchianum.
Desfontainea spinosa.
Deutzia gracilis.
Dracæna indivisa.
Echites atropurpurea.
„ splendens.
Escallonia macrantha.
„ organensis.
Eschscholtzia tenuifolia.
Eugenia Ugni.
Fitzroya Patagonica.
Fuchsia macrantha.
„ serratifolia.
„ spectabilis.
Hexacentris lutea.
„ mysorensis.
Hindsia longiflora.
„ violacea.
Hoya bella.
„ campanulata.
„ fraterna.
Impatiens Jerdoniæ.
Ixora javanica.
„ Lobbii.
„ salicifolia.
Lapageria rosea.
Lardizabala triternata.
Laurus aromatica.
Leptosiphon aureum.
„ luteum.
Lilium giganteum.
Limatodes rosea.
Lomaria magellanica.
Lomatia ferruginea.
Magnolia fragrantissima.
Mahonia Leschenaulti.
Manettia coccinea.
Medinilla magnifica.
„ speciosa.
Mitraria coccinea.
Nepenthes albo-marginata.
„ lævis.
„ lanata.
„ Phyllamphora.
„ sanguinea.
Pernettya mucronata speciosa.
Phalænopsis grandiflora.
„ Lobbii.
„ rosea.
Philesia buxifolia.
Picea bracteata.
Pleione humilus.
„ lagenaria.
„ maculata.
Pleroma elegans.
Podocarpus nubigena.
Quercus ægrifolia.
Rhododendron californicum.
„ jasminiflorum.
„ javanicum.
Rubus japonicus.
„ leucodermis.
Saccolabium Blumei major.
„ curvifolium.
„ miniatum.
Saxe-Gothea conspicua.

Sobralia dichotoma.
Sonerila margaritacea.
Telipogon obovatus.
Thuya gigantea.
Torreya myristica.
Tropæolum azureum.
„ Lobbii.
„ Smithanum.
Tropæolum speciosum.
Vanda cærulea.
„ suavis.
„ tricolor.
Veronica salicifolia.
Viola lutea.
Wellingtonia gigantea.
Whitlavia speciosa."

The son of Mr. James Veitch, Mr. James Veitch junior, conducted the nursery at Chelsea for some years, and created that at Coombe Wood. In reference to his death, the *Gardeners' Chronicle* contains the following notice in the issue of September 18th 1869:—

"In the person of Mr. Veitch, whose sudden death it was last week our melancholy duty briefly to announce, we have lost another of the horticultural notabilities of the last two decades; and though placed in a somewhat different sphere of action from such men as Lindley, Paxton, or Thompson, for example, it will be found that James Veitch the younger, as he was till quite recently best known, has left his mark upon the garden history of our time.

"James Veitch was born on May 24th 1815, in the neighbourhood of Exeter, where his grandfather (of Scotch extraction) and his father were at that time carrying on the business as nurserymen. When about eighteen years of age he was sent to London for two years for the purpose of acquiring that experience which in those days could only be gained in a London establishment. One year of this period was passed in the nursery of Mr. Chandler of Vauxhall, and the other in that of Messrs. Rollisson of Tooting. Returning again to Exeter, and resuming his routine of duties there, he at the same time, impelled by the new ideas and impulses acquired in London, devoted his energies to the gradual extension and improvement of the establishment at Mount Radford, making it eventually one of the first nurseries of the day. All this time he was acquiring the mastery over the mysteries of the nursery trade, in which, first (since 1838) as a partner in the firm of James Veitch & Son, and subsequently, on the death of his father, as the head of that of James Veitch & Sons, he was destined to raise himself to one of the very foremost positions. This prominent place amongst his compeers was won not less by his untiring zeal and energy, his keen perception, his clear-headed business habits, and his great personal influence, than by his thorough practical acquaintance with all professional details. It was on his return to Exeter from Tooting that, taking with him a collection of the Orchid gems of those days, he first started in the cultivation of those favourite plants; a taste which he always cultivated with the greatest possible zest, and which came in time to be ministered to by the introductions of his own collectors, and by his intimate personal friendship with men like Mr. G. Ure-Skinner and Colonel Benson, who had made acquaintance with orchids in their native homes. It should here be stated that Messrs. Rollisson, with whom young Veitch had been placed for the purpose of acquiring a knowledge of his business, declined to receive any adequate remuneration for the benefits conferred by them on their pupil; and the only method of acknowledgment open to the father of the subject of our notice was to commission the son to purchase orchids from the Messrs. Rollisson. These orchids became the nucleus of the collection for which Messrs. Veitch are now, and have for so long been renowned. It may be interesting to mention, as an instance of versatility, that about the year 1839 Mr. Veitch entered with great spirit upon the growth of Dahlias for competition, the Exeter Dahlia Shows, open to all England, furnishing at that period one of the most tempting arenas for the display of horticultural prowess.

INTRODUCTION

"In 1853, while still a partner in the nursery business at Exeter, which had then become famous as the first English home of multitudes of new plants, introduced directly by the agency of Messrs. Veitchs' collectors—the brothers Lobb, Mr. James Veitch removed to London, and took possession of the once famous establishment of Messrs. Knight & Perry at Chelsea. Here he was more directly brought into contact with all the leading horticulturists; and his estimable personal qualities, his sound sense, and his energetic manner, soon lifted him into a very influential position in the gardening world which he maintained for many years, until, as we may suppose, the foreshadowings of his fatal malady induced him gradually to withdraw from active participation in what may be called public life. All this time, however, the establishment at Chelsea, which still bore the name of the Royal Exotic Nursery, was being remodelled and improved, and a gigantic business, one of the most prominent in England, perhaps in Europe, was being worked up, sufficient of itself to form a striking monument of successful commercial skill and enterprise.

"Were we to attempt to show how far our gardens are indebted to the herculean and unflagging labours of Mr. Veitch, we should have to write a history of most of the new plants introduced during the last thirty years; for it was to his active superintendence of their importation, and to his discriminative choice of collectors, that we may largely attribute the success which was realized in this department. The later explorations of Pearce, Hutton and others, by which also many fine novelties have been acquired, were even more directly under his control; while in order to form some notion of all the services rendered to horticulture by Mr. Veitch in this direction, we must add to the foregoing the results of the two journeys of his eldest son, Mr. John Gould Veitch, to

Japan and the South Pacific, which have proved so prolific of first-class novelties. The pages of the garden periodicals bear witness to these facts in the number of first-class plants introduced through his intervention, such as Wellingtonia gigantea, Lapageria rosea and alba, Lilium auratum, Maranta Veitchii, Vanda suavis, tricolor, cœrulea and insignis; Phalænopsis grandiflora, Cypripedium caudatum, Rhododendron jasminiflorum, Pleroma elegans, Nepenthes (numerous species), Desfontainea spinosa, Thuya Lobbii, Abies bracteata, Begonia Veitchii, Masdevallia Veitchiana, Tropæolum azureum and speciosum, Calanthe vestita, Medinilla magnifica, Dipladenia splendens, Berberis Darwinii, *cum multis aliis*.

"It is, however, not only as an introducer and a dispenser of plants that Mr. Veitch's name must be boldly inscribed in the annals of horticulture, for he occupied a no less worthy position as a cultivator and exhibitor. Indeed Mr. Veitch was a thorough cultivator, as those who have seen the fine specimens sent from Exeter to Chiswick Shows will acknowledge when we say that many of them—Heaths and Orchids to wit—were the fruits of his own cultural manipulation, aided by his devoted and faithful servant and friend Dominy. Thus, when he desisted from the laborious task of potting his own plants, few knew better how to direct the labours of others. Then, as an exhibitor, Mr. Veitch has had a large share in making our shows the glorious monuments of cultural skill which they really are, despite all the grumbling concerning them. For many years he has been in the constant habit of bringing forward materials, the most excellent in quality, and these have been so abundant in quantity, and arranged with so much skill and taste, that it has been rare indeed to see the name of Veitch either absent from or occupying a secondary place in the award sheets. In all this, as well as in business transactions, he has for the last few years been ably assisted by his sons, Mr. J. G. Veitch and Mr. H. J. Veitch, who were admitted to partnership in 1865, and by whom the business will now be continued.

"We should not omit to state that Mr. Veitch was one of the best and most hard-working friends of the Gardeners' Royal Benevolent Institution, of which he was also a trustee.

"For several years—from 1856 to 1864—Mr. Veitch was a member of the Council of the Royal Horticultural Society, and took a very active part in the administration of its affairs.

"At his own hospitable board the interests of horticulture were often the subjects of a very free commentary; and many schemes which had been discussed or concocted there, were in due time realized. In particular it may be stated that the idea of the Fruit and Floral Committees of the Royal Horticultural Society had its origin in a happy thought of Mr. Veitch's, which was first broached and talked over, even into the small hours, in the parlour at the Royal Exotic Nursery, and at a subsequent gathering of a few kindred spirits was so thoroughly discussed as to smooth away the difficulties which at first seemed to stand in the way of its being realized. On the basis thus obtained a scheme was drawn out, virtually that adopted by the Council, which has not only worked well, but proved the horticultural mainstay of the Society. In Mr. Veitch's parlour, too, the particular mode in which the Great International Show of 1866 should be presented to the public was agreed to, and action taken thereupon. There had at that period sprung up amongst a small section of horticulturists a most unaccountable feeling antagonistic to the gentleman who afterwards so efficiently filled the office of chairman, and whose loss we have since had to deplore; and this feeling was battled with and overcome, chiefly by the influence of Mr. Veitch, and of a few others who supported him. In all movements for the advancement of horticulture he was ready to take a foremost part. Latterly, however, owing to his failing health, he has been less active in matters of this kind, but even so late as the occasion of the organization of the English Committee of the Hamburg International Show, those who attended the preliminary meetings were assisted by his advice.

"Some two years ago, owing to premonitory symptoms of heart disease, which have proved but too well founded, Mr. Veitch ceased to take so active a part as he had been wont, either in horticultural affairs or in matters of business; but latterly he had been in better health and spirits than usual, and even on the day before his decease had greatly enjoyed a visit from his old friend and collector, Thomas Lobb, so that his death on the morning of the 10th instant, at Stanley House, Chelsea, at the age of 54, came suddenly on his family and friends, although, under the circumstances, it can scarcely be said to have been wholly unexpected. His mortal remains have been deposited in the Brompton Cemetery, and there lie peacefully in the immediate vicinity of the scenes of the most active portion of his life. That he himself was not unprepared for the change that was to befall him is shown by the fact that only a few days before his death he selected, without the knowledge of any member of his family, a site for a family grave; and in its selection and attendant negotiations manifested those business habits so characteristic of him.

"Little remains for us to add. We have said enough to show that this was no ordinary man. Zeal and energy pervaded his every action. A quick temper and impatience of opposition were natural adjuncts to such a character; but at the same time it must be added there was thorough conscientiousness and straightforwardness, a disgust to all semblance, even, of meanness or underhandedness, and a warmth of friendly feeling that can be adequately gauged only by those who knew him."

In October 1869 a general movement throughout the country indicated the propriety of in some way perpetuating the memory of Mr. James Veitch junior, and resulted, after much discussion and many suggestions, in the Veitch Memorial Medal. For this purpose, after payment of expenses, the sum of £890 18*s.* 4*d.* was available, and was invested in the names of Trustees, the annual interest thereof being devoted to prizes for the advancement of horticulture. At the same time, through the liberality of Robert Crawshay, Esq., a portrait of the late Mr. James Veitch junior was hung in the Council Chamber of the Royal Horticultural Society.

Until August 1870 the two eldest sons of Mr. James Veitch junior—Mr. John Gould Veitch and Mr. Harry James Veitch—continued the work, when the death of the first-named had also to be recorded.

In reference to this unhappy event, the *Gardeners' Chronicle* of August 20th 1870 contained the following:—

"Scarcely twelve months have elapsed since we had to record the decease of Mr. James Veitch, the late indefatigable head of the firm, Veitch & Sons, of Chelsea, and one of the foremost horticulturists of our day. We have now the mournful task of chronicling the death of his eldest son, Mr. John Gould Veitch, who was, like his father, a devoted horticulturist, and who, although he has been taken from amongst us at an early age, will long be remembered as an intrepid voyager, and one to whom we are greatly indebted for many contributions to the enjoyment of our gardens in the introduction of valuable plants discovered in the course of his travels, and also as a young man full of zeal and enterprise in his profession—a worthy representative of his father's and grandfather's name.

"John Gould Veitch was born at Exeter in April 1839, and had, therefore, only reached his 32nd year. He was at an early age initiated in the mysteries of the nursery trade, and took an active part in the management of the establishment at Chelsea. It was in April 1860, almost as soon as he had attained his majority, that he started on his voyage to Japan and China, whence he proceeded to the Philippine Islands. The result of this journey was the enrichment of our collections with many choice plants, amongst which the lovely Primula cortusoides amœna would of itself form no mean monument to his memory. Various handsome Conifers, as Abies firma and Alcoquiana, Cryptomeria elegans, and other plants, as Lilium auratum, Ampelopsis tricuspidata (Veitchii) and japonica, &c., were, however, also obtained as the fruits of that first journey, and our volumes for 1860-1861 contain from his own pen the interesting records of his journeyings and discoveries during the two years which elapsed previously to his returning in the spring of 1862.

"The spirit of enterprise and the desire of making discoveries, which prompted him at first to set sail for Japan, then lately made accessible to Europeans, did but slumber for a season, for in 1864 we find him again *en route*, this time bound for Australia and the South Sea Islands, whence he returned in February 1866, after an absence of some eighteen or twenty months, bringing with him some of the most beautiful plants of modern introduction: witness the numerous richly-coloured forms of Croton and Dracæna which are only now becoming known. Of the Crotons alone no fewer than twenty-three distinct kinds (described in our volume for 1868, pp. 843, 943) were obtained; and of Dracænas, regina, magnifica, Mooreana, Chelsoni, Macleayi, and several others. To these must be added such distinct and popular subjects as Acalypha Wilkesiana (tricolor), Amaranthus melancholicus ruber, Coleus Veitchii, and Gibsoni; the more choice and valuable Pandanus Veitchii, Aralia Veitchii, and many others. During this journey Cape York in Northern Australia was visited, and here was obtained a new Palm, which has since been dedicated to his honour, under the name of Veitchia Johannis.

"The record of this second journey, specially interesting as referring to many little known and rarely visited islands, will be found in our volume for 1866.

"In the early part of 1867, Mr. Veitch, then recently married, was taken seriously ill with an affection of the lungs, and for some time his life was despaired of. He, however, rallied, under careful treatment and the potent aid of his indomitable spirit, and though subsequently obliged to winter in a warmer clime, his friends were not without hope that his life might have been for some time spared to them. But this was not to be. On the 9th inst. hæmorrhage from the lungs, under which he gradually sank, set in, and he expired on the evening of the 13th inst. at his residence at Coombe Wood, leaving behind him a widow and two boys. On Thursday last he was laid beside his father in the Brompton Cemetery, having been borne to his grave by the same trusty workmen: some of whom had, moreover, assisted to carry his grandfather to his last resting-place.

"So we part sorrowing from one of the most gifted and promising of our younger commercial horticulturists, one who, if his life had been spared and his health had permitted, would have worthily filled a prominent position in the world of horticulture; one, moreover, whose memory will continue to be cherished by those who had the pleasure to know him intimately as that of a manly, straightforward, single-hearted, earnest and sincere friend. The Veitch Memorial will now, in most people's minds, possess a double interest, as it will be henceforth impossible to dissociate the memory of the son from that of the father."

From 1870 for thirty years the responsibility of successful guidance has rested with Mr. Harry J. Veitch, for some few years assisted by his brother the late Mr. Arthur Veitch, and to his ceaseless watchfulness and ability is due the steady progress of the business.

The direct control Mr. Harry J. Veitch relinquished some few years since, when advantage was taken of the Company Law to convert the business into a Limited Liability Company, with Mr. James H. Veitch as Managing Director and Mr. John Gould Veitch as Secretary, the whole of the shares remaining in the hands of the family.

Amongst other efforts in the interests of Horticulture are a Manual of Coniferæ, the first edition long since exhausted, the Manual of Orchidaceous Plants, and various papers in the Journals of the Linnean and the Royal Horticultural Societies. The material for the following 422 plates in *Curtis's Botanical Magazine* is attributed by the editors to Messrs. Veitch:—

1. Gloxinia speciosa, var. macrophylla . . . t. 3934.
2. Alstrœmeria nemorosa . t. 3958.
3. Primula denticulata . t. 3959.
4. Habranthus pratensis, var. quadriflora . . . t. 3961.
5. Echites (Dipladenia) splendens . . . t. 3976.
6. Rondeletia longiflora . t. 3977.
7. Tropæolum azureum . t. 3985.
8. Begonia coccinea . . t. 3990.
9. Gesneria polyantha . . t. 3995.
10. Echites hirsuta . . . t. 3997.
11. Passiflora Actinia . . t. 4009.
12. Stigmaphyllon heterophyllum . . . . t. 4014.
13. Canavalia ensiformis. . t. 4027.
14. Hypocyrta strigillosa . t. 4047.
15. Clematis montana, var. grandiflora . . . t. 4061.

16. Gomphrena pulchella . t. 4064.
17. Houlletia Brocklehurstiana . . . . t. 4072.
18. Tropæolum Lobbianum . t. 4097.
19. Asclepias vestita . . t. 4106.
20. Hindsia violacea . . t. 4135.
21. Barbacenia squamata . t. 4136.
22. Calceolaria floribunda . t. 4154.
23. Calceolaria alba . . t. 4157.
24. Sida(Abutilon)pæoniæflora t. 4170.
25. Fuchsia serratifolia . . t. 4174.
26. Siphocampylos coccineus . t. 4178.
27. Tacsonia mollissima . . t. 4187.
28. Hebecladus biflorus . . t. 4192.
29. Cuphea cordata . . t. 4208.
30. Sida (Abutilon) vitifolium t. 4227.
31. Fuchsia macrantha . . t. 4233.
32. Cypripedium barbatum . t. 4234.
33. Æschynanthus purpurascens . . . . t. 4236.
34. Tropæolum crenatiflorum . t. 4245.
35. Collania andinamarcana . t. 4247.
36. Clematis smilacifolia . t. 4259.
37. Æschynanthus Lobbianus t. 4260.
38. Pleroma elegans . . t. 4262.
39. Æschynanthus pulcher . t. 4264.
40. Scutellaria incarnata . t. 4268.
41. Escallonia organensis . t. 4274.
42. Nepenthes Rafflesiana . t. 4285.
43. Calceolaria amplexicaulis . t. 4300.
44. Liebigia speciosa . . t. 4315.
45. Æschynanthus speciosus . t. 4320.
46. Medinilla speciosa . . t. 4321.
47. Tropæolum speciosum . t. 4323.
48. Æschynanthus longiflorus t. 4328.
49. Rhododendron javanicum t. 4336.
50. Tropæolum umbellatum . t. 4337.
51. Hoya cinnamomifolia . t. 4347.
52. Ceropegia Cumingiana . t. 4349.
53. Fuchsia spectabilis . . t. 4375.
54. Tropæolum Smithii . . t. 4385.
55. Cantua pyrifolia . . t. 4386.
56. Sonerila stricta. . . t. 4394.
57. Hoya bella . . . t. 4402.
58. Dipladenia urophylla . t. 4414.
59. Heterotrichum macrodon. t. 4421.
60. Macleania punctata . . t. 4426.
61. Loasa picta . . . t. 4428.
62. Vanda tricolor . . . t. 4432.
63. Lapageria rosea . . t. 4447
64. Mitraria coccinea . . t. 4462.
65. Escallonia macrantha . t. 4473.
66. Dendrobium tortile . . t. 4477.
67. Oxalis elegans . . . t. 4490.
68. Dipteracanthus spectabilis t. 4494.
69. Lardizabala biternata . t. 4501.
70. Hoya coriacea . . . t. 4518.
71. Hoya purpureo-fusca . t. 4520.
72. Ixora salicifolia . . t. 4523.
73. Rhododendron jasminiflorum . . . . t. 4524.
74. Stylidium saxifragoides . t. 4529.
75. Bolbophyllum Lobbii . t. 4532.
76. Medinilla magnifica . . t. 4533.
77. Hoya campanulata . . t. 4545.
78. Didymocarpus crinita . t. 4554.
79. Thibaudia macrantha . t. 4566.
80. Cantua buxifolia . . t. 4582.
81. Berberis Darwinii . . t. 4590.
82. Fitzroya patagonica . . t. 4616.
83. Eugenia Ugni . . . t. 4626.
84. Berberis wallichiana. . t. 4656.
85. Calanthe vestita . . t. 4671.
86. Rubus biflorus . . . t. 4678.
87. Hoya fraterna . . . t. 4684.
88. Dendrobium cretaceum . t. 4686.
89. Cantua bicolor . . . t. 4729.
90. Gilia (Leptosiphon) lutea. t. 4735.
91. Philesia buxifolia . . t. 4738.
92. Abies bracteata. . . t. 4740.
93. Wellingtonia gigantea . t. 4777.
94. Cerastostema longiflorum. t. 4779.
95. Torreya myristica . . t. 4780.
96. Desfontainia spinosa . t. 4781.
97. Hexacentris mysorensis . t. 4786.
98. Ceanothus floribundus . t. 4806.
99. Ceanothus Lobbianus . t. 4810.
100. Eschscholtzia tenuifolia . t. 4812.
101. Whitlavia grandiflora . t. 4813.
102. Ceanothus papillosus . t. 4815.
103. Befaria æstuans . . t. 4818.
104. Dipladenia Harrisii . . t. 4825.
105. Escallonia pterocladon . t. 4827.
106. Dipladenia acuminata . t. 4828.
107. Crawfurdia fasciculata . t. 4838.
108. Embothrium coccineum . t. 4856.
109. Rhododendron californicum . . . . t. 4863.

110. Leptodactylon californicum . . . . t. 4872.
111. Gilia dianthoides . . t. 4876.
112. Phygelius capensis . . t. 4881.
113. Delphinium cardinale . t. 4887.
114. Cœlogyne speciosa . . t. 4889.
115. Æschynanthus fulgens . t. 4891.
116. Tecoma fulva . . . t. 4896.
117. Aphelandra variegata . t. 4899.
118. Rhododendron moulmainense . . . . t. 4904.
119. Correa cardinalis . . t. 4912.
120. Coffea benghalensis . . t. 4917.
121. Ribes subvestitum . . t. 4931.
122. Rhododendron Brookeanum . . . . t. 5935.
123. Hypericum oblongifolium. t. 4949.
124. Adhatoda cydoniæfolia . t. 4962.
125. Hoya coronaria . . t. 4969.
126. Sonerila elegans . . t. 4978.
127. Befaria Mathewsii . . t. 4981.
128. Rhododendron Veitchianum . . . . t. 4992.
129. Forsythia suspensa . . t. 4995.
130. Viola pedunculata . . t. 5004.
131. Azalea occidentalis . . t. 5005.
132. Sonerila speciosa . . t. 5026.
133. Cosmanthus grandiflorus . t. 5029.
134. Eugenia Luma . . . t. 5040.
135. Calanthe Dominii (hybrida) . . . . t. 5042.
136. Pentstemon Jaffrayanus . t. 5045.
137. Clianthus Dampieri . . t. 5051.
138. Gesneria Donklarii . . t. 5070.
139. Cœlogyne Schilleriana . t. 5072.
140. Æsculus californica . t. 5077.
141. Nepenthes villosa . . t. 5080.
142. Thunbergia natalensis . t. 5082.
143. Fuchsia simplicicaulis . t. 5096.
144. Sonerila margaritacea . t. 5104.
145. Nepenthes ampullaria . t. 5109.
146. Ceanothus Veitchianus . t. 5127.
147. Dendrobium albo-sanguineum . . . . t. 5130.
148. Æschynanthus cordifolius . . . . t. 5131.
149. Dendromecon rigidum . t. 5134.
150. Spraguea umbellata . . t. 5143.
151. Calceolaria flexuosa . t. 5154.
152. Ceanothus velutinus . . t. 5165.
153. Chamæbatia foliolosa . t. 5171.
154. Richardia hastata . . t. 5176.
155. Ceanothus oreganus . . t. 5177.
156. Pteris quadriaurita . . t. 5183.
157. Phalænopsis grandiflora . t. 5184.
158. Vanda gigantea . . t. 5189.
159. Pentapterygium rugosum. t. 5198.
160. Phalænopsis rosea . . t. 5212.
161. Stenogaster concinna . t. 5253.
162. Mutisia decurrens . . t. 5273.
163. Verticordia nitens . . t. 5286.
164. Ceropegia Gardneri . . t. 5306.
165. Limatodes rosea . . t. 5312.
166. Saccolabium miniatum . t. 5326.
167. Nolana (Sorema) lanceolata . . . . t. 5327.
168. Philadelphus hirsutus . t. 5334.
169. Ourisia coccinea . . t. 5335.
170. Lilium auratum . . t. 5338.
171. Berberidopsis corallina . t. 5343.
172. Plumbago rosea, var. coccinea . . . . t. 5363.
173. Lycioplesium pubiflorum . t. 5373.
174. Calanthe Veitchii: hybrida . . . . . t. 5375.
175. Calceolaria punctata . t. 5392.
176. Homoianthus viscosus . t. 5401.
177. Eranthemum tuberculatum . . . . t. 5405.
178. Miconia pulverulenta . t. 5411.
179. Trichantha minor . . t. 5428.
180. Canscora Parishii . . t. 5429.
181. Alstrœmeria Caldasii . t. 5442.
182. Corylopsis spicata . . t. 5458.
183. Urceolina pendula . . t. 5464.
184. Cypripedium caricinum . t. 5466.
185. Eranthemum Cooperi . t. 5467.
186. Genethyllis fimbriata . t. 5468.
187. Linum Macraei . . . t. 5474.
188. Mimulus luteus, var. cuprea . . . . t. 5478.
189. Proustia pyrifolia . . t. 5489.
190. Manettia micans . . t. 5495.
191. Aglaonema marantæfolium, var. fol. mac. . t. 5500.
192. Cypripedium lævigatum . t. 5508.
193. Arum palæstinum . . t. 5509.
194. Hypœstes sanguinolenta . t. 5511.

195. Marianthus Drummondianus . . . . t. 5521
196. Bertolonia guttata . . t. 5524
197. Primula cortusoides, var. amœna . . . . t. 5528.
198. Alstrœmeria densiflora . t. 5531.
199. Calathea Veitchiana. . t. 5535.
200. Dendrobium Tattonianum t. 5537.
201. Dendrobium Johannis . t. 5540.
202. Calathea tubispatha . . t. 5542.
203. Begonia Pearcei . . t. 5545.
204. Peperomia marmorata . t. 5568.
205. Cymbidium Hookerianum t. 5574.
206. Ancylogyne longiflora . t. 5588.
207. Fremontia californica . t. 5591.
208. Sanchezia nobilis . . t. 5594.
209. Nierembergia Veitchii . t. 5599.
210. Bolbophyllum reticulatum t. 5605.
211. Nierembergia rivularis . t. 5608.
212. Vanda Bensoni. . . t. 5611.
213. Cattleya Dowiana . . t. 5618.
214. Curcuma Australasica . t. 5620.
215. Angræcum citratum . . t. 5624.
216. Myrtus Cheken. . . t. 5644.
217. Amaryllis pardinam . . t. 5645.
218. Stemonacanthus Pearcei . t. 5648.
219. Dendrobium macrophyllum Veitchianum . . t. 5649.
220. Gloxinia hypocyrtiflora . t. 5655.
221. Begonia boliviensis . . t. 5657.
222. Begonia Veitchii . . t. 5663.
223. Lilium Leichtlinii . . t. 5673.
224. Cymbidium Huttoni. . t. 5676.
225. Calceolaria pisacomensis . t. 5677.
226. Dendrobium Bensoniæ . t. 5679.
227. Begonia rosæflora . . t. 5680.
228. Saccolabium Huttoni . t. 5681.
229. Thunia Bensoniæ . . t. 5694.
230. Begonia falcifolia . . t. 5707.
231. Eranthemum aspersum . t. 5711.
232. Nasonia punctata . . t. 5718.
233. Epidendrum paniculatum. t. 5731.
234. Puya Whytei . . . t. 5732.
235. Masdevallia Veitchiana . t. 5739.
236. Aphelandra nitens . . t. 5741.
237. Agalmyla staminea . . t. 5747.
238. Thibaudia acuminata . t. 5752.
239. Cœlogyne (Pleione) Reichenbachiana . . . t. 5753.
240. Vanda insignis . . . t. 5759.
241. Dendrobium crassinode . t. 5766.
242. Saccolabium bigibbum . t. 5767.
243. Palava flexuosa . . t. 5768.
244. Odontoglossum Krameri . t. 5778.
245. Dipladenia boliviensis . t. 5783.
246. Aphelandra acutifolia . t. 5789.
247. Ærides japonicum . . t. 5798.
248. Blandfordia aurea . . t. 5809.
249. Vanda Denisoniana . . t. 5811.
250. Dendrobium lasioglossum t. 5825.
251. Orthosiphon stamineus . t. 5833.
252. Vanda cœrulescens . . t. 5834.
253. Vanda Cathcarti . . t. 5845.
254. Cymbidium canaliculatum t. 5851.
255. Beloperone ciliata . . t. 5888.
256. Begonia crinita . . t. 5897.
257. Darlingtonia californica . t. 5920.
258. Epidendrum pseudepidendrum . . . . t. 5929.
259. Gilia achilleæfolia . . t. 5939.
260. Cœlogyne lentiginosa . t. 5958.
261. Restrepia elegans . . t. 5966.
262. Dendrobium amethystoglossum. . . . t. 5968.
263. Phajus Blumei, var. Bernaysii . . . . t. 6032.
264. Odontoglossum vexillarium . . . . t. 6037.
265. Rhododendron malayanum . . . . t. 6045.
266. Sonerila Bensoni . . t. 6049.
267. Caraguata Zahnii . . t. 6059.
268. Campsidium chilense . t. 6111.
269. Blumenbachia (Caiophora) contorta . . . t. 6134.
270. Blumenbachia chuquitensis . . . . . t. 6143.
271. Odontoglossum maxillare. t. 6144.
272. Epidendrum syringothyrsis . . . . . t. 6145.
273. Wahlenbergia tuberosa . t. 6155.
274. Masdevallia Peristeria . t. 6159.
275. Odontoglossum (Miltonia) Warscewiczii. . . t. 6163.
276. Hemichæna fruticosa . t. 6164.
277. Balbisia verticillata . . t. 6170.
278. Cypripedium Argus . . t. 6175.
279. Masdevallia Davisii . . t. 6190.

280. Calathea leucostachys . t. 6205.
281. Pescatorea Dayana, var. rhodacra . . . t. 6214.
282. Viburnum dilatatum . t. 6215.
283. Episcia erythropus . . t. 6219.
284. Bouchea pseudogervaô . t. 6221.
285. Hypoestes aristata . . t. 6224.
286. Allium anceps . . . t. 6227.
287. Odontoglossum prænitens t. 6229.
288. Calceolaria tenella . . t. 6231.
289. Monopyle racemosa . . t. 6233.
290. Pescatoria lamellosa. . t. 6240.
291. Leucothœ Davisiæ . . t. 6247.
292. Lilium phillipinense . . t. 6250.
292. Lycaste lasioglossa . . t. 6251.
293. Begonia Davisii . . t. 6252.
294. Masdevallia inocharis . t. 6262.
295. Masdevallia triaristella . t. 6268.
296. Monardella macrantha . t. 6270.
297. Masdevallia attenuata . t. 6273.
298. Oncidium cheirophorum . t. 6278.
299. Cordia decandra . . t. 6279.
300. Boronia elatior. . . t. 6285.
301. Cypripedium Haynaldianum . . . . t. 6296.
302. Haplopappus spinulosus . t. 6302.
303. Lycaste Linguella . . t. 6303.
304. Notylia albida . . . t. 6311.
305. Dendrobium crystallinum t. 6319.
306. Oncidium Euxanthinum . t. 6322.
307. Arthropodium neo-caledonicum . . . . t. 6326.
308. Æchmea (Chevalliera) Veitchii. . . . t. 6329.
309. Calceolaria lobata . . t. 6330.
310. Lilium neilgherrense . t. 6332.
311. Alloplectus peltatus. . t. 6333.
312. Stenospermatium Wallisii t. 6334.
313. Eranthemum laxiflorum . t. 6336.
314. Pleroma Gayanum . . t. 6345.
315. Crossandra guineensis . t. 6346.
316. Dendroseris macrophylla . t. 6353.
317. Spathoglottis Petri . . t. 6354.
318. Ardisia Oliveri . . . t. 6357.
319. Magnolia stellata . . t. 6370.
320. Eurygania ovata . . t. 6393.
321. Watsonia densiflora . . t. 6400.
322. Burbidgea nitida . . t. 6403.
323. Calceolaria deflexa . . t. 6431.
324. Cypripedium Lawrenceanum . . . . t. 6432.
325. Pachystoma Thomsonianum . . . . t. 6471.
326. Conandron ramondioides. t. 6484.
327. Cypripedium Spicerianum t. 6490.
328. Mormordes Ocanæ . . t. 6496.
329. Chionographis japonica . t. 6510.
330. Stelis Brückmülleri . . t. 6521.
331. Crinum purpurascens . t. 6525.
332. Wormia Burbidgei . . t. 6531.
333. Jasminum gracillimum . t. 6559.
334. Schismatoglottis crispata . . . . t. 6576.
335. Primula proculiformis (obconica) . . . . t. 6582.
336. Drosera capensis . . t. 6583.
337. Abelia spathulata . . t. 6601.
338. Zephyranthes citrina . t. 6605.
339. Globba atro-sanguinea . t. 6626.
340. Columnea Kalbreyeri . t. 6633.
341. Celmisia spectabilis . . t. 6653.
342. Utricularia Endresii. . t. 6656.
343. Hamamelis japonica. . t. 6659.
344. Fraxinus Mariesii . . t. 6678.
345. Saxifraga cortusifolia . t. 6680.
346. Hoya linearis . . . t. 6682.
347. Rogersia podophylla. . t. 6691.
348. Cephælis tomentosa . . t. 6696.
349. Eranthemum Borneense . t. 6701.
350. Sarmienta repens . . t. 6720.
351. Medinilla Curtisii . . t. 6730.
352. Rhododendron multicolor t. 6769.
353. Berberis congestifolia, var. hakeoides . . . t. 6770.
354. Corylopsis himalayana . t. 6779.
355. Caryopteris mastacanthus . . . . t. 6799.
356. Eucomis bicolor . . t. 6816.
357. Odontoglossum Œrstedii. t. 6820.
358. Rhododendron javanicum, var. tubiflora. . . t. 6850.
359. Ixora macrothyrsa . . t. 6853.
360. Myrmecodia Beccarii . t. 6883.
361. Tillandsia chrysostachys . t. 6906.
362. Amasonia calycina . . t. 6915.
363. Pleurothallis insignis . t. 6936.
364. Phalænopsis Mariæ . . t. 6964.
365. Anthurium Veitchii . . t. 6968.

366. Heloniopsis japonica . t. 6986.
367. Masdevallia gibberosa . t. 6990.
368. Styrax Obassia . . . t. 7039.
369. Sarcochilus luniferus . t. 7044.
370. Licuala Veitchii . . t. 7053.
371. Enkianthus campanulatus . . . . . t. 7059.
372. Eucryphia pinnatifolia . t. 7067.
373. Heliamphora nutans . . t. 7093.
374. Zamia Wallisii . . . t. 7103.
375. Abies brachyphylla . . t. 7114.
376. Caraguata angustifolia . t. 7137.
377. Nepenthes Curtisii . . t. 7138.
378. Scaphosepalum pulvinare . . . . . t. 7151.
379. Musa Basjoo . . . t. 7182.
380. Impatiens mirabilis . . t. 7195.
381. Didymocarpus lacunosa . t. 7236.
382. Æschynanthus obconica . t. 7336.
383. Dendrobium atroviolaceum . . . . t. 7371.
384. Trochodendron aralioides. t. 7375.
385. Angræcum Kotschyi. . t. 7442.
386. Scutellaria formosana . t. 7458.
387. Hypocyrta pulchra . . t. 7468.
388. Celmisia Munroi . . t. 7496.
389. Didymocarpus malayana . t. 7526.
390. Zamia obliqua . . . t. 7542.
391. Odontoglossum retusum . t. 7569.
392. Rhododendron yunnanense . . . . t. 7614.
393. Rhododendron rubiginosum . . . . t. 7621.
394. Elæagnus macrophylla . t. 7638.
395. Dryandra calophylla . t. 7642.
396. Rhododendron dilatatum. t. 7681.
397. Corylopsis pauciflora . t. 7736.
398. Lhotskya ericoides . . t. 7753.
399. Cœlogyne Veitchii . . t. 7764.
400. Manettia bicolor . . t. 7776.
401. Rubus palmatus . . t. 7801.
402. Corydalis thalictrifolia . t. 7830.
403. Astilbe Davidii . . . t. 7880.
404. Hamamelis mollis . . t. 7884.
405. Senecio clivorum . . t. 7902.
406. Senecio tanguticus . . t. 7912.
407. Lysimachia crispidens . t. 7919.
408. Corydalis Wilsoni . . t. 7939.
409. Dicentra chrysantha . t. 7954.
410. Lysimachia Henryi . . t. 7961.
411. Loropetalum chinense . t. 7979.
412. × Zygocolax Veitchii . t. 7980.
413. Jasminum primulinum . t. 7981.
414. Dendrobium bellatulum . t. 7985.
415. Nepenthes Rajah . . t. 8017.
416. Lycaste Locusta . . t. 8020.
417. Meconopsis integrifolia . t. 8027.
418. Primula tangutica . . t. 8043.
419. Sciadopitys verticillata . t. 8050.
420. Primula Veitchii . . t. 8051.
421. Lonicera pileata . . t. 8060.
422. Lonicera tragophylla . t. 8064.

JAMES VEITCH

JAMES H. VEITCH

JAMES VEITCH, JUNIOR

JOHN VEITCH

HARRY J. VEITCH

FAMILY TREE

John Veitch.
Born 1752, at Jedburgh, Scotland.
Died 1839.
Rented land at Lower Budlake, near Killerton, near Exeter, in 1808.
Acquired additional land in 1810.
Moved to Mount Radford in 1832.
Succeeded in 1837 by

James Veitch.
Born 1792.
Died 1863.
Nurseryman at Mount Radford.
James Veitch and James Veitch junior, as James Veitch & Son, of Exeter, acquired the business and rented the land of Messrs. Knight & Perry, Chelsea, in 1853.
James Veitch remained at Mount Radford, Topsham Road, Exeter.
James Veitch junior moved to Chelsea in 1853.

James Veitch junior.
Born 1815.
Died 1869.
Nurseryman at Chelsea, Coombe Wood, &c., as James Veitch & Son.
Ceased all interest in the Exeter firm in 1864.

Robert Toswill Veitch.
Born 1823.
Died 1885.
Nurseryman at New North Road and High Street, Exeter, in 1864, on the death of James Veitch in 1863, as Robert Veitch & Son.

Peter C. M. Veitch.
Nurseryman at Exeter as Robert Veitch & Son.

John Gould Veitch.*
Born 1839.
Died 1870.

Harry James Veitch.*

Arthur Veitch.*
Born 1844.
Died 1880.

*Nurserymen at Chelsea, Coombe Wood and Langley, as James Veitch & Sons.

James H. Veitch. **John Gould Veitch.**
Nurserymen at Chelsea, Coombe Wood, Langley and Feltham, as James Veitch & Sons, Ltd.

JOHN GOULD VEITCH ROBERT T. VEITCH P. C. M. VEITCH
ARTHUR VEITCH JOHN G. VEITCH

CONTENTS

LIST OF ILLUSTRATIONS

LIST OF ILLUSTRATIONS

SCOTLAND
AUSTRIA
BELGIUM

RUSSIA

ITALY
ENGLAND
FRANCE

MESSRS. VEITCHS' GOLD MEDALS

LIST OF REFERENCES

NAME IN FULL	ABBREVIATION
Curtis's Botanical Magazine	Bot. Mag.
Dictionnaire Iconographique des Orchidées, Bruxelles	Dict. Ic. des Orchidées.
Die Gartenwelt, Leipsig.	
Flora and Sylva.	
Floral Magazine	Fl. Mag.
Florist and Pomologist	Fl. and Pom.
Garden and Forest	Gard. and For.
Gordon's Pinetum.	
Hooker's Icones Plantarum	Hooker's Ic. Pl.
Hooker's Species Filicum	Hooker's Sp. Fil.
Journal of Horticulture	Jour. of Hort.
Journal of the Linnean Society of London . .	Jour. Linn. Soc.
Journal of the Royal Horticultural Society . .	Jour. R.H.S.
La Belgique Horticole, Liège	La Belg. Hort.
La Flore des Serres et des Jardins de l'Europe, Gand.	Fl. des Serres.
l'Illustration Horticole, Gand	l'Illus. Hort.
Lindenia, une Iconographie des Orchidées . . .	Lindenia.
Lindley's Botanical Register	Lindl. Bot. Reg.
Lindley and Paxton's Flower Garden . . .	Paxt. Fl. Gdn.
Loddige's Botanical Cabinet	Lodd. Bot. Cab.
l'Orchidophile.	
Loudon's Encyclopædia of Plants	Loudon's Ency. Pl.
Nicholson's Illustrated Dictionary of Gardening .	Nich. Dict. Gard.
Orchid Review.	
Paxton's Magazine of Botany	Paxt. Mag. Bot.
Plantæ Delavayanæ.	
Proceedings of the Royal Horticultural Society .	Proc. R.H.S.

LIST OF REFERENCES

Name in Full	Abbreviation
Regel's Gartenflora, Erlangen.	
Reichenbachia, Orchids Illustrated and Described .	Reichenbachia.
Reichenbach's Xenia Orchidacea	Rchb. Xen. Orch.
Revue Horticole, Paris	Rev. Hort.
Sargent's Trees and Shrubs, Illustrations of New or Little-known Ligneous Plants, &c., Boston, U.S.A.	Sargent's Trees and Shrubs.
The Gardeners' Chronicle	Gard. Chron.
The Orchid Album	Orchid Album.
The Orchid Grower's Manual	Williams' Orch. Man.
Veitchs' Manual of Coniferæ	Man. Con.
Veitchs' Manual of Orchidaceous Plants . . .	Veitchs' Man. Orch. Pl.
Veitchs' Catalogue of New and Rare Plants . .	Veitchs' Catlg. of Pl.
Veitchs' Catalogue of Trees and Shrubs . . .	" " " Trees and Shrubs.
Warner's Select Orchidaceous Plants	Warner's Selec. Orch.

LIVES OF TRAVELLERS

LIST OF MESSRS. VEITCHS' TRAVELLERS

William Lobb . . .	1840—1857
Thomas Lobb . . .	1843—1860
Richard Pearce . . .	1859—1866
John Gould Veitch . .	1860—1870
David Bowman . . .	1866
Henry Hutton . . .	1866—1868
Carl Kramer . . .	1867—1868
Gottlieb Zahn . . .	1869—1870
George Downton . . .	1870—1873
J. Henry Chesterton . .	1870—1878
A. R. Endres . . .	1871—1873
Gustave Wallis . . .	1872—1874
Walter Davis . . .	1873—1876
P. C. M. Veitch . . .	1875—1878
Guillaume Kalbreyer . .	1876—1881
Christopher Mudd . .	1877
F. W. Burbidge . . .	1877—1878
Charles Maries . . .	1877—1879
Charles Curtis . . .	1878—1884
David Burke . . .	1881—1897
James H. Veitch . . .	1891—1893
E. H. Wilson . . .	1899—1905

LIVES OF TRAVELLERS

WILLIAM LOBB.

Collector in California and South America.

1840—1857.

William Lobb was born in the eastern division of Cornwall. The place is unknown, nor is anything known of his early life. When a young man he applied himself to gardening, and was in 1837 employed in the nursery of Mr. James Veitch senior at Exeter. In this same year Mr. Veitch sent him to be gardener to Mr. Stephen Davey, of Redruth, whose horticultural establishment appears to have been on a modest scale, but which, under Lobb's management, became thoroughly efficient.

For a long time William Lobb had cherished an ardent desire for travel and adventure: he was quick of observation, ready in resources, and practical in their application; he had devoted much of his leisure to the study of botany, in which considerable proficiency had been acquired. Accordingly, after three years in Mr. Davey's service, he gladly accepted the proposal of Mr. James Veitch senior to go on a mission to various parts of South America for the purpose of collecting plants, and he sailed from Plymouth in 1840 for Rio Janeiro. On his arrival in Brazil he first proceeded to the Orgãos Mountains, and met with several beautiful and notable orchids at that time extremely rare in English gardens; he then left for Chili, crossing the great Pampas of the Argentine Republic and the Chilian Andes. Continuing his journey southwards, Lobb penetrated the great Araucaria forests, where he collected a large quantity of seeds of Araucaria imbricata, and was thus instrumental in bringing this remarkable Conifer into general use for ornamental planting.

He returned to England in 1844, renewed his engagement, and sailed again for Brazil in April of the following year.

After sending home from Rio Janeiro a consignment of plants collected in Southern Brazil, he proceeded to Valparaiso for the purpose of exploring Southern Chili, at that time but little known to Europeans, except along the coast. Here a rich harvest awaited him.

Amongst his earliest successful introductions from this region were Lapageria rosea, Escallonia macrantha, Embothrium coccineum, Philesia buxifolia, and Desfontainea spinosa.

Following up these brilliant achievements, he continued explorations in Valdivia, Chiloe, and Northern Patagonia, where he collected seeds and plants of Libocedrus tetragona, Fitzroya patagonica, Saxe-Gothæa conspicua, and Podocarpus nubigena, "four most interesting Conifers for this country, after Araucaria imbricata, that South America produces." Nor must mention be omitted of Berberis Darwinii, first introduced to British gardens during this interesting expedition. Lobb returned to England in 1848.

The wonderful Conifers discovered by Douglas in California and Oregon were then still very scarce in England, and young plants of most of the species could scarcely be bought : it was therefore decided that Lobb should proceed to California with a view of obtaining seeds of all the most important kinds known, and, if possible, discover others.

He landed at San Francisco in the summer of 1849, and at once made arrangements for exploring Southern California.

THUYA GIGANTEA (LOBBII)

One of the first fruits of the expedition was the successful introduction of Abies bracteata.

During the years 1850-1851 consignments were sent home of cones and seeds of Pinus radiata, P. muricata, P. Sabiniana, P. Coulteri and P. tuberculata; and also of many shrubs and flowering plants, quite new to British gardens.

In the autumn of 1851 his operations were pushed further north, and cones and seeds collected of the Redwood (Sequoia sempervirens), Pinus Lambertiana, P. monticola, and others.

In 1852 Lobb made an excursion to the Columbia River and Oregon, where he succeeded in obtaining seeds of Abietia (Pseudotsuga) Douglasii and Abies nobilis, still rare at that time in England, and the beautiful Thuia provisionally named after him.

Returning through North California, Lobb collected seeds of Abies grandis, A. magnifica, which he sent home under the name of A. amabilis, believing it to be the A. amabilis of Douglas, A. concolor Lowiana, the first received in England of that fine Fir, Juniperus californica, and of Pinus ponderosa.

In 1853 the Sierra Nevada was explored, the traveller tempted by the reports of the discovery of trees of extraordinary magnitude, which he had the good fortune to find, and to secure the first cones and seeds of Sequoia gigantea received in England.

Lobb brought these home at the end of the year, and with them two living plants, afterwards planted out at Exeter, where they survived but three or four years, nor was there at any time much hope of their living.

Lobb returned to California in the autumn of 1854, and from that time up to the end of 1856 continued to send home consignments of plants and seeds.

In 1857 his engagement with Mr. Veitch terminated. He remained, however, in California, and sent collections of seeds to England from time to time. In 1863, seized with paralysis, Lobb lost the use of his limbs; he died at San Francisco in the autumn of the same year, and was buried in Lone Mountain Cemetery.

Reference :—Man. Con. 1900, ed. 2, p. 243.

THOMAS LOBB.

Collector in India and Malaya.

1843—1860.

Thomas Lobb was a native of Cornwall, but scarcely anything is known of his early life till he entered the nursery of Mr. James Veitch senior, of Exeter, by whom he was engaged as a collector. From an agreement dated January 11th 1843 it appears that "Thomas Lobb agrees to proceed to the British Settlement of Singapore, in the employ of James Veitch & Son as botanical collector, to make collections of living plants, seeds, and dried specimens of plants, and to collect for the said James Veitch & Son and for no other person. The understanding of this agreement is that the said Thomas Lobb's principal destination is to be China, should that country be open to admit a botanical collector, and in the absence of any definite instructions from James Veitch & Son, Thomas Lobb is to use his own discretion and be guided by existing circumstances as to what parts of China he proceeds to, and if on arrival at Singapore he finds circumstances are not favourable for his proceeding to China, he shall be at liberty to proceed to such of the oriental islands as may appear to him most desirable; but next to China the island of Java appearing to offer the greatest advantages to a botanical collector (if facilities offer for exploring the same with safety), he is directed to proceed thither, but it is left to his own discretion." Apparently Lobb did not find China ready to receive a botanical collector, and he adopted the alternative, and visited Java and the adjacent islands.

By a second agreement Thomas Lobb agreed to go to India to collect plants, seeds and other objects of Natural History for three years, and left England for Calcutta on December 25th 1848. During the twenty years or upwards he travelled for the Veitchian firm he visited the Khasia Hills, Assam, and other parts of North-East India, and subsequently Moulmein and parts of Lower Burmah, sending home from those districts most of the finest Orchids found there, many previously known to science, but introduced by him to cultivation for the first time.

Worthy of mention among these are Vanda cœrulea, Cœlogyne (Pleione) lagenaria, C. maculata, Aërides Fieldingi, A. multiflorum Lobbii, A. m. Veitchii, Dendrobium infundibulum, Calanthe (Limatodes) rosea, and Cypripedium villosum.

Lobb afterwards visited the southern parts of the Malay peninsula, North Borneo (Labuan and Sarawak) and other Eastern Isles, when he discovered and introduced the ancestral forms of the superb and useful race of Rhododendrons, known in gardens as the Javanico-jasminiflorum hybrids; the original forms being Rhododendron javanicum, R. Lobbii, R. jasminiflorum and R. Brookeanum.

From this region, too, he successfully introduced some of the first Nepenthes cultivated in British gardens, including Nepenthes Rafflesiana, N. Veitchii, N. sanguinea, and N. ampullaria; and among the very many Orchids he sent home were Vanda tricolor and its variety suavis, of which he was also the discoverer; Cœlogyne speciosa, Calanthe vestita, Cypripedium barbatum, and others. Lobb subsequently went to the Philippine Islands, and collected the best Orchids found in the neighbourhood of Manila, among the Phalænopsis

CYPRIPEDIUM VILLOSUM

being P. intermedia, the first natural hybrid subsequently proved by artificial means.

It was greatly in Lobb's favour that much of the region explored was virgin soil for a collector, but his discrimination equalled his energy, and he sent home but few plants that proved unworthy of cultivation. It is not saying too much to assert that during the long period Lobb collected in the East, British gardens were enriched with more beautiful plants of Indo-Malayan origin than by any single collector of his own or any other time.

Lobb also collected many herbarium specimens, a list of which is given by Planchon in *Hooker's London Journal of Botany*, 1847-1848, vols. vi., vii., where the following notice appears regarding them :—

"Mr. Heward, Young Street, Kensington, is charged with the distribution of the sets of the exquisitely beautiful and rare specimens of the mountains of Java, collected by Mr. Thomas Lobb. The number of sets is but small, and the amount of species in each varies from 100 to 200, or nearly so.

"More perfect specimens have never been offered for sale."

And again in another volume :—

"The early numbers of this valuable collection were made by Mr. Thomas Lobb in Java. The beauty and variety of the specimens gave such satisfaction to the subscribers that Mr. Lobb has extended his researches: some of the present list are from Singapore (a very large collection having been lost by shipwreck); and we are now happy to say that a further addition has arrived of extremely interesting and beautiful specimens from Moulmein, which are placed in the hands of Mr. Heward for distribution.

"This set contains, amongst others of great rarity and novelty, the singular Barclaya longifolia, Wall. (*Linn. Trans.* v. 15,

p. 443, t. 18), with flowers resembling in structure those of Euryale, and the leaves in shape like those of Scolopendrium vulgare; a plant so scarce that Dr. Wallich is not aware that a good specimen is to be found in any herbarium."

The genus Lobbia, of the order Aristolochiaceæ, founded by Planchon on a specimen collected by Thomas Lobb in Singapore, was named by him Lobbia dependens, the generic name being given to commemorate the labours of the two brothers, Thomas and William Lobb.

In manner Lobb was modest and retiring, of few words, and it was difficult to get him to describe a plant, but if he ventured on calling it "very pretty," it was quite sufficient to induce extra care.

As the result of exposure in his work, he had the misfortune to lose one of his legs, a circumstance which induced him to settle at Devoran in Cornwall, where he remained for the rest of his life, and it is rather remarkable that the only time he was induced to leave his home was to pay a visit to his last employer, Mr. James Veitch junior, with whom he was staying on the occasion of his sudden death in 1869.

Thomas Lobb died on April 30th 1894, at Devoran in Cornwall, at a very advanced age.

References:—Hooker's London Journal of Botany, 1847-1848, p. 145; Journal of Botany, 1894, vol. xxxii. p. 191; Cottage Gardener, xiii. p. 274; Gard. Chron. 1894, vol. xv. p. 636.

RICHARD PEARCE.

Collector in Chili, Peru and Bolivia.

1859—1866.

The name of this collector is indissolubly connected with the history of the Tuberous Begonia, for it was to his energy and daring as a traveller that we are indebted for the introduction of the early species from Bolivia and Peru.

Richard Pearce was born at Stoke Devonport, and was first employed in the nursery of Mr. Pontey in the town of Plymouth, where he stayed till about the year 1858, when he entered the nursery at Mount Radford, near Exeter.

It appears by an agreement drawn up between James Veitch & Son, of the Mount Radford Nursery, Exeter, and Richard Pearce, in February 1859, that the latter agreed to go out to South America for three years as collector of plants, seeds, land-shells and other objects of Natural History.

Pearce was instructed to proceed to Valparaiso in South America, and collect in Chili and Patagonia. His particular attention was directed to the collection of seeds of Libocedrus tetragona, at that time supposed to be the tree which produced the famous Alerze timber; the Lapageria rosea and L. alba; the Chilian Pine (Araucaria imbricata), and other hardy trees and shrubs; secondly, to procure such plants as require a greenhouse temperature; and thirdly, Orchidaceæ and stove and greenhouse flowering plants.

Pearce carried out these instructions, and besides the above-named plants obtained and introduced Prumnopitys elegans, Podocarpus nubigena, Eucryphia pinnatifolia, several Bomareas, Thibaudia acuminata, Ourisia coccinea, O. Pearcei and quantities of Ferns. At the same time, through his researches, the true Alerze-producing tree was found to be Fitzroya patagonica and not Libocedrus tetragona as had been generally supposed.

During 1860 Pearce made many journeys to the Cordilleras and the interior of the country, to Los Baños, the Baths of Chillan, and to Los Luganos, the Lakes.

Of the scenery of that part of the Cordilleras which he explored he writes :—" It is of the most charming description—gently undulating meadows covered with a carpet of short grass, placid lakes reflecting from their smooth surface the mountains around, foaming cataracts and gentle rivulets, deep gorges and frightful precipices, over which tumble numerous dark, picturesque waterfalls reaching the bottom in a cloud of spray, high rocky pinnacles and lofty peaks, surround one on every side.

" Nor is the vegetation less beautiful and interesting. At an elevation of 4,000 ft. the vegetation exhibits a totally different character from that of the coast. Here one finds Antarctic Beeches (Fagus antarctica and F. betuloides), which constitute with Fitzroya patagonica the large forest trees. The Embothrium coccineum, Desfontainea spinosa, Philesia buxifolia, three species of Berberis, Pernettya and Gaultheria are the most abundant of the flowering shrubs, whilst the numerous pretty little rock-plants meet one at every step with their various forms and colours."

Early in 1862 Pearce left Chili, travelled north to Peru and Bolivia in search of stove and greenhouse plants, and plants with fine foliage. From Cuença he sent a good number of

EUCRYPHIA PINNATIFOLIA

seeds, including Befaria ledifolia, Lisianthus magnificus, Calceolaria ericoides, and several good Tacsonias, and from Guayaquil despatched some six large cases of plants, amongst which was the handsome Maranta Veitchii.

It appears from a second agreement drawn up between James Veitch & Son, of the Mount Radford Nursery, Exeter, and Richard Pearce, in January 1863, that the latter again agreed to go to South America for three years to collect plants, seeds and other objects of Natural History. He was to proceed to Lima, thence to Muña and Veloc, and afterwards to such parts of South America as by written instructions to him should be determined. Particular attention was to be directed to the Province of Teukamon, should there be facilities for reaching that country.

Amongst other fine additions from Muña were Aphelandra nitens, Gymnostachys Pearcei, and Sanchezia nobilis—three well-known and much-prized stove plants.

Pearce succeeded in reaching Teukamon, where he collected Nierembergia rivularis and N. Veitchii, Begonia boliviensis, Palava flexuosa, Mutisia decurrens and several Peperomias.

The next journey was to La Paz, and in November 1865 were sent home, Begonia Pearcei, B. Veitchii, a number of good Hippeastrums, such as H. pardinum and H. Leopoldii, the progenitors of that magnificent race of garden varieties (Amaryllis) so general to-day, and two or three excellent species of Ecremocarpus subsequently lost.

On returning from La Paz in 1866 Pearce's engagement with James Veitch & Son terminated, and he returned to his home at Plymouth, where he married.

In 1867 he came to London and entered into an engagement with the late Mr. William Bull to travel in South America.

On arrival in Panama, he was taken ill on July 13th,

and died on the 17th of that month, of a bilious remittent fever.

As a botanical collector Pearce was one of the best, and his untimely death was a great loss to the world of Horticulture.

References :—"The Tuberous Begonia," by W. Wynne, ed. 1, 1888, with sketch of Pearce's travels and portrait; Journal of Botany, 1868, pp. 320, 134; Gard. Chron. 1868, pp. 874, 893.

JOHN GOULD VEITCH.

Collector in Japan, South Sea Islands, and Australia.

1860—1870.

John Gould Veitch was born at Exeter in 1839. He was at an early age initiated in the working of the nursery business, and took an active part in the management of the Veitchian establishment after its removal to Chelsea in 1853, at that time rapidly acquiring prestige through the introduction of new plants, a prestige it was determined to maintain, and if possible enhance.

His majority scarcely attained, an opportunity offered by the opening of the ports of Japan to foreigners, and in April 1860 John Gould Veitch started on a voyage to the Far East, arriving at Nagasaki in the July following.

He remained in Japan about a year, collecting plants, many previously unknown in British gardens.

Attached to the suite of Sir Rutherford Alcock, the British Envoy to Japan, he was enabled to make the ascent of Fujiyama, and was one of the first Europeans to reach the summit of the "sacred mountain."

After despatching the collections to Europe he proceeded to the Philippine Islands on a similar mission, but with the especial object of obtaining plants of various species of Phalænopsis, natives of the islands, at that period extremely rare in European gardens. The mission proved successful.

The result of the voyage to Japan was the enrichment of European gardens with many choice coniferous trees, several beautiful evergreen and deciduous trees and shrubs, various herbaceous and other plants and bulbs. Amongst the first named are Abies firma, A. microsperma, Cryptomeria japonica elegans, Cupressus obtusa varieties, Juniperus chinensis aurea, Larix leptolepis, Picea Alcockiana, P. ajanensis, P. polita, Pinus densiflora, P. parviflora, P. Thunbergii, Tsuga diversifolia, and the introduction in quantity of the rare Sciadopitys verticillata; among the latter especial mention should be made of several beautiful forms of Acer palmatum, Ampelopsis Veitchii (Vitis inconstans), Lilium auratum, Primula japonica and P. cortusoides.

The spirit of enterprise and the desire of further important discoveries induced him again to undertake a long voyage to the East, and in 1864 John Gould Veitch sailed for Australia and the South Sea Islands, returning to England in February 1866.

Among the most enduring results of the second voyage was the introduction of many richly coloured Crotons and Dracænas (varieties of Codiæum and Cordyline), the forerunners of the handsome races now so constantly in request for decorative purposes; the beautiful Pandanus Veitchii, the elegant Aralia Veitchii, and other plants of merit were also added to our stoves.

In the early part of 1867 this most successful traveller was taken ill with an affection of the lungs, from which, however, under careful treatment he rallied for a time, but in August 1870 hæmorrhage set in, and he died shortly afterwards, at the early age of thirty-one.

In reference to the introduction of several of the most distinct Japanese conifers the following letter appeared in the *Times* from Mr. Veitch junior :—

"To the Editor of the *Times*.

"Sir,—A paragraph under the head of 'Expeditious Journey to Japan' appears in your paper of to-day, which, by inference, gives the credit of first finding and forwarding to England seeds of the Sciadopitys verticillata, or umbrella pine, to Mr. Robert Fortune.

"I have not the slightest desire to detract from Mr. Fortune in any way, and I hope his researches in China and Japan will prove valuable to the botanical public, but as Mr. John Gould Veitch arrived in Japan on the 20th of July last (which was before Mr. Fortune had left England), and obtained and forwarded seeds of that magnificent tree, and of the scarcely less beautiful one, Thuiopsis dolobrata, on the 22nd of September, nearly a month before Mr. Fortune reached Japan, I feel bound, in simple justice to Mr. John Gould Veitch, to state the fact, that he may have the full credit due to him of first introducing into England so fine a tree: and I shall therefore be obliged by your inserting this letter in your paper of to-morrow.

"Mr. John Gould Veitch has also discovered and sent home seeds of numerous other fine trees and shrubs, one of which, a beautiful new fir, he has named Abies Alcockiana, in compliment to Mr. Rutherford Alcock, Her Majesty's Minister at Jeddo, to whose kind assistance he has been much indebted ; and another entirely new pine has been named by Professor Lindley Abies Veitchii, in compliment to the discoverer.

"These seeds, with a number of others forwarded by Mr. John Gould Veitch, are now being raised at our establishments at Chelsea and Exeter.

"Full particulars of Mr. J. G. Veitch's arrival in Japan, with the particulars of his first collection sent home, have appeared in the *Gardeners' Chronicle*, of which I enclose you extracts.

"The fact of Mr. John Gould Veitch being my son will plead my apology for troubling you.

"I am, Sir, your most obedient servant,

"JAMES VEITCH JUNIOR.

"Royal Exotic Nursery, King's Road, Chelsea, S.W.
"February 12th 1861."

References.—Gard. Chron. 1860-1862, *passim*, "Extracts from Mr. John Gould Veitch's Letters on Japan"; *id*. 1866, *passim*, "Extracts from the Journal of Mr. John Gould Veitch during a Trip to the Australian Colonies and the South Sea Islands;" Man. Con. 1900, ed. 2, p. 543.

SCIADOPITYS VERTICILLATA
THE DELL, EGHAM

DAVID BOWMAN.

COLLECTOR IN BRAZIL.

1866.

DAVID BOWMAN was born at Arniston near Edinburgh, where his father was a gardener in a small position, on September 3rd 1838.

His gardening career commenced at Arniston under the father, and subsequently he went to the gardens at Dalhousie Castle, Archerfield, and Dunmore Park, in Scotland, coming later to England to enter the gardens of the Royal Horticultural Society at Chiswick as foreman.

Bowman left England early in 1866 on our behalf for Rio de Janeiro, Brazil, whence he sent, with other plants, Dieffenbachia Bowmani, which perpetuates his name, and Paullinia thalictrifolia, a pretty stove climber with beautiful fern-like foliage.

He subsequently sent home plants to the Royal Horticultural Society, to Mr. Wilson Saunders, and other amateurs, but when, having collected a large number in the vicinity of Bogota, and preparing to sail for England, he suffered heavy losses through robbery, and his stay had to be prolonged.

Shortly after this Bowman contracted a violent attack of dysentery, from which he died on June 25th 1868.

He is buried in the British cemetery at Bogota.

References:—Gard. Chron. 1868, pp. 924, 942.

HENRY HUTTON.

Collector in Java and the Malay Archipelago.

1866—1868.

Henry Hutton, a son of a former head-gardener to Lord Houghton, already, in young days, an ardent student and promising explorer, was sent on a collecting mission to the East in 1866.

After twelve months' residence in Java, when it was hoped he had become sufficiently inured to the climate to extend his explorations, his health broke down and he fell a victim to his enthusiasm.

In commemoration of his zealous services and early death, his name is associated with a new species of Cymbidium* he was the means of introducing to this country from Java, together with Saccolabium (Aërides) Huttoni from the same country, and Dendrobium Huttoni, which he sent from the island of Timor in the Malay Archipelago. He rediscovered and successfully introduced the very rare Vanda insignis.

* Cymbidium Huttoni, Hook. f. Bot. Mag. t. 5676.

CARL KRAMER.

Collector in Japan and Costa Rica.

1867—1868.

Carl Kramer, son of the late Herr Kramer, gardener to Senator Janisch of Flotbeck near Hamburg, the well-known possessor of a once fine collection of Orchids, was sent on a collecting mission to Japan to supplement the collection made there by the late John Gould Veitch, but his mission was a failure.

He was afterwards sent to Costa Rica and Guatemala for Orchids, but he again sent home little of note. Kramer proved entirely unsuitable for the work he had undertaken, and apparently had not that adaptability and resource essential to successful exploration.

His name is associated with a species of Odontoglossum* allied to O. citrosmum, introduced from Costa Rica in 1868.

* Odontoglossum Krameri, Rchb. f. Gard. Chron. 1868, p. 98.

GOTTLIEB ZAHN.

Collector in Central America.

1869—1870.

A German who collected during 1869 and 1870 in Central America, sending several consignments of plants, for the most part Orchids and Ferns.

The main object of Zahn's journey was the introduction of the rare Odontoglossum Warscewiczii (Miltonia Endresii), discovered by Warscewicz about the year 1849, but which hitherto had resisted all attempts at introduction; Zahn also failed, and it was left for Endres (*q.v.*) two years later to accomplish this often attempted task.

Zahn arrived in Panama in September 1869, collected in the neighbourhood of Chiriqui, and was proceeding to Costa Rica, when he perished by drowning.

In gardens his name is perpetuated by the beautiful Bromeliaceous plant, distributed as Tillandsia Zahnii, but now correctly named Caraguata Zahnii.

Reference :—Bot. Mag. sub t. 6059.

GEORGE DOWNTON.

Collector in Central and South America and the Islands of Juan Fernandez.

1870—1873.

George Downton received part of his early training in Horticulture in the gardens of Wilton House, Salisbury, under Mr. T. Challis, V.M.H., and afterwards in the Royal Horticultural Society's Gardens at Chiswick, where he distinguished himself as a student, winning in 1870 both the Royal Horticultural Society's and the Society of Arts' first prizes for Floriculture and Fruit and Vegetable culture.

In 1870 he was engaged as a plant-collector, and sent for Orchids to Central America, whence he forwarded several consignments to Chelsea; he was subsequently instructed to join Endres, plant-collecting on our behalf in Costa Rica.

Downton succeeded in joining Endres, and brought to England shortly afterwards the result of their joint enterprise, principally Orchids, and a few plants (the bulk, unfortunately, succumbed *en route*) of the long-desired and much-sought-for Odontoglossum Warscewiczii (Miltonia Endresii).

In October 1871 he started on a mission to Chili, to collect a further supply of seed of Embothrium coccineum, Tropæolum azureum, T. tricolor, and other plants of horticultural interest, and to introduce any new plants he might be fortunate enough to discover.

Downton visited the little-known islands of Juan Fernandez in 1873, and sent home many fine ferns, including Dicksonia berteroana, several half-hardy shrubs, such as Tricuspidaria dependens (Crinodendron Hookeri), and various species of Berberis, Eugenia, and Citharexylon.

On the termination of his engagement in 1873 he entered the service of an English firm of coffee-planters, and continued to reside in Central America until his death, which took place suddenly about 1895.

J. HENRY CHESTERTON.

Collector in South America.

1870—1878.

Very little is known of the early life of this traveller, who was afterwards such a very successful orchid-collector over a wide area in South America.

When this collector first came to notice he was serving as valet to a gentleman who travelled much in foreign lands, and as Chesterton wished to bring home with him some of the floral treasures that he met with during his travels, he applied to Messrs. Veitch for information as to the best means of packing.

Such information and advice as was likely to prove useful was readily given, and opportunities offered him of seeing plants packed for long journeys, and of examining those newly imported, by which he was enabled to form an idea of the great care required to safely convey living plants thousands of miles by sea and through various climates and greatly varying temperatures.

Nothing more was heard of Chesterton for some time, until, on arriving in England from a visit to South America, he came to Chelsea with a collection of Orchids, so carefully packed and well looked after, that they arrived in the best possible condition.

In return for the information given, he made Messrs. Veitch the first offer of purchase, which was accepted, and the collection passed into our hands. Subsequently he entered our service as a traveller, and made several journeys to South America in search of Orchids.

The special object for which Chesterton was engaged was the introduction of the much-talked-of and long-desired "scarlet Odontoglossum" (Miltonia vexillaria), the existence of which was made known through Bowman, afterwards by Wallis and Roezl, all of whom made unsuccessful attempts at its introduction.

Provided with but the scantiest information as to the native habitat, long kept secret and shrouded in mystery, Chesterton started, and not only succeeded in discovering the plant, but safely introduced it to Chelsea, where it flowered for the first time in 1873. Some of the finest forms of Odontoglossum crispum were sent home by Chesterton, one named Chestertonii by Professor Reichenbach in compliment to its discoverer: some fine Masdevallias were also sent home, including the beautiful Masdevallia coccinea Harryana.

After his engagement with us terminated Chesterton continued to collect plants on his own behalf and for several other firms until his death, which took place in South America in 1883.

The following obituary note was contained in the Shipping List of January 30th 1883 :—

"Mr. J. H. Chesterton, the botanist, died at Puerto Berrio on the 26th. He had been quite ill, but left the hotel 'San Nicholas,' thinking that he had sufficiently improved to be able to make his trip up the river. Sad mistake! He continued to decline, and was barely put on shore at Puerto Berrio ere he died. Poor Chesterton's reckless spirit rendered him very efficient as a plant-collector."

Reference:—"Veitchs' Man. Orch. Pl." pt. viii. p. 113.

ODONTOGLOSSUM CRISPUM VEITCHIANUM

THE DELL, EGHAM

A. R. ENDRES.

COLLECTOR IN COSTA RICA.

1871—1873.

THIS traveller was a half-caste named to us by Mr. James Bateman, who had employed him to collect Orchids in Guatemala through Mr. G. Ure-Skinner.

On the untimely death of Zahn by drowning, Endres, at that time in Costa Rica, was engaged to continue Zahn's work and search for rare plants known to exist in that country, more especially Odontoglossum Warscewiczii (Miltonia Endresii), Cattleya Dowiana, and Anthurium Scherzerianum, all at that time valuable plants.

Endres commenced collecting in 1871, and was later joined by Downton (*q.v.*), who brought the collection to England.

Many Orchids were subsequently sent home, but few of horticultural merit, with the exception of Cattleya Dowiana and Miltonia Endresii; the mission, which terminated in April 1873, was expensive and scarcely a success.

Epidendrum Endresii, Odontoglossum Warscewiczii (Miltonia Endresii) and Utricularia Endresii commemorate this traveller's name.

References :—Bot. Mag. sub tt. 7855 and 6656. "Veitchs' Man. Orch. Pl." pt. viii. p. 101.

GUSTAVE WALLIS.

COLLECTOR IN BRAZIL, NEW GRENADA, AND VARIOUS PARTS OF TROPICAL SOUTH AMERICA.

1872—1874.

GUSTAVE WALLIS was born on May 1st 1830, at Lüneburg, Hanover, where his father was an advocate.

Deaf and dumb until he was six years of age, it was not till 1836 that he could articulate. About this time the father died, leaving the mother a widow with six children. Her means of support gone, she was compelled to leave Lüneburg and retire to Detmold, her native town. In this romantic and picturesque country, surrounded by mountains and forests, young Wallis spent his schooldays, and developed that love of Nature and of Botany which excited in maturer years such a strong desire to see foreign lands, and above all the tropics.

The youth possessed an indomitable energy, and in spite of his defective speech acquired considerable proficiency in foreign languages, an accomplishment which always stood him in good stead during his career.

At the age of sixteen Wallis was apprenticed to a goldsmith, but, disliking the work, it was abandoned, and he became apprenticed to a gardener at Detmold.

After the term of apprenticeship had terminated he obtained employment at Münich, and during this period he made several excursions to the Alps, for the purpose of collecting

ANTHURIUM VEITCHII

PALMEN GARTEN, FRANKFURT-AM-MAIN

and studying in their native habitats the plants belonging to those rugged regions.

In 1856 Wallis went to Southern Brazil, and in connection with a German house started a horticultural establishment, but owing to the failure of the parent firm the branch ceased to exist, and Wallis was left practically penniless.

In 1858 he offered his services as a plant-collector to the late M. Linden of Brussels, who accepted them, and Wallis then commenced his remarkable journey across the continent of South America, from the mouth to the source of the Amazon, exploring that great river as well as some of the more important tributaries.

In 1870 he entered Messrs. Veitchs' service and proceeded to the Philippines to obtain as his principal object plants of various species of Phalænopsis known to inhabit the Islands. —— Seyfarth, a young German, was sent to Manila to bring the collection home.

The mission proved very expensive, was practically a failure, and Wallis had to be recalled.

In December 1872 he was sent to New Grenada, a country already known to him, and returned in 1874, with many fine tropical plants, including Anthurium Veitchii, A. Warocqueanum, and several interesting Orchids.

After his engagement terminated he still continued to collect plants in South America, and commenced his last journey at the end of the summer of 1875, when he left to explore the north and central regions of South America.

Wallis was next heard of at Panama, dangerously ill with fever, from which he, however, recovered, and again commenced work, but a second attack of the malady, combined with dysentery, soon proved fatal. His last letter was dated Cuença, March 24th 1878, where, according to

Mr. Edward Klaboch, he died in the hospital on June 20th of that year.

The specific names of the following plants were given at various times by botanists in commemoration of his services to Botany and Horticulture:—

Anthurium Wallisii, Batemannia Wallisii, Curmeria Wallisii, Dieffenbachia Wallisii, Epidendrum Wallisii, Maranta Wallisii, Masdevallia Wallisii, Stenospermation Wallisii.

References:—"La Belg. Hort." 1879, p. 1, with portrait; "Hamburger Garten und Blumenzeitung," 1878, p. 433.

WALTER DAVIS.

Collector in South America.

1873—1876.

Walter Davis was born at Amport, a small village in the county of Hampshire, and inherited from his father a taste for Natural History and outdoor pursuits, which later took the form of a love of gardening.

Davis began his horticultural career in the gardens of the Marchioness of Winchester at Amport House, at a time when these were being remodelled. From Amport House he was sent to Wilton Park Gardens, where he stayed four years, ultimately becoming departmental foreman, and thence he went to the gardens of C. Ryder, Esq., Slade, and to those of the late T. W. Evans, Esq., at Allestree Hall, Derby.

In 1870 Davis came to Chelsea, served under John Dominy in the New Plant Department, eventually becoming foreman of the Nepenthes and Fine Foliage Plants.

In 1873 an opportunity occurring to send a collector to South America, Davis was selected, and he sailed on August 2nd 1873, with the special object to secure a quantity of Masdevallia Veitchiana, introduced to this country in 1867 by Pearce, but still very scarce.

In this undertaking Davis was successful, and in addition to Masdevallia Veitchiana met with several other species of this interesting genus, one of which* was named by Professor Reichenbach in compliment to the discoverer.

During his stay in South America Davis crossed the Cordilleras of the Andes in Peru and Bolivia no less than twenty times, at elevations of 14,000-17,000 ft., and he traversed that vast continent from one side to the other, along the whole length of the Amazon valley.

On his return to England in 1877 he was selected to conduct the botanical analysis of the herbage on the experimental plots at Rothamstead, and, returning to Chelsea on the termination of this engagement, is still in the employ of Messrs. Veitch.

* Masdevallia Davisii, Rchb. in Gard. Chron. 1874, vol. ii. p. 710.

PETER C. M. VEITCH.

COLLECTOR IN AUSTRALIA, SOUTH SEA ISLANDS, AND BORNEO.

1875—1878.

PETER C. M. VEITCH, for some years the chief of the firm of Robert Veitch & Son, of Exeter, travelled in the service of the Chelsea house for several years in Australia, the Fiji and the South Sea Islands, Borneo and the neighbouring islands, and he sent several new plants, as well as many rare in the British Isles at that time.

P. C. M. Veitch's earliest connection with Chelsea commenced in 1867, when he entered Coombe Wood as an assistant, to acquire a knowledge of Trees and Shrubs and of the working of that intricate branch of the nursery business. Subsequently transferred to the New Plant Department at Chelsea, he stayed till 1869, when he was sent to a seed-growing establishment in Germany.

From Germany he proceeded to a seed-house in France for some six months, and again returned to Chelsea.

In the spring of 1875, deemed advisable he should take a sea-voyage, it was arranged that he should visit, on behalf of the firm, the clients in Australasia, and, at the same time, introduce to England any plants likely to be of value for horticultural purposes.

With this object P. C. M. Veitch left England in 1875 for Sydney, by the long sea route, and almost immediately after landing left for the Fiji Islands, having an offer to sail in H.M. cruising schooner *Renard.*

Several months were spent in visiting the various islands of the Fiji group and in collecting plants.

In February 1876 a trading vessel having called at Fiji, P. C. M. Veitch secured a passage and proceeded to the South Sea Islands, where he remained until the following September. The whole of the collection of plants made in the Fiji Islands was lost in a gale, but that from the South Sea Islands was despatched to England in 1877.

From September to December 1876 excursions were made to various parts of the Australian Colonies, and useful plants found in cultivation in gardens sent home, amongst these Lomaria discolor, L. bipinnatifida and Microlepia hirta cristata.

During the early part of the year 1877 a visit was made to New Zealand, and a special journey to Mount Cook, where seeds of the beautiful Ranunculus Lyalli (or, as it is more commonly called in New Zealand, Mountain Lily or Rookwood Lily) were gathered and sent to Chelsea, from which plants were raised and flowered. From other parts of New Zealand several species of Celmisia, Veronica and the beautiful and somewhat difficult Notospartium Carmichæliæ were introduced.

In June 1877 P. C. M. Veitch again visited Australia, but in August, as he was making his way to New Guinea, had the misfortune to be shipwrecked off the north coast of Australia, and for a second time the collections were lost

Instructed to join F. W. Burbidge, plant-collecting for us in Borneo, he started for that country, stopping in Penang, Sumatra and Singapore *en route*, and arriving at Labuan in November 1877.

During his stay in Borneo he accompanied Burbidge on several excursions to the mainland and adjacent islands, and also undertook the journey to Kina Balu, the Sugar Loaf Mountain of Borneo.

In the spring of 1878 P. C. M. Veitch returned to Chelsea, and with him the collection made in Borneo in company with Burbidge. In 1880 he left for Exeter, to enter the firm of Robert Veitch & Son, of which he is now the head.

GUILLERMO KALBREYER.

Collector on the West Coast of Africa and Colombia, S. America.

1876—1881.

Guillermo Kalbreyer, a promising young man, twenty-nine years of age, entered Messrs. Veitchs' service as a plant-collector in 1876, and his first trip was to the West Coast of Africa in search of tropical flowering and foliage plants, very popular at that time.

Kalbreyer left Liverpool in November, arrived at Fernando Po on Christmas Eve, and reached Victoria a week later.

He proceeded to Old and New Calabar, Bonny, and the Cameroon Mountains and River, collecting many plants, sent to Chelsea.

At that time travelling in Africa was difficult, and, owing to the hostility of native traders, foreigners were unable to penetrate far into the country.

In July 1877 Kalbreyer returned to England, bringing with him a small collection of plants obtained on the West Coast, including five species of Mussænda, Gardenia Kalbreyeri, and two new Orchids:—Brachycorythis Kalbreyeri, a terrestrial species named by Reichenbach in compliment to its discoverer, and Pachystoma Thomsoniana, an epiphyte, named, at Kalbreyer's request, in honour of a Rev. George Thomson, for many years an earnest missionary in that unhealthy region.

The great heat endured and the frequent attacks of malaria from which he suffered on the West Coast seriously affected Kalbreyer's health, and it was decided to send him to a more healthy country to collect.

The next journey was to Colombia, where the climate is delightful on the highlands and mountain-slopes, though in the lowlands and along the coast almost as hot as in parts of Africa.

He left England in October 1877, collected in the Eastern Cordillera near Ocaña, La Cruz and Sierra Palado, the results being principally Odontoglossum Pescatorei and O. triumphans.

In February 1878 he left Ocaña to proceed to England, but owing to the River Magdalena being very low, the passage to the coast was difficult, and a journey which usually occupies from seven to ten days required nearly a month, so that by the end of April, when Kalbreyer arrived in England, more than half his collection was useless.

The third trip was again to Ocaña, but further east. A start was made in July 1878, and he proceeded to the Eastern Cordillera, through the towns of San Pedro, Salazar and Pamplona; here he again met with some extraordinary forms of Odontoglossum Pescatorei, including O. P. Veitchii, and O. triumphans, O. tripudians, O. hastilabium, O. coronarium, O. crocidipterum and O. blandum, a difficult species to import alive, and, until Kalbreyer's consignment arrived, rare in this country. After sending several very fine consignments of orchids, Kalbreyer returned to England, bringing with him a large and choice collection.

The fourth journey was commenced in September 1879, when he again left England for Colombia, on this occasion travelling down the River Magdalena to the Central and

Western Cordillera, and as far west as the Valle de Atrato or Llanos de Murry. Passing from the water-shed of the Atrato to the plains, he was particularly struck by the richness of the vegetation. Here Anthurium Veitchii, with leaves over 6 ft. in length, climbed trees more than twenty yards in height, and growing luxuriantly were a great many palms, of which he collected specimens of more than 100 species, and seed of many. Kalbreyer traversed the towns of Rio Negro, Medellin, Antioquia, Sopetran, Frontino, Rio Verde, and many others, and, later on, to the north, he passed Santa Rosa Amalfi, and on the south, Concordia.

Orchids were the principal plants collected, amongst others Odontoglossum ramosissimum, O. sceptrum, Miltonia vexillaria, Cattleya aurea, C. gigas, Cypripedium Roezlii, C. Schlimii alba, and several species of the curious large-flowered Masdevallias.

Several consignments of these were sent to Chelsea, and in September 1880 Kalbreyer returned to England, bringing with him many living plants and a great collection of dried Ferns, comprising some 360 species, of which eighteen were new to science. These were described by Mr. Baker in the *Journal of Botany* for July 1881.

The last journey as a plant-collector was commenced by Kalbreyer in December 1880; on this occasion he again proceeded to Ocaña, where he arrived in January 1881, and sent home a consignment of Orchids. Leaving Ocaña at the end of the month, he went southwards to Andinamarca and Bogota, on the high plains of the Eastern Cordillera In this neighbourhood he made a collection of Orchids, consisting principally of Odontoglossum crispum, brought safely to England in June of the same year. His engagement with Messrs. Veitch then terminated, and after a short stay he returned to Columbia, commenced business in Bogota as a nurseryman and exporter of Orchids, in which occupation he is still engaged.

The following new Ferns were discovered by Kalbreyer in New Grenada during the summer of 1880, and were determined by Mr. Baker of Kew:—

Acrostichum (Polybotrya) botryoides.
„ (Gymnopteris) suberectum.
„ „ polybotryoides.
„ „ juglandifolium.
Alsophila podophylla.
„ hispida.
„ ? late-vagans.
Asplenium (Euasplenium) filicaule.
„ (Diplazium) longisorum.
Danæa serrulata.
Dicksonia pubescens.
Gymnogramme vellea.
Nephrodium (Lastrea) longicaule.
„ „ valdepilosum
„ (Sagenia) antioquoianum.
Polypodium (Phegopteris) sylviculum.
„ (Eupolypodium) antioquoianum.
Trichomanes Kalbreyeri.
Selaginella longissima.

CHRISTOPHER MUDD.

Collector in South Africa.

1877.

This collector, son of a former curator of the Cambridge Botanic Gardens, went on an expedition to South Africa in 1877, and great things were expected to result from the undertaking. These expectations, however, were not realized, for Mudd, who seemed to have no special aptitude for collecting, and entirely lacked the explorer's instinct, sent home little of horticultural value, and the mission, which was practically a failure, had to be recalled.

Mudd subsequently settled in New Zealand.

F. W. BURBIDGE.

Collector in Borneo.

1877—1878.

F. W. Burbidge, a native of Wymeswold, Leicestershire, undertook in our service a collecting expedition in Borneo during the years 1877—1878, the special object, the introduction of certain Pitcher Plants known to inhabit that island, and to be accessible.

Burbidge was well equipped with both practical and scientific knowledge to undertake such a mission: he had distinguished himself as a student in the Royal Horticultural Society's Gardens at Chiswick and in the Royal Gardens at Kew, and had journalistic experience.

The story of his travels and adventures is told in his book, *The Gardens of the Sun*, written on his return from Borneo, from which the following particulars are abstracted.

The object of the journey was the collection and introduction of beautiful new plants, as well as of birds and other objects of Natural History, and he was fortunate in adding about fifty new species of Ferns to the list of those already collected in Borneo, and of this number about twenty were new to science.

Perhaps the greatest good fortune which attended Burbidge's work was the introduction of the Giant Pitcher Plant of Kina Balu—Nepenthes Rajah.

This wonderful plant and its geographical allies were discovered in 1851 by Sir Hugh Low, who made repeated journeys from Labuan to Kina Balu, but failed in his endeavours to introduce specimens to European gardens.

Thomas Lobb, the most successful of all Eastern plant-hunters, also endeavoured to find the habitat of these plants in 1856, and had actually reached the foot of the mountain on which they grew, but was prevented by the hostility and extortion of the natives from completing the ascent. Burbidge was successful in introducing living plants and seeds, but unfortunately Nepenthes Rajah has proved of such very difficult culture that few specimens exist in gardens.

Another curious species also discovered and introduced by Burbidge is Nepenthes bicalcarata, remarkable in having two long spurs projecting over the mouth of the pitcher.

The three native courts of Jahore, Brunei and Sulu were visited, and Burbidge was enabled to make extended excursions into the interior of Sulu itself.

The collection made on the last-named island comprised new Ferns, rare Mosses and several beautiful Orchids, including Phalænopsis Mariæ, Dendrobium Burbidgei and Aërides Burbidgei, all rare plants at the present day.

The first expedition to Kina Balu, the Sugar Loaf Mountain of Borneo, was made in company with P. C. M. Veitch (*q.v.*), who joined Burbidge on his return from an extended tour through Australia and the Fiji Islands.

This journey was a critical and tedious one, and the entire route from Gaya Bay to the mountain and back to the coast through the villages of Kuong, Kalawat and Bawang, had to be accomplished on foot.

These labours were rewarded by finding all the large species in one locality, and in addition a distinct form of Nepenthes Edwardsiana not previously collected.

On his return from Borneo in 1879 Burbidge was appointed

NEPENTHES BICALCARATA

Curator of the Botanical Gardens at Trinity College, Dublin, and in 1894 became keeper of the College Park.

In 1889 he had conferred on him, *honoris causa*, the degree of Master of Arts of Dublin University, and in 1897 was a recipient of the Victoria Medal of Honour granted by the Royal Horticultural Society in recognition of his services to Horticulture.

His much-lamented death, of a serious heart trouble, occurred in Dublin on Sunday, December 24th 1905, in his fifty-eighth year. On the 27th day of that month the *Times* had the following appreciation:—"Mr. Burbidge had the academic as well as the horticultural mind; he filled his office with distinguished success, and made many important contributions to the literature of his subject, on which he was a recognized authority."

Burbidge was the author of several works on gardening, probably the most important his *Cultivated Plants* and *The Narcissus*, whilst the *Gardens of the Sun* already mentioned contains an account of his travels as a plant-collector.

The sixty-sixth volume of *The Garden*, to the editorial staff of which he was attached from the year 1873 to 1877, is dedicated to him by the Editor.

The following is a list of Ferns discovered by Burbidge in the neighbourhood of Labuan and Kina Balu, and which were first described by Mr. Baker of Kew from Burbidge's specimens:—

Alsophila Burbidgei.
Asplenium (Diplazium) porphyrorachis.
" (") xiphophyllum.
Davallia Veitchii.
Lindsaya crispa.

Lindsaya Jamesonioides.
Nephrodium (Sagenia) nudum.
Polypodium (Eupolypodium) minimum.
„ („) Burbidgei.
„ („) streptophyllum.
„ („) taxodioides.
„ (Phymatodes) stenopteris.
„ („) holophyllum.

In the Sulu Archipelago, a group of small islands lying off Borneo and the Philippines, Burbidge discovered the following new species:—

Cyathæa suluensis.
Pteris Treacheariana.
Polypodium (Phegopteris) oxyodon.
„ (Eupolypodium) Leysii.

Of the plants introduced at the same time, probably the following are the best known:—

Alocasia guttata, A. pumila, A. scabriuscula, Aërides Burbidgei, Bulbophyllum Leysianum, B. mandibulare, Burbidgea nitida, the type of a new genus; Cryptocoryne caudata, Cypripedium Dayanum, C. Lawrenceanum, Dendrobium Burbidgei, D. cerinum, Jasminum gracillimum, a beautiful stove plant with fragrant white flowers; Nepenthes Rajah, N. bicalcarata, a species new to science; N. Rafflesiana nivea. Phalænopsis grandiflora, the Bornean variety, P. Mariæ, Pinanga Veitchii, Pothos celatocaulis, and Wormia Burbidgei, a new species named in compliment to the discoverer by Sir Joseph Hooker.

Reference:—Gard. Chron. 1905, vol. xxxviii. p. 460, obituary note and portrait.

CHARLES MARIES.

Collector in Japan and China.

1877—1879.

Charles Maries was born at Stratford-on-Avon in the county of Warwick. He was educated at the Grammar School at Hampton Lucy during the years 1861 to 1865, where the present Professor Henslow was Head-master, and to whom Maries was greatly indebted for his knowledge of Botany.

From school Maries proceeded to Lytham in Lancashire, where his brother had a small nursery, in which he worked for the next seven years, afterwards obtaining employment in the houses at Chelsea. He proved to be an industrious and steady workman and was eventually selected to undertake an exploring expedition to the Far East, the object of which was to obtain seeds of the coniferous trees of Japan, and to explore the great Yangtsze valley of China, rightly believed to possess a rich arboreal vegetation, and many plants that would prove hardy in this country.

Maries left London on February 1st 1877, visiting Hong Kong, Ningpo, the snowy mountains of which he explored, and Shanghai in China, arriving at Nagasaki, Japan, on April 20th. Remaining a short time to note the various cultivated plants, many of which had been introduced to Western gardens by Fortune, he left Nagasaki, for Shimenosiki, and, by way of the Inland Sea, Ozaka and Kioto, reached Yokohama.

At Yokohama he visited the nurseries of which Fortune had written in such glowing terms, but, disappointed with what he saw, yet found several plants rare in cultivation.

From Yokohama and Tokio Maries proceeded overland to Nikko, the great shrine of Japan, and thence to Awomori, the northernmost port of the main island. Whilst waiting at Awomori for a steamer to convey him to Hakodate, Maries noticed a Conifer new to him growing in a garden, and learnt that it could be found in quantity on a neighbouring mountain. He went in search, and had reached a height of 3,500 ft., when it became obvious that the bamboo scrub formed an impassable barrier on that side of the mountain, and he reluctantly had to turn back, although the object of his search could plainly be seen. The following day he again made the ascent, but this time from the north side, and he succeeded in procuring cones of a new species, since named by Dr. Masters, Abies Mariesii.

On the same trip, what was at first thought to be a variety of Abies Veitchii, but which eventually proved to be A. sachalinensis, Mast., was re-discovered. It had previously been met with by one Friedrich Schmidt, a German botanical traveller, in the island of Saghalien in 1866, but not introduced.

Leaving Awomori, Maries arrived in Hakodate, in Yezo, the northern island, on June 20th, and was much struck with the beauty of Azalea Rollisoni (Rhododendron indicum balsaminæflorum) found growing in masses of Kæmpfer's Iris on the banks of streams. He was able to procure seed which he sent to Chelsea. From this district he also sent Styrax × obassia, common on the volcanic slopes of that island, the racemes of pure white flowers and very large light-green leaves objects of great beauty.

Continuing the journey, Maries reached Sapporo, passing through swamps swarming with wild-fowl, and swollen rivers, noting by the way large masses of Platycodon grandiflorum Mariesii and Lilium Thunbergianum. From the thickly wooded and mountainous districts in the neighbourhood of Sapporo Maries sent home seeds of Abies sachalinensis, A. yessoensis, Daphniphyllum glaucescens, many Maples and climbers, including Schizophragma hydrangeoides and Actinidia Kolomikta, the "Cat's Medicine" of the Japanese. From Sapporo he travelled by way of Chitose and Yubetsu, visiting Uragawa and Shamani, to Horidzumi: near Shamani in a sandy plain he met with masses of the pretty little Dracocephalum Ruyschiana, and obtained seed.

Making Horidzumi, on the south-west cape, his head-quarters, Maries stayed in the country from June to October 1877, exploring the mountains and making extensive entomological and botanical collections. These collections were shipped in a vessel laden with sea-weed bound for Hakodate, but which was wrecked the following morning; the sea-weed, wet and swollen, had burst open the vessel and the captain ran her ashore. The box containing the seeds was rescued and put into another boat which immediately capsized and sank. It was not too late, however, to still gather seeds of the Conifers, and Maries lost no time in replacing the loss by a fresh collection. He left Yezo by H.M.S. *Modeste*, arrived at Niigata, on the south-west coast of the main island, in December 1877, and travelled overland to Yokohama. On Christmas Day 1877 Maries left Yokohama for Hong Kong, arrived on January 2nd 1878, and sailed a few days later for the island of Formosa.

He landed at Sia-wau-fu on the 16th, left shortly afterwards for the interior, but did not penetrate far into the country. After a distant view of Mount Morrison he returned to the port, having obtained, amongst other material, seed of a new species of Lilium.

An attempt to enter the island from the north side met with a like failure and he returned to Shanghai.

In the spring of 1878 Maries visited Chin-kiang and Kui-kiang, collecting *en route* the pretty Daphne Genkwa, found growing in quantity with Exochorda grandiflora, Spiræas, Hypericums, Deutzias, Weigelas and the stunted Pinus sinensis, and, in more sheltered situations, Forsythias, Loniceras, Akebias and Wild Cherries were common.

On the Looshan Mountains a white form of Daphne Genkwa with Rhododendron Fortunei and Loropetalum sinense were discovered, seeds of all being obtained and successfully introduced.

On an excursion to the "Teen Cha" Temple, known also as the "Yellow Dragon" and "Heavenly Pool" Temple, Maries saw magnificent trees of Larix Kæmpferi, Cryptomeria japonica, and the Chinese form of Liriodendron tulipiferum, as well as that beautiful Lily since named by Baker, Lilium lancifolium formosanum.

On this trip Maries suffered severely from sunstroke and returned to the coast.

The summer of 1878 he again spent in Japan, collecting seeds of plants, especially of Conifers, of which he had made notes on his former trip.

In December he left Japan and went to Han-kow on the Yangtsze, starting early in the spring of 1879 for Ichang, 800 miles higher up the river.

Amongst the gorges of Ichang, where the Great River rushes out of the mountains, Maries found Primula obconica, and sent seed to Chelsea.

ABIES VEITCHII

PENCARROW, CORNWALL

With the natives of China Maries did not succeed so well as with the Japanese, he was not sufficiently gentle, and was often threatened and occasionally robbed of his baggage; in the summer of 1879 he returned to Japan.

On this trip seeds of many Japanese Oaks were gathered and the beautiful dwarf Bamboos, including the square Bamboo, which he successfully introduced. Altogether about 500 living plants were sent home, and large quantities of seed of various Conifers and other fine trees; thirty-eight new plants are recorded by Bretschneider as being first discovered in China by Maries.

In British gardens he will always be remembered by the following, which he was fortunate to be able to introduce to this country:—

Abies Veitchii, A. sachalinensis, Daphniphyllum glaucescens, Acer polymorphum varieties, several new and distinct forms of Hydrangea rosea, Styrax obassia, Lilium auratum gloriosoides, L. a. platyphyllum, Spiræa palmata alba, Conandron ramondioides, Primula obconica, Platycodon grandiflorum Mariesii, Iris Kæmpferi (many varieties), Davallia Mariesii, Osmunda japonica corymbifera, and many others hitherto very scarce in Europe.

Maries returned to England in February 1880, when his herbarium was sent to Kew, the Conifers brought from Japan being dealt with by Dr. Masters, *Conifers of Japan*, *Linn. Soc.'s Jour.* xviii. 1881, 473-541, and in Veitchs' *Manual of Coniferæ*, ed. 2, 1900. A collection of insects was accepted by the British Museum.

In 1882 Maries was recommended by Sir Joseph Hooker for the post of Superintendent of the gardens of the Maharajah of Durbhungah, in India, where he laid out the very extensive grounds which surround the palaces.

Subsequently entering the service of the Maharajah Scindia of Gwalior, he again laid out the palace gardens, of which, and of the Gwalior State Gardens, he remained superintendent until his death, which took place in India on October 11th 1902.

Maries had enthusiasm, but lacked "staying" power: he was musical, much to the delight of the Japanese peasants, and doubtless this must often have helped the work: he was a skilled shot, as the buck on the domains of the Maharajah Scindia learnt to their cost when Maries was living in the country of the Mahrattas.

Maries was a Fellow of the Linnean Society, an original recipient of the Victoria Medal of Honour, granted by the Royal Horticultural Society in recognition of his services to Horticulture, and a frequent contributor to the Horticultural Press.

References :—Bretschneider's "History of European Botanical Discoveries in China," vol. ii. p. 741; Maries in The Garden, 1881-1882, vols. xx.—xxii. *passim*; "Rambles of a Plant Collector"; Burbidge in The Garden, 1883, vol. xxiv. p. 444, with portrait; Gard. Chron. 1902, vol. xxii. p. 360, obituary note and portrait.

CHARLES CURTIS.

Collector in Madagascar, Borneo, Sumatra, Java, and the Moluccas.

1878—1884.

Charles Curtis had been employed some four years at Chelsea when, in 1878, an opportunity occurring for sending a plant-collector to the East, he was selected to undertake the important mission.

The first trip was to Mauritius and Madagascar, whence he sent the handsome Pitcher, Nepenthes madagascariensis, and various tropical plants. This occupied rather over a year. In 1879 Curtis returned to England, and was sent in 1880 to Malaysia, where he explored Borneo, Sumatra, Java and the Moluccas, and collected many interesting Stove plants, Palms, and Orchids, subsequently sent to Chelsea. The special object of the journey was to collect specimens of Miss North's Pitcher-plant (Nepenthes Northiana), the existence of which had been made known through a drawing by that lady in Borneo, now in the North Gallery at Kew. The precise locality where this plant grew was unknown, but after much search Curtis was successful and introduced it.

A full description of Nepenthes Northiana is given in this work in the special chapter devoted to Nepenthes: there is no Pitcher more striking.

On the trip to Borneo Curtis was accompanied by the young gardener, David Burke, who returned with the collection made in Sarawak, and who himself afterwards became a plant-collector.

The plants brought home by Burke on this occasion included large consignments of Cypripedium Stonei, C. Lowii, many Vandas, Rhododendrons, and the beautiful Stove-foliage plant, Leea amabilis, was also found.

After seeing Burke and the collection safely shipped at Singapore, Curtis proceeded to Pontianak in Dutch Borneo, with the special object of obtaining a consignment of Phalænopsis violacea, known in England but still rare. In this again successful, but, owing to a mishap with the boat, a month's collections and all his clothes and instruments were lost, and he narrowly escaped with life.

The beautiful Rhododendron Teysmanni and R. multicolor, with the red variety Curtisii, were introduced through Curtis: in themselves most gorgeous plants, they are quite eclipsed by the valuable hybrids which have since been derived from their cross fertilization.

On terminating his engagement early in 1884, Curtis was appointed Superintendent of the Botanic Gardens at Penang, a post which he held until December 1903, when retirement was necessary, and he returned to England, settling in his native town of Barnstaple, in the county of Devon.

The following plants received the specific name of Curtis in commemoration of his services to Botany and Horticulture:—Nepenthes Curtisii, Cypripedium Curtisii, Medinella Curtisii, Gastrochilus Curtisii, Rhododendron multicolor Curtisii, and several others, enumerated in the body of this work.

CYPRIPEDIUM CURTISII

DAVID BURKE.

Collector in the East Indies, Burmah and Colombia.

1881—1897.

This traveller crossed a greater area of the earth's surface and covered more miles in search of plants than any other Veitchian collector, with the possible exception of the two brothers Lobb.

Born in Kent in 1854, Burke entered the houses at Chelsea as a young gardener, and, having a wish to travel, was sent on a trial trip to Borneo with Curtis, and brought home a collection of plants, the result of their joint work.

His next mission, commenced in 1881, was to British Guiana, where he re-discovered the interesting insectivorous plant Heliamphora nutans, which had not been seen since its discovery on Mount Roraima in 1839 by the two brothers Schomburgk, and successfully introduced it to England. Amongst Orchids from British Guiana was the rare Zygopetalum Burkei which perpetuates his name, and from that country he also sent the handsome Amasonia punicea (calycina), the brilliant scarlet bracts of which are so effective in stoves during the winter months.

Subsequent journeys included two to the Philippine Islands for Phalænopsis, two to New Guinea, and one in 1891 to the then newly annexed provinces of Upper Burmah for Orchids. During the years 1894 to 1896 he made three trips to

Colombia for Cattleya Mendelii, C. Schröderæ, C. Trianæ and Odontoglossum crispum (Alexandræ), and finally in 1896, having spent a short time in England, Burke left for the Celebes Islands and the Moluccas, and in the island of Amboina, belonging to the last-named group, he died on April 11th 1897.

Burke was one of those curious natures who live more or less with natives as a native, and apparently prefer this mode of existence; his early death is partly due to this cause, and there is little doubt that this sad event occurred in a lonely hut far from any European settlement, and, had it not been for information sent by a German commercial traveller, apparently the only white man in that part of the island at the time, Messrs. Veitch would not have heard of Burke's ultimate fate.

JAMES HERBERT VEITCH.

TRAVELLER THROUGH INDIA, MALAYSIA, JAPAN, COREA, THE AUSTRALIAN COLONIES AND NEW ZEALAND.

1891—1893.

JAMES H. VEITCH sailed from Naples at the end of October 1891 on a somewhat extended tour, passing through Ceylon to Cape Tuticorin, the most southern port of India, and northwards overland to Lahore; southwards from Calcutta to the Straits Settlements and to Java to visit the very noble gardens at Buitenzorg:—and to eat the Mangosteen. It is necessary to eat the Mangosteen grown within three or four degrees of latitude of the equator to realize at all the attractive and curious properties of this fruit.

From Java he proceeded north to Japan, where several months were spent in the woody districts of that very extended country, the district of Nikko and the neighbourhood of Sapporo offering the widest field of interest, and possessing, of their kind, the richest flora known.

From Japan to Corea, by way of the Sea of Japan, is a short voyage, and several weeks were spent in Cho-sen, "the Land of the Morning Calm." Prior to the Japanese-Chinese War, the isolation of the country and of the people was remarkable, and a journey of 600 miles—on two occasions crossing the peninsula from coast to coast—proved unusually interesting.

The land is thinly populated; there is no scattered peasantry, and the people congregate in villages and towns—the only means of communication a limited and sorry lot of ponies.

On approaching a village or small town, uniformed runners from the local Yamen were sent to meet the traveller, and woe betide any unfortunate pedestrian who did not make way on the public road for the officially recognized personage.

The official gates of the Yamen were opened for the inevitable reception, held in spacious courtyards, with buildings barbaric in design flanking the boundary walls. The lesser officials were much interested in the general accoutrements, and apparently now saw most of these for the first time—women, when met, turned their heads away, and enveloped the face and hands in the long, loose, spotlessly white cotton robes which the entire nation wear; children, in the outlying districts, fled as the white man rode down the village street.

There is no rice in the Far East to equal the rice of the Coreans, the seed is larger and has more weight than the varieties Japan produces; nor can Japan grow sufficient for her own population, and is dependent for much of her supply on Corea.

It was known that the flora of the peninsula was of little value, that the nature of the country and climatic conditions were such, that it would not be possible to find any plants new and suitable to English gardens; hill-slopes covered with Platycodon grandiflorum Mariesii were pretty, and solitary specimens of Pinus koraiensis, never found wild in Japan, of interest.

This pine was introduced to Europe from a Japanese nursery in 1861 by the late John Gould Veitch.

Continuing the journey, James H. Veitch sailed southwards, spent several months in Australia and New Zealand, and returned to England in 1893.

The chief results to gardens from this long tour were, the distribution of the large-fruited Winter-cherry (Physalis Francheti), the variety of Cerasus pseudo-cerasus known as James H. Veitch, and the re-introduction of the large-leaved Vine (Vitis Coignetiæ) from Japan.

Rhododendron Schlippenbachii, discovered some forty years earlier on the Corean coast by a Russian collector, was also introduced, and other plants rare in cultivation.

References :—Gard. Chron. 1892-1894, vols. xi.—xvi. *passim*, " A Traveller's Notes " (these notes were collected, revised and fully illustrated from photographs by the author, and published in 1896 under this name); Sargent's "Forest Flora of Japan," 1894; Bretschneider's "History of European Botanical Discoveries in China," 1898, p. 767; Man. Con. 1900, ed. 2, p. 335.

E. H. WILSON.

Collector in Central and Western China and on the Tibetan Frontier.

1899—1905.

E. H. Wilson, the most recent of our successful travellers and collectors, was born in Birmingham, and received part of his early training in Botany and Horticulture in the Botanic Gardens of that town.

He afterwards entered the Royal Gardens at Kew as a young gardener, and when an application was made to Sir W. T. Thiselton-Dyer, the late Director of the Gardens, for a man likely to prove suitable to undertake a prolonged journey in China, the late Director suggested Wilson for the post.

The object in sending a traveller to Central China was to obtain seed of species likely to prove hardy in Great Britain, and living representatives of certain plants only known to exist from dried specimens in the herbaria of various European countries.

Wilson sailed from Liverpool in April 1899. Travelling by way of America, he visited Professor Sargent, the well-known authority on ligneous plants at Boston, and consulted him respecting the trees and shrubs likely to be found in China.

The desired information obtained, Wilson proceeded, and arrived at Hong Kong on June 3rd 1899.

Before leaving for the interior it was considered advisable

that Wilson should consult Dr. Henry and benefit by his unrivalled knowledge of the Chinese flora.

Dr. Henry was at that time in the Chinese Customs Service, stationed at Szemao in the south-west corner of the province of Yunnan, and Wilson, who left Hong Kong on June 14th to find him, travelled *viâ* the French colonial settlement of Tonkin.

At Laokai he was detained owing to a native rising at Mengtsze, which made it dangerous or impossible for Europeans to travel in the interior.

After several weeks' delay, during which Messrs. Veitch had lost hope of his being able to proceed, the disturbance quieted down, and Wilson was allowed to continue his journey unmolested.

Szemao was reached on September 24th, and a cordial welcome from Dr. Henry awaited the young traveller.

That gentleman freely imparted important information regarding the plants Wilson was in search of, and the ways and means of reaching them. The information was valuable, as the district Wilson was instructed to explore was practically a closed book to all but a few, amongst whom was Dr. Henry.

On his return from Szemao Wilson collected plants of the beautiful Jasminum primulinum, afterwards successfully introduced, and which flowered for the first time in this country at Coombe Wood in October 1901.

Returning to Hong Kong again in December, he left immediately for Shanghai, made at once preparations for a journey to Ichang in the Yangtsze valley, and left fully equipped in February 1900.

In the usual type of house-boat of the Yangtsze, a dwelling-house and a conveyance at the same time, many months were spent. On arrival he commenced exploring the mountain-ranges south and south-west of Ichang, and in

April 1900 discovered the wonderful Davidia involucrata, the principal object of the journey.

The country generally was disturbed by a revolt in the north, known in England as the "Boxer" rising, but after an anxious period of several months the trouble subsided and the work of plant-collecting continued uninterruptedly.

As a result of this year's work Wilson obtained seeds of 671 different species of plants, herbarium specimens numbering 1,764 species, and great quantities of bulbs and roots of herbaceous plants.

During the year 1901, the third of his mission, Wilson explored the high mountain-ranges on the Hupeh-Szechuan boundary, north-west and south of Ichang, and collected quantities of seed, though the season was exceptionally wet and cold.

Davidia involucrata was again met with, growing in large quantities ;—a striking feature in the landscape.

The collection this year consisted of 305 varieties of seeds, of many herbs, trees and shrubs, and of herbarium specimens numbering 906 species, in addition to thirty-five cases of bulbs and living roots and rhizomes of herbaceous plants, all shipped to England.

Among the best finds collected on this, his first mission, and successfully introduced to our gardens, are :—

Davidia involucrata, Astilbe Davidii, Clematis montana rubens, Senecio clivorum, Buddleia variabilis Veitchiana, Brandisia racemosa, Actinidia chinensis, numerous Vines, Acers, Viburnums, Spiræas, Roses, and Magnolias.

On the whole, Wilson succeeded remarkably well with the natives, and, though the country was disturbed by political risings and riots, met with no serious mishap, and lost no part of his collection.

SENECIO CLIVORUM

Wilson returned to England in April 1902, spent the summer at Coombe Wood, and left for the second journey to the extreme west of China, to the border of Tibet, a thousand miles further west beyond the former field of exploration, in January 1903.

On arrival at Shanghai he followed the former route as far as Ichang, reaching Kiating, which was to be his base, on June 19th 1903.

The mountainous country west of the Min river to the Yalung river, about 100 miles west of the border town of Tatien-lu, was explored, as was Mount Omi, a sacred mountain of the Chinese. Specimens of the flora were obtained and some few seeds.

In the neighbourhood of Tatien-lu was discovered the principal object of the search—the magnificent yellow Poppy, Meconopsis integrifolia, and Wilson was successful in securing seeds from which plants were raised and flowered at Langley in September 1904.

In July 1904 Wilson left Kiating for Sungpan in the extreme north-west corner of Szechuan, a trip which occupied some fifty-two days. The experience was arduous, owing to the severity of the weather, the bad roads and the scarcity of food, but on the whole, from the plant-collector's standpoint, a brilliant success. The country is very mountainous and possesses a rich flora. Ranking next to the truly magnificent Rhododendrons, are several species of Primula, one of the most beautiful, P. vittata, growing in enormous quantities in moist Alpine meadows, and by the sides of streams.

Besides securing a further supply of seeds of Meconopsis integrifolia, Wilson discovered the scarlet-flowered species, M. punicea, and, from seed successfully introduced, plants were raised and flowered at Langley in September 1904. Another remarkable find was a new, striking, and most promising Incarvillea.

This ended work for the season, and Wilson at once commenced preparations for his return home. Leaving Kiating on December 8th, he arrived at Chung-king on the 14th, and early in 1905 reached Ichang. The collections were here repacked, and with them he finally left China in January of that year, arriving in England in the month of March.

The five years' collections comprised some 25,000 dried specimens, representing some 5,000 species: these were distributed amongst the principal herbaria in Europe and America; and in addition seeds of 1,800 species, some 30,000 bulbs of new and rare species of Lilium, and living roots and rhizomes of various herbs and shrubs were sent to Messrs. Veitch.

Wilson was unusually sympathetic to the Chinese temperament, always prepared to yield a point, but firm when necessary, a contrast to his predecessor, Charles Maries, who, when in the Ichang region of the Yangtsze valley in 1879, could have sent many plants, subsequently found by Wilson, and which even at that time were undeniably accessible. Maries was *difficile;* and the natives, naturally resenting this, destroyed the collection, and he returned to the coast reporting the people hostile.

References:—James H. Veitch in Jour. R.H.S. vol. xxviii. pt. i.; Gard. Chron. 1905, vol. xxxvii. p. 113, with portrait; *id.* "Leaves from my Chinese Note-book," *passim.*

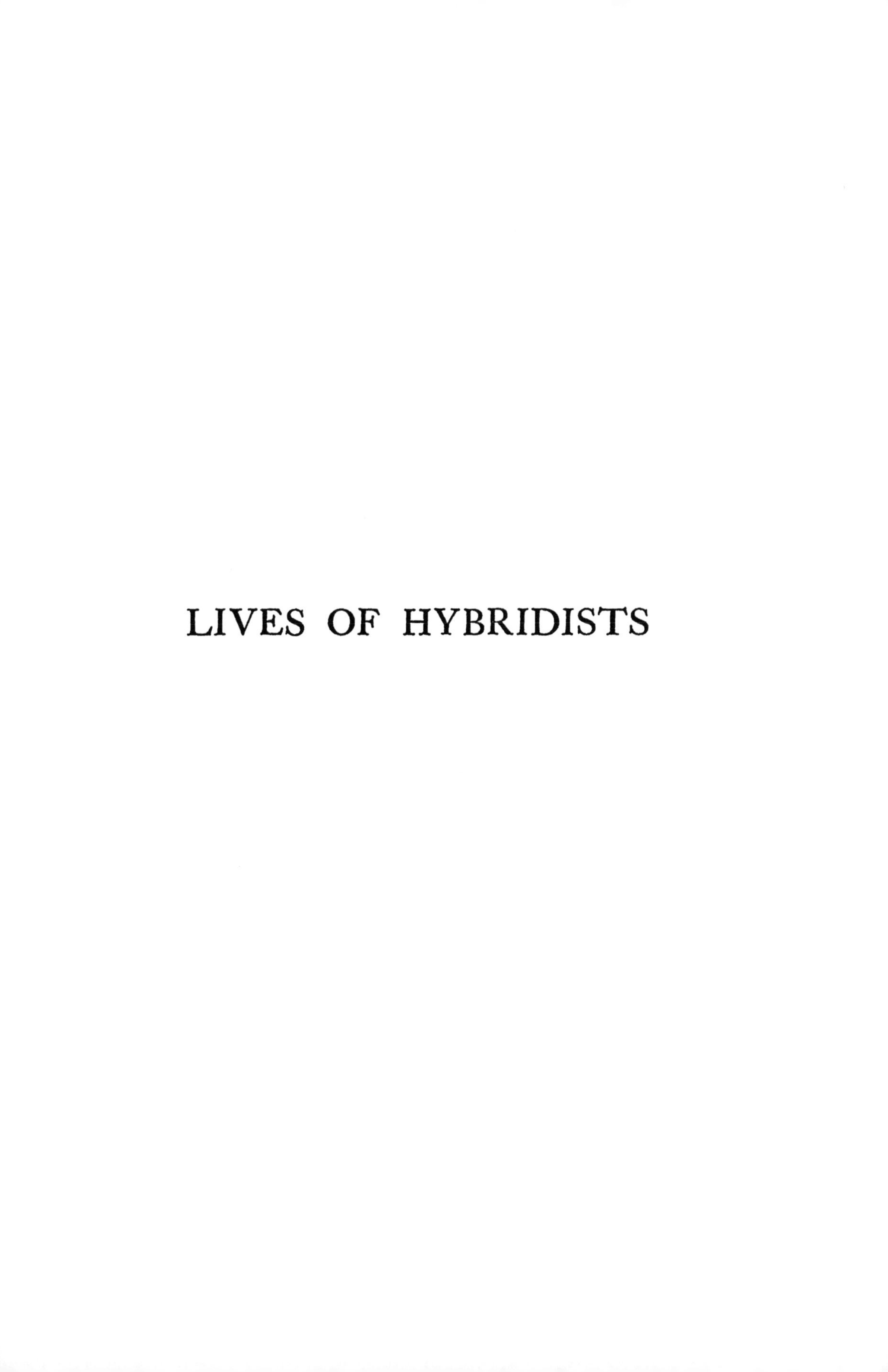

LIVES OF HYBRIDISTS

LIST OF MESSRS. VEITCHS' LEADING HYBRIDISTS

John Dominy.

William Court.

John Seden.

John Heal.

George Tivey.

LIVES OF HYBRIDISTS

JOHN DOMINY.

John Dominy was born at Gittisham, Devon, in 1816, and early in life adopted gardening as a profession. In 1834, after completing his term of apprenticeship in a private garden, he entered the nursery of Messrs. Lucombe, Pince & Co., of Exeter, where he stayed for two or three months; he then joined Messrs. Veitch, who at that time possessed only the Exeter establishment.

Here Dominy remained until 1841, in which year he accepted an appointment as head gardener to J. P. Magor, Esq., of Redruth, with whom he stayed nearly five years, after which he again entered the nursery at Exeter, and continued with the firm, both at Exeter and Chelsea, till 1880, when failing strength compelled retirement.

Dominy was an excellent cultivator of Stove and Greenhouse Plants, but it was his skill as a hybridizer of Orchids and Nepenthes that won for him the high position his name holds in the list of practical horticulturists of the last century.

Mr. John Harris, a surgeon of Exeter, who possessed an acquaintance with Botany, first suggested to Dominy the possibility of obtaining hybrid orchids, and explained to him the structure of the orchid flower and the process of pollination. As soon as an opportunity presented itself Dominy lost no time in turning the suggestion to practical account, and Calanthe × Dominii,* which flowered in 1856, was the first of his successes. This resulted from crossing

* Bot. Mag. t. 5042.

Calanthe Masuca with the pollen of C. furcata, and the seedling took two years to flower.

Considered a great cultural feat by the gardeners of the day, botanists were less enthusiastic in welcoming the new plant, and the exclamation of Dr. Lindley, the leading botanist and systematist of his time—"You will drive the botanists mad," is well known, and expressed the feelings of many scientists regarding hybrids, or, as they were then called, "mules."

Calanthe × Dominii was soon followed by others, an account of which will be found in other portions of this work, but mention may here be made of some of the more important hybrid orchids, which, in spite of the great advance made in Orchid culture since Dominy's day, still hold a favourable position in collections.

Læliocattleya exoniensis, Dominy's principal success from a cultivator's point of view; Calanthe × Veitchii, long since recognized as one of the handsomest and most useful of winter-flowering hybrids, and a potent agent in the production of many new and beautiful forms of recent times; Phaiocalanthe irrorata, a bigeneric hybrid, the first to be raised; and Cypripedium × vexillarium, the forerunner of a group of handsome Cypripedes in which the beautiful Cypripedium Fairieanum has participated in the parentage, are all due to Dominy.

It was not, however, to orchids alone that Dominy devoted his attention; Nepenthes and Fuchsias gained much from his efforts, and some very successful results were obtained in these two genera, notably Nepenthes × Dominii, N. × hybrida and Fuchsia × Dominiana.

To the high estimation in which John Dominy was held in horticultural circles the following testifies:—On leaving Devon in 1864 to accompany the late James Veitch to Chelsea, the Exeter Horticultural Society presented him with a piece of plate "in recognition of the value of his experiments in hybridization carried on by him whilst a member of their association."

On his retirement in 1880 the Council of the Royal Horticultural Society presented the famous hybridizer with the Gold Flora Medal "for his successful labours as a raiser of hybrid Orchids, Nepenthes, and other garden plants," and a few years later his friends presented him, through the President of the Society, Sir Trevor Lawrence, Bart., with a handsome gold watch and a purse of 200 guineas.

The twenty-first volume of *The Garden* is dedicated by the founder to "John Dominy, of Exeter and Chelsea, in recognition of his long and useful work in the improvement and hybridization of garden plants, especially Orchids, and his general excellence as a cultivator."

After his retirement Dominy still retained an interest in horticultural pursuits, and was a constant attendant at the meetings of the R.H.S. Floral and Orchid Committees, of which he was a member.

He died on Thursday, February 12th 1891, after a short illness, and was buried at Exeter on the 17th of that month.

References:—Gard. Chron. 1880, vol. xiii. p. 752; *id.* vol. xiv. p. 112; *id.* 1881, vol. xv. p. 728, vol. xvi. pp. 405, 509; *id.* 1891, vol. ix. pp. 240, 278, obituary note, portrait p. 277; The Garden, 1882, vol. xxi. portrait frontispiece; *id.* 1891, vol. xxxix. p. 179, obituary note; Veitchs' Man. Orch. Pl. pt. x.

WILLIAM COURT.

William Court, well known during the seventies of the past century as a talented propagator and cultivator of stove plants, was born at Alphington, near Exeter, in 1843, and commenced his gardening career in the nurseries of Messrs. Lucombe & Prince, of that city.

In 1863 he entered the Exeter branch of the Veitchian firm, and shortly afterwards was transferred to Chelsea.

A successful hybridizer of Nepenthes and Sarracenias, Court raised several fine hybrids as a result of his experiments, some of the best being Nepenthes × Mastersiana, N. × intermedia, N. × Courtii, N. × rufescens, Sarracenia × melanorhoda, S. × Chelsoni, S. × Courtii, and several others.

Court made several journeys to North America, and introduced to that country many novelties sent home by our collectors from various parts of the world.

He died at Chelsea on September 17th 1888, after a short illness, and is buried in Brompton Cemetery, where a monument is erected to his memory by his American friends.

References :—Gard. Chron. 1888, vol. iv. p. 338, obituary note; The Garden, 1888, vol. xxxiv. p. 287, obituary note.

JOHN SEDEN.

John Seden is probably the best known of all hybridizers, and in connection with his retirement in 1905 the *Gardeners' Chronicle* of December 31st 1904 contains the following notice :—

"John Seden, V.M.H.

"To every lover of Orchids the name of Seden is familiar, as probably no other person now living has enriched our collections with so many fine hybrids or practised in so wide a field.

"John Seden was born at Dedham in Essex, July 6th 1840, and early in life commenced his career as a gardener, working in several private gardens before he came to Chelsea in January 1861. In the autumn of 1861 he was transferred to Exeter, under Dominy, amongst the Orchids and stove plants, and it was here that he was first initiated into the practice of hybridization, which he has since so persistently followed with such good results. The autumn of 1862 saw Seden again at Chelsea in charge of the Orchids, some of the stove plants, and the Nepenthes, and amongst these he commenced experiments in hybridization and cross-fertilization.

"Caladium × Chelsoni, Alocasia × Sedeni (which received a Gold Medal from the Horticultural Society), A. × Chelsoni, A. × intermedia, Nepenthes × Sedeni, N. × Chelsoni, Amaryllis (Hippeastrum) Brilliant, Chelsoni and maculata, the three first seedlings to be raised at Chelsea, are some of the results obtained from his early experiments.

"About a dozen varieties of Gloxinia, progenitors of the fine strain since developed at Chelsea, were distributed from seedlings raised by Seden from intercrossing the best existing forms.

"In 1867 the tuberous Begonia was taken in hand, several of the original species being then available through introductions of the firm's collector Pearce and, later on, of Davis. For a number of years hybrids and varieties were regularly distributed, and these laid the foundation on which have been built the fine strains existing at the present day. The first variety with pure white flowers was raised at this period, originating in a batch of seedlings of Begonia rosæflora, and the first double-flowered variety was obtained by fertilizing a flower of B. × Sedeni with its own pollen.

"About the same time Seden commenced hybridizing Orchids. Since his first hybrid, Cypripedium × Sedenii, flowered in 1873, Seden has raised 150 hybrid Cypripediums, 140 Lælio-Cattleyas, 65 Cattleyas, 40 Dendrobiums, 25 Lælias, 16 Phalænopsis, 20 Epidendrums, 12 Masdevallias, 9 Calanthes, 8 Sophro-Cattleyas, 5 Phaio-Calanthes, 6 Disas, 4 Zygopetalums, besides miscellaneous hybrids, such as Chysis × Chelsoni, C. × Sedenii, C. × langleyensis, Thunia × Veitchii, Sobralia × Veitchii, Cymbidium × eburneo-Lowianum, Phaius × amabilis, P. × maculato-grandifolius, × Epilælia radico-purpurata, E. × Eros, Leptolælia Veitchii, Angræcum × Veitchii, Miltonia × Bleuana splendens, Odontoglossum × excellens, Anguloa × intermedia, and many others.

"The following hybrids other than Orchids have also been obtained:—Echeveria glauca metallica, from E. secunda glauca and E. metallica; Veronica Purple Queen, from V. Hendersoni and V. Traversii; Escallonia × langleyensis, from E. philippinensis and E. macrantha sanguinea, a very valuable gain; Althæa Primrose Queen, from A. ficifolia and A. rosea; Hemerocallis × luteola, from H. Thunbergii and H. aurantiaca majus; Rose Queen Alexandra, from Crimson Rambler and Rosa multiflora simplex; Electra, from

R. multiflora simplex and W. A. Richardson; Myra, from Rosa Wichuriana and Crimson Rambler.

"In 1889 Seden was transferred to Langley, and since that time has devoted much of his attention to the improvement of hardy fruit. Amongst other varieties raised by Seden, most of the following have been distributed:—Strawberry Veitchs' Perfection, obtained from Waterloo and British Queen; Veitchs' Prolific, from Empress of India and British Queen; Lord Kitchener, from British Queen and Waterloo; The Khedive, from Lord Suffield and British Queen; President Loubet, from Waterloo and Lord Napier; the Alake, from Frogmore Late Pine and Veitchs' Perfection. Apple Langley Pippin, from Mr. Gladstone and Cox's Orange Pippin; Mrs. John Seden, from Transcendent Crab and King of Pippins; Mr. Leopold Rothschild, from John Downie and Cox's Orange Pippin; Middle Green, from Frogmore Prolific and Blenheim Pippin; Rev. W. Wilks, from Peasgood's Nonsuch and Ribston Pippin; Crab The Langley, from John Downie and King of Pippins; Veitchs' Scarlet, from Red Siberian Crab and King of Pippins; Bullace The Langley, from Damson Farleigh Prolific and Plum Black Orleans; The Mahdi, the product of a cross between the common Blackberry and Raspberry Belle de Fontenay; Gooseberry Langley Beauty, from the varieties Railway and Yellow Champagne; Langley Gage, from Pitmaston Green Gage and Telegraph; Golden Gem, from Whitesmith and Antagonist; Raspberry Yellow Superlative, from Superlative and Autumn Yellow; November Abundance, from Catawissa and Superlative; and Queen of England, from Superlative and Rubus laciniatus.

"In 1897 Seden was chosen as one of the original recipients of the Victoria Medal of Honour by the Royal Horticultural Society."

JOHN HEAL.

Born at Barnstaple in North Devon, this well-known cultivator and hybridizer commenced his gardening career in the Westacott nurseries near that town, and was trained in the usual routine work of a country nursery. In 1863 sent to Coombe Wood, he was after two years transferred to the houses at Chelsea.

His first charge at Chelsea was a large and representative collection of Ferns, but he afterwards had the care of the New Plant department, and, from 1873 onwards, of the houses devoted to Greenhouse Florists' Flowers.

Since 1873 Heal has devoted much attention to the improvement of various races of garden plants by hybridization and selection, with notable results. The Hippeastrum or Amaryllis has been greatly improved and a strain of high merit obtained, the Royal Horticultural and Royal Botanic Societies having awarded no less than 200 certificates to meritorious varieties.

The Streptocarpus hybrids, first produced by Mr. W. Watson, Curator of Kew, have been further improved, and entirely new forms and colours created, of which the achemeniflorus strain is a noteworthy example.

The gorgeous-flowered Phyllocacti, so much in favour with our forefathers, have also had attention, and by cross-fertilization and selection new colours and forms obtained.

Good work has also been done with the Greenhouse Rhododendrons, the fine modern varieties being derived from some seven species, natives of Java, Malaya and adjacent

RHODODENDRON BALSAMINÆFLORUM

islands, most of them introduced through Messrs. Veitchs' travellers Thomas Lobb and Curtis.

George Taylor, who had charge of these species prior to Heal's day, commenced their hybridizing, and raised several fine varieties; Heal continued the work and produced a large number of excellent forms remarkable for the pure, rich and varied colours of their flowers, the increased size of the individual blooms, and the large compact trusses in which they are produced.

A distinct race in this section of the genus was created by self-pollinating a flower the stamens of which showed a tendency to petaloidy.

The flowers produced by the varieties of this race, known as the balsaminæflorum hybrids, are double, of great substance, and have the same rich colours characteristic of the javanico-jasminiflorum hybrids.

The latest class of plants evolved by this skilful hybridizer are the winter-flowering Begonias. These valuable green-house plants were obtained by crossing varieties of the summer-flowering tuberous-rooted Begonias with Begonia socotrana, a species from the island of Socotra, discovered by Professor Balfour of Edinburgh. A dozen or more forms are now in cultivation, some with single, others with double or semi-double flowers of bright rose or rose-carmine shades of colour.

Useful work has also been done amongst the Clivias (Imantophyllums), Kalanchoes, Cinerarias, and fine-foliage Begonias.

The Veitch Medal was awarded to Heal in 1892 by the Veitch Memorial Trustees in recognition of his services to Horticulture, and in 1897 he received the Victoria Medal of Honour from the Royal Horticultural Society.

Reference:—Gard. Chron. 1892, vol. xi. p. 812, with portrait.

GEORGE TIVEY.

George Tivey has had unusual success in the hybridization of Nepenthes, and has for many years, and to an exceptional degree, thoroughly understood their culture.

The first notable result was Nepenthes × Chelsoni excellens, followed by N. × mixta, N. × Tiveyi, N. × Balfouriana, N. × cylindrica, N. × Sir William T. Thiselton-Dyer, N. × picturata, and the very bold, striking N. × F. W. Moore, all of which are improvements on the species or varieties from which they were derived.

ORCHID SPECIES

ORCHID SPECIES

A LIST OF THE PRINCIPAL ORCHID SPECIES INTRODUCED BY MESSRS. VEITCH

ACANTHOPHIPPIUM CURTISII, *Rchb. f.*

Rchb. in Gard. Chron. 1881, vol. xv. p. 169.

Introduced from the Malay Archipelago through Curtis, after whom it is named.

The five keels between the side laciniæ distinguish the species from Acanthophippium bicolor and A. sylhetense, to which it is closely allied.

AËRIDES FIELDINGI, *Lindl.*

Orchid Album, vii. t. 309; Veitchs' Man. Orch. Pl. pt. vii. p. 69.

Introduced through Thomas Lobb in 1850, and named in honour of Colonel Fielding, an officer in the Indian army. It is known as the "Fox Brush" Aërides, a name evidently given in reference to the appearance of the inflorescence.

AËRIDES FIELDINGI, *Lindl.*, var. WILLIAMSII.

Veitchs' Man. Orch. Pl. pt. vii. p. 69; Warner's Selec. Orch. i. t. 21.

This variety of the type was also introduced through Thomas Lobb, and is extremely rare.

AËRIDES JAPONICUM, *Lindl. & Rchb. f.*

Bot. Mag. t. 5798; l'Illus. Hort. 1883, t. 461; Veitchs' Man. Orch. Pl. pt. vii. p. 70.

Aërides japonicum was originally introduced from Japan by Mr. Linden of Brussels in 1862 and subsequently by ourselves. From a plant flowering at Chelsea in June 1869 the plate in the Botanical Magazine was prepared.

AËRIDES MULTIFLORUM, *Roxb.*, var. LOBBII, *Veitch.*

Syns. *A. Lobbii*, Hort. Veitch.

Veitchs' Man. Orch. Pl. pt. vii. p. 75, fig.; l'Illus. Hort. 1868, xv. t. 559.

This variety, the handsomest and most generally cultivated of all the multiflorum forms, was introduced from Moulmein through Thomas Lobb. As compared with the type the stem is shorter, the leaves much crowded, almost lying one upon the other, and the flowers, produced on longer peduncles, are more numerous and more richly coloured.

AËRIDES MULTIFLORUM, *Roxb.*, var. VEITCHII, *Morren.*

La Belg. Hort. 1881, p. 123; Les Orchidées, t. 4; Veitchs' Man. Orch. Pl. pt. vii. p. 75.

A form closely resembling the variety Lobbii, but with less crowded leaves, and flowers lighter in colour: the sepals and petals are white dotted with rose and the lip light rose-purple. It was introduced from Moulmein through Thomas Lobb with the variety which bears his name.

AËRIDES PACHYPHYLLUM, *Rchb. f.*

Gard. Chron. 1880, vol. xiv. p. 230.

A fine species imported from Burmah in a consignment of Aërides crassifolium, but now apparently lost to cultivation. The leaves are short, thick, fleshy and unequally bilobed at the apex; the flowers, in short racemes, are light crimson-lake, with white spur and column and lip painted with purple.

ANGRÆCUM CITRATUM, *Thouars.*

Bot. Mag. t. 5624; l'Illus. Hort. 1886, xxxiii. t. 592; Veitchs' Man. Orch. Pl. pt. vii. p. 125, fig.

First discovered by the French botanist Du Petit Thouars towards the end of the eighteenth century in Madagascar, but subsequently lost sight of until a plant, which we believe we obtained through Mr. Ellis, flowered at Chelsea in 1865.

At that time Angræcum citratum was exceedingly rare in British orchid collections, and continued to be so till the opening of the Suez Canal afforded facilities for the more rapid transmission of plants from Madagascar.

ANGRÆCUM FALCATUM, *Lindl.*

Lindl. Bot. Reg. t. 283; Bot. Mag. t. 2097; Veitchs' Man. Orch. Pl. pt. vii. p. 128; *id.* Catlg. of Pl. 1869, p. 23.

An unpretending little orchid of great botanical and horticultural interest from the fact that it was the first Angræcum cultivated in the glass-houses of Europe, and one of the earliest of the Japanese orchids ever introduced. It was first sent to this country about the year 1813 by Dr. Roxburgh, but probably lost to cultivation until we re-introduced it from Japan about the year 1868.

ANGRÆCUM HYALOIDES, *Rchb. f.*

Rchb. in Gard. Chron. 1880, vol. xiii. p. 264; l'Orchidophile, 1889, p. 347, col. pl.; Veitchs' Man. Orch. Pl. vii. p. 132.

Introduced in 1879 through Curtis, who discovered it in North-East

Madagascar growing on small shrubs, which form the undergrowth of the dense forest along the swampy coast.

The small white flowers of a delicate semi-transparent texture suggested the specific name, which is from the Greek, meaning "crystal."

ANGRÆCUM KOTSCHYI, *Rchb. f.*

Rchb. in Gard. Chron. 1880, vol. xiv. pp. 456 and 693, fig.; *id.* 1884, vol. xxii. p. 712; Veitchs' Man. Orch. Pl. pt. vii. p. 133; Bot. Mag. t. 7442.

Discovered in 1838 by Theodor Kotschy, after whom it is named, and subsequently met with by several other travellers.

In 1876 it was found by the German Hildebrandt on the coast of Zanzibar, and three years later living plants were sent by Sir John Kirk, the British Consul at Zanzibar, to Mr. Gerald Walker, from whom we acquired them. It flowered for the first time at Chelsea in the autumn of 1880.

ARACHNANTHE CATHCARTII, *Benth.*

Syns. *Vanda Cathcartii*, Lindl.; *Esmeralda Cathcartii*, Rchb. f.

Bot. Mag. t. 5845; Gard. Chron. 1870, p. 1409, with fig.; Fl. Mag. n.s. t. 66; Veitchs' Man. Orch. Pl. pt. vii. p. 7, fig. opposite p. 8.

A native of shady valleys in the Eastern Himalayas, where it was first detected by Sir Joseph Hooker, by whom it was sent to the Calcutta Botanic Gardens. Repeated attempts were made to introduce plants to England with more or less success, and it flowered the first time in this country in our houses in March 1870.

The species is dedicated to the memory of Mr. James F. Cathcart, of the Indian Civil Service, an ardent amateur naturalist, and one of the earliest explorers of the rich flora of the Eastern Himalayas.

ARACHNANTHE LOWII, *Benth.*

Syns. *Vanda Lowii*, Lindl.; *Renanthera Lowii*, Rchb. f.

Gard. Chron. 1858, p. 175; *id.* 1847, p. 239; Bot. Mag. t. 5475; Veitchs' Man. Orch. Pl. pt. vii. p. 12, fig.; Orchid Review, 1904, vol. xii. p. 283.

Discovered in Sarawak by Sir Hugh Low, in whose honour it was named by Dr. Lindley, but first flowered in this country from specimens sent by Thomas Lobb in 1858. It was later collected by Curtis in the low swampy forests near the coast of Sarawak.

The plant is remarkable for the enormous length of its racemes and for the occurrence of two kinds of flowers on the same inflorescence.

BRASSIA ARCUIGERA, *Rchb. f.*

Gard. Chron. 1869, p. 389.

A Peruvian orchid described by Professor Reichenbach from material introduced by us: an insignificant species of botanical interest only, and apparently not now in cultivation.

BRASSIA THYRSODES, *Rchb. f.*

Gard. Chron. 1868, p. 842.

Introduced from Peru, and apparently not now in cultivation. The flowers are greenish-yellow with purplish spots, borne in panicles.

BULBOPHYLLUM LOBBII, *Lindl.*

Lindl. Bot. Reg. 1847, sub t. 29; Bot. Mag. t. 4532; Veitchs' Man. Orch. Pl. pt. iii. p. 97.

Sent from Java by Thomas Lobb in 1846, and one of the few members of the genus that have flowers sufficiently showy to gain admission into the orchid collections of amateurs.

BULBOPHYLLUM MANDIBULARE, *Rchb. f.*

Rchb. in Gard. Chron. 1882, vol. xvii. p. 366; Gard. Chron. 1899, vol. xxvi. p. 293, fig. 99.

Introduced from North Borneo through F. W. Burbidge, and first flowered at Chelsea in 1882.

The flowers are curious, of botanical interest only. The sepals and petals are greenish-yellow striped with brown, and the curious lip has a mass of purple hairs on a pale ground.

BULBOPHYLLUM RETICULATUM, *Batem.*

Bot. Mag. t. 5605; Veitchs' Man. Orch. Pl. pt. iii. p. 97.

Discovered in North Borneo by Thomas Lobb, and through him introduced to Exeter about the year 1852.

Its handsome leaves and singular flowers render it a very interesting species. The former are pale-green reticulated with deep-green veins; the whitish flowers striped with red-purple are sometimes spotted.

CALANTHE CURTISII, *Rchb. f.*

Rchb. f. in Gard. Chron. 1884, vol. xxii. p. 262.

One of the many plants sent from the Sondaic regions by Curtis and not now in cultivation. The flower is interesting, but more curious than pretty.

CALANTHE LABROSA, *Rchb. f.*

Syns. *Limatodes labrosa*, Rchb. f.

Gard. Chron. 1883, vol. xix. p. 44; Veitchs' Man. Orch. Pl. pt. vi. p. 63; Gard. Chron. 1879, vol. xi. p. 202.

A fine species with rose-purple flowers, sent to us by a correspondent from Burmah, the precise locality being unknown.

It appears to be fast disappearing from cultivation, but is of importance as having produced a distinct race of hybrid Calanthes—Calanthe × porphyrea, C. × lentiginosa, and varieties.

ORCHID SPECIES

CALANTHE PLEIOCHROMA, *Rchb. f.*

Gard. Chron. 1871, p. 938; Veitchs' Man. Orch. Pl. pt. vi. p. 65.

Introduced from Japan and flowered for the first time at Chelsea in May 1871.

It is possibly only a geographical form of Calanthe sylvatica, a native of Mauritius and Bourbon, a species not now in cultivation in this country.

CALANTHE PROBOSCIDEA, *Rchb. f.*

Rchb. f. in Gard. Chron. 1884, vol. xxi. p. 476.

Imported from the Sunda Islands, this species is nearest to Calanthe furcata, Bat., but has a distinctive feature in that the anterior part of the column is bent down in a curve, like the snout of certain insects.

CALANTHE ROSEA, *Benth.*

Syns. *Limatodes rosea*, Lindl.

Veitchs' Man. Orch. Pl. pt. vi. p. 65; Lindl. in Paxt. Fl. Gdn. 1852, vol. iii. t. 81; Bot. Mag. t. 5312; Fl. des Serres, xxii. t. 2294.

First discovered by Thomas Lobb in Moulmein early in the year 1850, and sent by him to Exeter, where it flowered in the winter of that year. It has now been superseded in gardens by the beautiful Calanthe × Veitchii and other hybrids, in the parentage of which it participated largely to their benefit.

CALANTHE TEXTORI, *Miquel.*

Veitchs' Man. Orch. Pl. pt. vi. p. 67.

Introduced from Japan through Charles Maries in 1877, and probably only a form of the widely distributed Calanthe veratrifolia.

CALANTHE TRICARINATA, *Lindl.*

Veitchs' Man. Orch. Pl. pt. vi. p. 67.

First discovered by Wallich in Nepaul and later by Maximowicz in Japan, it was introduced from the latter country with Calanthe Textori in 1879. The absence of a spur to the flowers distinguishes it from the other members of the genus in cultivation.

CALANTHE VESTITA, *Wall.*

Bot. Mag. t. 4671; Paxt. Fl. Gdn. vol. i. p. 106, fig. 72; Veitchs' Man. Orch. Pl. pt. vi. p. 71, fig.; Fl. des Serres, 1858, tom. iii. 2me serie, p. 33; Paxt. Mag. Bot. vol. xvi. p. 129.

First introduced into European gardens by Dr. Kane, who sent the type species and a variety from Moulmein to Exeter in 1848.

Shortly after Thomas Lobb sent the same two forms, with another variety since named Turneri, from the same locality.

It has entered largely into the production of artificial hybrids, the best known being Calanthe × Veitchii, one of Dominy's earliest efforts, and one of the most largely cultivated of all hybrid orchids.

CALANTHE VESTITA, *Wall.*, var. TURNERI, *Veitch.*

Syns. *C. Turneri*, Hort.

Veitchs' Man. Orch. Pl. pt. vi. p. 71, fig.

A very charming variety introduced with the type from Moulmein through Thomas Lobb.

The flowers resemble those of the variety rubro-oculata, having a red-purple blotch on the disk of the lip, but they appear later in the season.

It was named in compliment to Mr. J. A. Turner of Pendlebury, near Manchester, one of the most ardent orchid amateurs of his time.

CATASETUM SACCATUM, *Lindl.*, var. PLICIFERUM, *Rchb. f.*

Gard. Chron. 1869, p. 1182.

Imported from Peru in 1869, but long since lost to cultivation. It is described by Reichenbach as having sepals and petals dull olive-green marbled with numerous brown spots and a green lip with numerous cinnamon-brown blotches.

CATTLEYA BOWRINGIANA, *Hort. Veitch.*

Syns. *C. autumnalis*, Hort.

Gard. Chròn. 1885, vol. xxiv. p. 683; Veitchs' Man. Orch. Pl. pt. ii. pp. 31, 32, with figs.; Veitchs' Catlg. of Pl. 1886, p. 10, figs. p. 3.

Sent to us in 1884 from British Honduras in Central America, by a correspondent who stated that the plant grows on cliffs by a rapid stream flowing over a succession of waterfalls, where the atmosphere is always highly charged with moisture. As a species Cattleya Bowringiana is close to C. Skinneri, but differs in its flowering season, which is during the dull months of October and November.

It is dedicated to the late Mr. J. C. Bowring of Forest Farm, near Windsor, for many years a well-known amateur.

CATTLEYA DOWIANA, *Batem.*

Syns. *C. Lawrenceana*, Warsc.; *C. labiata*, var. *Dowiana*, Veitch.

Gard. Chron. 1866, p. 922; Bot. Mag. t. 5618; Fl. des Serres, tt. 1709-1710; l'Illus. Hort. t. 525; The Garden, 1877, vol. xii. t. 99; Veitchs' Man. Orch. Pl. pt. ii. p. 16.

This superb orchid flowered for the first time in this country at Chelsea in the autumn of 1865. The plants were obtained through Mr. G. Ure-Skinner, whose collector Mr. Arce, a zealous naturalist, had obtained them in Costa Rica. Plants had previously been sent to this country in 1850, but arriving in a bad condition, had all died without flowering.

It was the wish of Warscewicz, the original discoverer, that his plant should bear the name Lawrenceana, in compliment to Mrs. Lawrence of Ealing, a generous patroness of Horticulture, but as his specimens miscarried, this fact was not made known until after Bateman had named it in compliment to Captain J. M. Dow of the American Packet Service, to whose kindness orchidists and men of science owe so much. It has proved potent as a parent for hybridizing, many fine seedlings now in cultivation being due to its influence.

CATTLEYA IRICOLOR, *Rchb. f.*

Gard. Chron. 1874, p. 162; Veitchs' Man. Orch. Pl. pt. ii. p. 40; Orchid Review, 1893, vol. i. p. 63.

The only known plant of this interesting orchid was obtained at one of the orchid sales at Stevens's Rooms, where it was sold without any specific name or intimation of its origin.

Flowered at Chelsea in 1874, the flowers proved cream-white in colour with W-shaped yellow markings, on the lip on either side of which are maroon-purple stripes.

CATTLEYA SKINNERI, *Batem.*, var. ALBA, *Rchb. f.*

Gard. Chron. 1877, vol. vii. p. 810; Veitchs' Man. Orch. Pl. pt. ii. p. 47.

Discovered in Costa Rica by Endres, and sent to us.

It has ever since its introduction been acknowledged one of the loveliest white orchids in cultivation; the pure white flowers, with a yellowish blotch on the disk of the lip, are produced in the same manner as those of the type.

CIRRHOPETALUM RETUSIUSCULUM, *Rchb. f.*

Syns. *Bulbophyllum retusiusculum*, Rchb. f.

Gard. Chron. 1869, p. 1182.

Introduced through Colonel Benson, by whom it was discovered in Moulmein. A botanical species not now in cultivation.

CIRRHOPETALUM ROBUSTUM, *Rolfe.*

Gard. Chron. 1895, vol. xvii. p. 771, fig. 116; Veitchs' Catlg. of Pl. 1896, p. 4, fig. reproduced.

This Cirrhopetalum, probably the largest species known, and one of the most remarkable yet introduced, was sent from New Guinea by David Burke.

The flowers are produced on a short scape, almost sub-umbellate. The sepals are yellowish-green, tinged with red in the centre, and have a varnished surface; the petals are small, brown-coloured, whilst the mobile lip is reddish chocolate.

CŒLOGYNE DAYANA, *Rchb. f.*

Rchb. in Gard. Chron. 1884, vol. xxi. p. 826; Veitchs' Man. Orch. Pl. pt. vi. p. 37; Orchid Album, t. 247.

Imported from Borneo through Curtis, and dedicated by Prof. Reichenbach, at our request, to Mr. John Day, of Tottenham.

It flowered for the first time in this country at Chelsea in 1884.

As a species it resembles Cœlogyne Massangeana in its long pendulous racemes, but in the colour of the flowers and more especially in the vegetative organs is abundantly distinct.

CŒLOGYNE FLAVIDA, *Hook. f.*

Veitchs' Man. Orch. Pl. pt. vi. p. 39.

Discovered by Thomas Lobb on the Khasia Hills, and afterwards by Cathcart on the Sikkim Himalayas.

Closely allied to Cœlogyne barbata and C. elata, it is inferior in a horticultural sense to both these species.

CŒLOGYNE (PLEIONE) HUMILIS, *Lindl.*

Bot. Mag. t. 5674; Paxt. Fl. Gdn. vol. ii. p. 65, t. 51; Gard. Chron. 1883, vol. xix. p. 46 (in W. B. Hemsley's List of Garden Orchids); Veitchs' Man. Orch. Pl. pt. vi. p. 55, figs.

Originally discovered by Dr. Buchanan Hamilton, and afterwards by Griffiths, this plant was first introduced into British Gardens in 1849 through Thomas Lobb, who found it at Sanahda on the Khasia Hills.

The lip is beautifully fringed and the flowers vary much in colour.

CŒLOGYNE (PLEIONE) LAGENARIA, *Lindl.*

Bot. Mag. t. 5370; Veitchs' Man. Orch. Pl. pt. vi. p. 56, fig.; Paxt. Fl. Gdn. 1851, vol. ii. t. 39.

Introduced through Thomas Lobb, who sent plants from the Khasia Hills, Northern India, to Exeter in 1849, and always a great favourite with orchid amateurs on account of its beautiful flowers, freely produced in October and November.

CŒLOGYNE LENTIGINOSA, *Lindl.*

Bot. Mag. t. 5958; Veitchs' Man. Orch. Pl. pt. vi. p. 43.

Introduced in 1847 through Thomas Lobb, who discovered it in Moulmein.

There are two forms, that figured in the Botanical Magazine above quoted being inferior in the beauty of its flowers. It has been imported at various times from the same locality.

CŒLOGYNE (PLEIONE) MACULATA, *Lindl.*

Gard. Chron. 1850, p. 710; Bot. Mag. t. 4691; Paxt. Fl. Gdn. ii. p. 5, t. 39, fig. 1.

Introduced from the Khasia Hills in 1849 by Thomas Lobb, who sent plants to Exeter. In 1852 it was sent from Assam to the Royal Gardens, Kew, by Simons, and from this plant was made the figure which appears in the Botanical Magazine. It was exhibited by us November 5th, 1850.

CŒLOGYNE (PLEIONE) MACULATA, var. ARTHURIANA, *Veitch.*

Syns. *C.* (*Pleione*) *Arthuriana*, Rchb.

Veitchs' Man. Orch. Pl. pt. vi. p. 57; Rchb. in Gard. Chron. 1881, vol. xv. p. 40.

The variety Arthuriana was sent in 1881 by a correspondent at Rangoon, and dedicated by Professor Reichenbach to the memory of the late Mr. Arthur Veitch.

CŒLOGYNE PELTASTES, *Rchb. f.*

Rchb. in Gard. Chron. 1880, vol. xiv. p. 296; Lindenia, vi. t. 258.

Introduced from Borneo, this remarkable Cœlogyne has peculiar pseudobulbs, produced at different levels on the rhizome; these are somewhat crescent-shaped, flattish, convex on one side and concave on the other, closely pressed against the surface over which they grow, forming reservoirs for water.

CŒLOGYNE (PLEIONE) POGONIOIDES, *Rolfe.*

Rolfe in Kew Bulletin, 1896, p. 196; Orchid Review, 1903, vol. xi. p. 291.

Introduced from the province of Hupeh, Central China, through Wilson, and cultivated at Chelsea, but not flowered at the present date.

The bulbs are used by the Chinese as a drug, under the name of "Pei-mu."

It is closely allied to Cœlogyne (Pleione) humilis, which it resembles in the flowering stage when the leaves are partly developed.

CŒLOGYNE (PLEIONE) REICHENBACHIANA, *Moore.*

T. Moore in Gard. Chron. 1868, p. 1210; Bot. Mag. t. 5753; Veitchs' Man. Orch. Pl. pt. vi. p. 59.

Discovered by Colonel Benson, of Rangoon, on the mountains of Arracan, and by him introduced to the Royal Gardens, Kew, and to Chelsea, but now very rarely seen in collections. In both places the plants flowered simultaneously for the first time in November 1868.

CŒLOGYNE SCHILLERIANA, *Rchb. f.*

Bot. Mag. t. 5072; Fl. des Serres, tom. xxii. t. 2302; Veitchs' Man. Orch. Pl. pt. vi. p. 49.

Introduced through Thomas Lobb from Moulmein in 1857, and dedicated by Professor Reichenbach to Consul Schiller of Hamburg, at that time one of the most prominent amateur orchidists in Europe.

CŒLOGYNE SPECIOSA, *Lindl.*

Syns. *C. salmonicolor*, Rchb. f.

Bot. Mag. t. 4889; Lindl. Bot. Reg. 1847, t. 23; Veitchs' Man. Orch. Pl. pt. vi. p. 50.

Imported from Java through its discoverer Thomas Lobb, and first flowered in 1846.

The hairs that fringe the crest of the lip are among the most beautiful microscopic objects possible.

CŒLOGYNE VEITCHII, *Rolfe.*

Rolfe in Kew Bulletin, November 1895, p. 282; Gard. Chron. 1895, vol. xvii. p. 248 (Report of R.H.S. Orchid Committee); Veitchs' Catlg. of Pl. 1896, p. 6, fig.; Bot. Mag. t. 7764.

A very distinct little species, introduced from Western New Guinea through David Burke.

The flowers, of the purest white, are produced in racemes almost as long as those of its near allies, the beautiful Cœlogyne Dayana and C. Massangeana.

CRYPTOPHORANTHUS GRACILENTUS, *Rolfe.*

Syns. *Masdevallia gracilenta*, Rchb. f.

Rolfe in Gard. Chron. 1887, vol. p. 693; *id.* Orchid Review, 1903, vol. xi. p. 304; Rchb. in Gard. Chron. 1875, vol. iv. p. 98.

One of the Costa Rica discoveries of M. Endres, a curious "window-bearing" species, the flowers of which are closed at the apex, the only access to the interior being by slits or "windows" at the sides.

CYCNOCHES PENTADACTYLON, *Lindl.*

Gard. Chron. 1842, p. 190; *id.* 1843, p. 319; Lindl. in Bot. Reg. 1843, vol. xxix. p. 18 (misc.); *id.* t. 22; Paxt. Fl. Gdn. iii. sub t. 75; Veitchs' Man. Orch. Pl. pt. ix. p. 143, figs.; Rolfe in Gard. Chron. 1889, vol. vi. p. 188, fig. 26; Gard. Mag. 1893, p. 77, with plate; Orchid Review, 1893, vol. i. p. 74.

Introduced from Rio de Janeiro to Exeter through William Lobb in 1841, this species produced a seven-flowered raceme in March 1842, from which material Dr. Lindley wrote his description.

The flowers of the two sexes differ much in size and appearance and offered considerable difficulty to the botanist before the phænomenon was thoroughly understood.

It is a handsome species, the flowers greenish-yellow, sometimes white, barred and blotted with chocolate brown; parts of the lips are white, spotted with red.

ORCHID SPECIES

CYMBIDIUM CANALICULATUM, *R. Br.*

Bot. Mag. t. 5851; Veitchs' Man. Orch. Pl. pt. ix. p. 12.

This species was first discovered by Robert Brown in the beginning of the last century, near Cape York, in North-East Australia, where in 1865 it was re-discovered by the late John Gould Veitch and by him introduced to our gardens.

It flowered at Chelsea for the first time in April 1870, and from this plant, the figure in the Botanical Magazine was prepared.

CYMBIDIUM GRANDIFLORUM, *Griff.*

Syns. *C. Hookerianum*, Rchb. f.

Gard. Chron. 1892, vol. xi. p. 267, with fig.; Veitchs' Man. Orch. Pl. pt. ix. p. 18; Rchb. in Gard. Chron. 1866, p. 7; Bot. Mag. t. 5574.

Introduced to Exeter through Thomas Lobb, the first plant flowering in 1866 at Chelsea.

Reichenbach at once described it, and named it in compliment to Sir Joseph Hooker, who had just succeeded his father as Director of the Royal Gardens, Kew. The plant had, however, previously been named by Griffith its discoverer, a fact overlooked by Reichenbach at the time.

CYMBIDIUM HUTTONI, *Hook. f.*

Bot. Mag. t. 5676; Orchid Review, 1900, vol. viii. p. 232; Gard. Chron. 1905, vol. xxxviii. p. 63, figs. 21, 22.

A rare and remarkable species sent from Java by the unfortunate Henry Hutton, in commemoration of whose labours and early death it is named. Apparently it was soon lost to cultivation, until re-imported to Kew, and flowered in the Gardens in 1900.

The flowers, densely spotted all over with dusky brown on a light yellow ground, are almost purple at the apex of the petals and lip.

CYMBIDIUM WILSONI, *Hort.*

Gard. Chron. 1904, vol. xxxv. p. 157, with fig.; The Garden, 1904, vol. lxv. p. 189, with fig.; Orchid Review, 1904, vol. xii. p. 79.

A remarkable species introduced from the province of Yunnan, South China, through E. H. Wilson in 1901, and first flowered at Chelsea in February 1904.

The species is allied to Cymbidium giganteum, but differs in being much smaller in all its parts.

The fragrant flowers have brownish-green sepals and petals obscurely marked with reddish dots at the base, the lip is cream-white with irregular reddish-brown blotches and markings, and the tip of the rostellum maroon purple.

CYPRIPEDIUM (PAPHIOPEDILUM) ARGUS, *Rchb. f.*

Rchb. f. in Gard. Chron. 1873, p. 608; *id.* 1874, p. 710; Fl. Mag. t. 220; Bot. Mag. t. 6175; La Belg. Hort. xxxii. (1882), p. 241; Veitchs' Man. Orch. Pl. pt. iv. p. 11.

Discovered by Gustav Wallis in 1872 in Luzon, one of the Philippine Islands, and introduced through him immediately afterwards, this Cypripedium flowered for the first time in Europe in April 1873.

It was named Argus by Professor Reichenbach in allusion to the warty eye-like spots on the petals, which form its most striking characteristic.

CYPRIPEDIUM (PAPHIOPEDILUM) BARBATUM, *Lindl.*

Lindl. Bot. Reg. 1842, t. 17; Bot. Mag. t. 4234; Veitchs' Man. Orch. Pl. pt. iv. p. 12, fig.

Discovered by Cuming in 1840 on Mount Ophir, near Malacca in the Malay Peninsula, and sent by him to Messrs. Loddiges of Hackney, with whom it first flowered.

Thomas Lobb collected it three years later in the same locality, and from his importation the plant became generally distributed.

CYPRIPEDIUM BOISSIERIANUM, *Rchb.*

Syns. *C.* (*Selenepedium*) *reticulatum*, Rchb.

Rchb. in Gard. Chron. 1887, vol. i. p. 143, fig.; *id.* vol. xviii. 1882, p. 520; Veitchs' Man. Orch. Pl. pt. iv. p. 57.

This first became known in British gardens through Walter Davis, who found it, unknown to himself at the time, with Cypripedium caudatum, near Muña in the Huanuco district of Peru in 1875-1876, in the same locality in which the latter had been collected by William Lobb in 1847. Presumably both species were found here by Ruiz and Pavon sixty years previously.

CYPRIPEDIUM (PHRAGMOPEDILUM) CARICINUM, *Lindl.*

Syns. *C. Pearcei*, Batem.

Lindl. in Paxt. Fl. Gdn. 1850, vol. i. t. 9; Bot. Mag. t. 5466; Fl. des Serres, tom. xvi. t. 1648; Veitchs' Man. Orch. Pl. pt. iv. p. 59, fig.

Introduced in 1863 through Pearce and flowered for the first time in this country at Chelsea in May of the following year. The specific name, from carex, "a sedge," is in allusion to the sedge-like leaves.

CYPRIPEDIUM (PHRAGMOPEDILUM) CAUDATUM, *Lindl.*

Lindl. in Paxt. Fl. Gdn. 1850-1851, i. p. 37, t. 9; Veitchs' Man. Orch. Pl. pt. iv. p. 60, fig.; Orchid Review, 1895, vol. iii. p. 355, frontispiece (the variety Wallisii).

Although previously known to science, Cypripedium caudatum remained unknown to Horticulture till introduced by William Lobb in 1847, from the Huanuco district of Peru, where thirty years afterwards it was collected by Davis, who at the same time sent the variety Wallisii. Pearce also sent a few plants to Chelsea in 1862, having met with them in the Caupolica district, on the Andes of Ecuador, at 5,000-6,000 ft. elevation.

CYPRIPEDIUM LAWRENCEANUM

ORCHID SPECIES

CYPRIPEDIUM (PAPHIOPEDILUM) CURTISII, *Rchb. f.*

Rchb. in Gard. Chron. 1883, vol. xx. p. 8; Orchid Album, iii. t. 122.

Discovered in Sumatra in 1882 and introduced by Curtis, whose name it bears. It inhabits the great mountain range that stretches almost through the entire length of the island, at elevations of 3,000-4,000 ft.

CYPRIPEDIUM (PAPHIOPEDILUM) HAYNALDIANUM, *Rchb.*

Gard. Chron. 1877, vol. vii. p. 272; Bot. Mag. t. 6296; Veitchs' Man. Orch. Pl. pt. iv. p. 28.

Introduced in 1893 from the Philippine Islands, through Gustav Wallis, who had discovered it at San Isidro near Manila, and dedicated to Cardinal Haynald, Archbishop of Kaloesa in Hungary.

CYPRIPEDIUM (PAPHIOPEDILUM) JAVANICUM, *Rwdt.*

Lindl. in Paxt. Fl. Gdn. 1850-1851, vol. i. p. 38; Fl. des Serres, 1851, vol. vii. t. 703; Veitchs' Man. Orch. Pl. pt. iv. p. 35.

First discovered by the Dutch botanist, Reinwardt, on the mountains of Eastern Java in 1826, though not introduced to European gardens till 1840, when Thomas Lobb sent plants to Exeter.

CYPRIPEDIUM (PAPHIOPEDILUM) LAWRENCEANUM, *Rchb. f.*

Rchb. in Gard. Chron. 1878, vol. ii. p. 748; *id.* 1880, vol. xiii. p. 780, fig. 134; Bot. Mag. t. 6432; Fl. des Serres, tom xxiii. t. 2372; l'Illus. Hort. 1883, vol. xxx. t. 478; Veitchs' Man. Orch. Pl. pt. iv. p. 36, fig. opposite.

Discovered by F. W. Burbidge in 1878 on the left bank of the Lawas River, near Meringit, North Borneo, at an altitude of 1,000-1,500 ft. above sea-level, growing in company with the dwarf palm, Pinanga Veitchii.

It was dedicated by Professor Reichenbach to Sir Trevor Lawrence, the President of the Royal Horticultural Society, and the owner of a very rare collection of orchids.

CYPRIPEDIUM (PHRAGMOPEDILUM) LINDLEYANUM, *Schomb.*

N. E. Brown in Gard. Chron. 1885, vol. xxiv. p. 262; Rchb. in Gard. Chron. 1886, vol. xxv. p. 680; Veitchs' Man. Orch. Pl. pt. iv. p. 64.

First discovered by Schomburgk, during his exploration of British Guiana, on the southern slopes of the Roraima mountain at 6,000 ft. elevation, it was re-discovered in 1881 by David Burke, who brought plants to Chelsea in the autumn of that year. None flowered till January 1886.

In the meantime it had been sent to Kew, where it produced its flowers for the first time in 1885.

CYPRIPEDIUM (PAPHIOPEDILUM) NIVEUM, *Rchb. f.*

Rchb. in Gard. Chron. 1869, p. 1038; *id.* 1883, vol. xix. p. 16, fig.; Bot. Mag. t. 5922; The Garden, 1876, vol. ix. t. 23; Veitchs' Man. Orch. Pl. pt. iv. p. 39, fig.; Orchid Review, 1903, vol. xi. p. 273, fig. 44.

The first appearance of Cypripedium niveum was a surprise. In 1868 we received from Moulmein a consignment of plants of a Cypripedium, supposed to be C. concolor, but which, on flowering in the spring of the following year, proved to be the very beautiful white species now known as C. niveum.

It is not a native of Moulmein, but of the Tambilan Islands, situate midway between Singapore and Sarawak and the Langkawi Islands, a few miles north of Penang, from which locality our plants were presumably obtained.

CYPRIPEDIUM (PAPHIOPEDILUM) PHILIPPINENSE, *Rchb.*

Syns. *C. lævigatum*, Batem.

Bot. Mag. t. 5508; Fl. des Serres, tt. 1760-1761; La Belg. Hort. 1867, t. 6; Gard. Chron. 1865, p. 914; Rev. Hort. Belge, 1881, p. 121; Veitchs' Man. Orch. Pl. pt. iv. p. 43, fig.

This species, discovered by the late John Gould Veitch in the Philippine Islands and by him sent to Chelsea in 1861, bloomed for the first time in March 1865.

It was found established on the roots of Vanda Batemanni, to obtain which was the object of the voyage, and for which the traveller long sought in vain, but once happily running the boat ashore in a bay of a small island, he was delighted and astonished to find the neighbouring rocks covered with the plant of which he was in quest.

CYPRIPEDIUM (PAPHIOPEDILUM) SUPERBIENS, *Rchb. f.*

Fl. des Serres, 1861, vol. xiv. p. 161, t. 1453; Veitchs' Man. Orch. Pl. pt. iv. p. 51, fig.

Only two plants of this species have ever been introduced, and all now growing in orchid collections have been derived from the two originally imported.

Messrs. Rollison introduced the first plant either from Java or Assam, and sold it in 1855 to Consul Schiller of Hamburg.

The second plant appeared in an importation of Cypripedium barbatum collected in 1857 by Thomas Lobb on Mount Ophir, near the southern extremity of the Malay Peninsula.

CYPRIPEDIUM TIBETICUM, *King.*

Orchid Review, 1905, vol. xiii. p. 194.

A hardy species introduced from Western China through Wilson.

The flowers resemble those of the Siberian Cypripedium macranthon but are larger; the sepals and petals have numerous broad blackish-purple lines on a greenish-yellow ground; the pouch is large, blackish-purple in front with a greenish area at the base and a purplish reticulation where the two colours meet.

Plants flowered for the first time at Coombe Wood in June 1905.

CYPRIPEDIUM (PAPHIOPEDILUM) TONSUM, *Rchb. f.*

Rchb. in Gard. Chron. 1883, vol. xx. p. 262; Veitchs' Man. Orch. Pl. pt. iv. p. 53.

Discovered in the mountains of Sumatra by Curtis, who collected it, mixed with Cypripedium Curtisii.

The specific name, tonsum, "shorn," refers to the absence of black marginal hairs that fringe the petals of closely allied species.

CYPRIPEDIUM (PAPHIOPEDILUM) VILLOSUM, *Lindl.*

Lindl. in Gard. Chron. 1854, p. 135; l'Illus. Hort. iv. (1857), pl. 126; Fl. des Serres, xiv. t. 1475; Veitchs' Man. Orch. Pl. pt. iv. p. 54; The Garden, 1891, vol. xxxix. p. 568, pl. 810.

First discovered by Thomas Lobb on the mountains near Moulmein at 4,000-5,000 ft. elevation, and introduced through him in 1853, Cypripedium villosum has proved to be one of the most potent of hybridizing agents, and has entered largely into the composition of some of the finest seedlings yet raised.

DENDROBIUM ACROBATICUM, *Rchb. f.*

Rchb. in Gard. Chron. 1871, p. 802.

Introduced from Moulmein, and named acrobaticum by Professor Reichenbach on account of the peculiar growth of the pseudo-bulbs, which reminded him of acrobatic contortions.

DENDROBIUM ALBOSANGUINEUM, *Lindl.*

Lindl. in Paxt. Fl. Gdn. vol. ii. t. 5; Bot. Mag. t. 5130; Fl. des Serres, tom. vii. p. 209; Veitchs' Man. Orch. Pl. pt. iii. p. 16.

Introduced in 1851 through Thomas Lobb, who found it on the hills near the Atran river. It occurs in several parts of Burmah, always on the tops of the highest trees. The flowers are white with a reddish maroon blotch on the sides of the lip.

DENDROBIUM AMETHYSTOGLOSSUM, *Rchb. f.*

Rchb. in Gard. Chron. 1872, p. 109; Bot. Mag. t. 5968; Veitchs' Man. Orch. Pl. pt. iii. p. 16.

Introduced in 1872 from the Philippines through Gustave Wallis, who sent a single plant mixed with Dendrobium taurinum. Though since imported in restricted numbers, it still remains a rare species in collections. The specific name is in allusion to the rich amethyst-purple of the lip.

DENDROBIUM ANNULIGERUM, *Rchb. f.*

Gard. Chron. 1871, p. 675; Gard. Chron. 1881, vol. xvi. p. 625 (W. B. Hemsley's List of Garden Orchids).

Introduced from Marisa. The limits of the internodes along the attenuated stems are marked with obscure bars, from which peculiarity the specific name was derived.

DENDROBIUM ANTELOPE, *Rchb. f.*

Rchb. in Gard. Chron. 1883, vol. xix. p. 656.

A curious species sent from the Moluccas by Curtis, and named antelope by Professor Reichenbach, from a fancied resemblance between the erect antenniform petals and the straight horns of such antelopes as Antelope Oreas.

DENDROBIUM ARACHNOSTACHYUM, *Rchb. f.*

Rchb. in Gard. Chron. 1877, vol. vii. p. 334; *id.* 1877, vol. viii. p. 38.

Sent to Chelsea by Peter C. M. Veitch. The flowers were thought by Professor Reichenbach to resemble green spiders; hence the specific name. Of botanical interest only, it does not now appear to be in cultivation.

DENDROBIUM ATRO-VIOLACEUM, *Rolfe.*

Rolfe in Gard. Chron. 1890, vol. vii. p. 512; *id.* 1894, vol. xv. p. 113, f. 12; Bot. Mag. t. 7371; Orchid Album, t. 444; Jour. of Hort. 1894, vol. xxviii. p. 65, f. 10; Orchid Review, 1895, vol. iii. p. 305, fig. 12.

Introduced from Eastern New Guinea, and flowered for the first time in Europe in April, 1890.

Sir Joseph Hooker writes of this species in the Botanical Magazine as follows:—

"Of all Dendrobes known to me I cannot recall amongst recent discoveries one so strikingly unlike its congeners in coloration, and at the same time so beautiful in this respect, as Dendrobium atro-violaceum."

The sepals and petals are primrose-yellow with numerous dusky brown spots, and the lip inside deep violet-purple, with a few paler radiating lines near the margin. Outside the lip is green, with a large dark violet irregular blotch on either side.

DENDROBIUM BELLATULUM, *Rolfe.*

Orchid Review, 1903, p. 103; *id.* 1904, vol. xii. p. 135; Gard. Chron. 1904, vol. xxxv. p. 258; Bot. Mag. t. 7985.

This beautiful little plant, much like a miniature Dendrobium formosum, was originally discovered by Dr. A. Henry in Yunnan, and afterwards introduced to cultivation through Wilson, who sent home living plants in

1900. The sepals and petals are white, and the front lobe of the lip reddish-orange.

It flowered for the first time in this country in the Royal Gardens, Kew

DENDROBIUM BENSONÆ, *Rchb. f.*

Bot. Mag. t. 5679; Fl. Mag. t. 355; Veitchs' Man. Orch. Pl. pt. iii. p. 22, with fig.; Orchid Review, 1903, vol. xi. p. 241, fig. 40.

Sent to us in 1866 from British Burmah by Colonel Benson, and named after Mrs. Benson at the Colonel's request; one of the best white-flowered Dendrobes in the section to which it belongs.

There are several varieties referable to this species, differing mainly in robustness of habit, size of flower, and lip-marking.

DENDROBIUM BIGIBBUM, *Lindl.*, var. SUPERBUM, *Rchb. f.*

Rchb. in Gard. Chron. 1878, vol. x. p. 748; Fl. Mag. n.s. pl. 229; Veitchs' Man. Orch. Pl. pt. iii. p. 23.

A variety, with larger flowers more brightly coloured than those of the type, discovered by the late John Gould Veitch, in 1865, on Mount Adolphus, near Torres Strait, and through him introduced. It flowered for the first time in December 1878.

At the same time the discoverer sent the first large importation of Dendrobium bigibbum ever received in this country.

DENDROBIUM BINOCULARE, *Rchb. f.*

Rchb. in Gard. Chron. 1869, p. 785; Veitchs' Man. Orch. Pl. pt. iii. p. 24.

Sent to us from British Burmah in 1868 by Colonel Benson, who found plants growing on hills eastward of Prome: it is now but rarely seen in British collections. The specific name refers to the two "eyes" or blotches on the labellum.

DENDROBIUM CANALICULATUM, *R. Br.*

Syns. *D. Tattonianum*, Batem.

Gard. Chron. 1865, p. 890; Bot. Mag. t. 5537; Veitchs' Man. Orch. Pl. pt. iii. p. 26.

Introduced in 1865 by the late John Gould Veitch, who discovered it at Endeavour Creek, York Peninsula, in North-East Australia, and very noticeable on account of its pseudo-bulbous stem, its deep-coloured lip, and the fragrance of its flowers.

DENDROBIUM CERINUM, *Rchb. f.*

Rchb. in Gard. Chron. 1879, vol. xii. p. 554.

Collected in the Malayan Archipelago and introduced to cultivation by F. W. Burbidge.

Flowered in July, 1879, for the first time it was described by Professor Reichenbach from material supplied by us.

The Professor says of it, "The lip is just alarming, it mimics that of Dendrobium sanguinolentum, but is oblong not three-lobed, and shows numerous minute teeth on its anterior edge. The whole flower is of very firm texture and shining as if made of wax."

It is not now in cultivation.

DENDROBIUM CRASSINODE, *Rchb. f.*

Rchb. in Gard. Chron. 1869, p. 164; Bot. Mag. t. 5766; Veitchs' Man. Orch. Pl. pt. iii. p. 31.

Sent to the Royal Gardens, Kew, and to Chelsea, by Colonel Benson in 1868, from the mountains of Arracan, near Moulmein, India, it flowered simultaneously in both establishments in January 1869, but had previously been made known to science by the Rev. C. Parish, who sent a sketch of it to Sir W. J. Hooker, of Kew, prepared from material obtained in the Siamese province of Kiong-Koung.

The specific name refers to the swollen joints on the pseudo-bulbs, by which this specimen can be readily distinguished.

DENDROBIUM CRETACEUM, *Lindl.*

Lindl. Bot. Reg. 1847, t. 62; Bot. Mag. t. 4686; Veitchs' Man. Orch. Pl. pt. iii. p. 33.

Sent to Exeter in 1846 by Thomas Lobb, who detected it in the Moulmein district.

The specific name, from creta, "chalk," refers to the colour of the flowers, which appear in May and June.

DENDROBIUM CRYSTALLINUM, *Rchb. f.*

Gard. Chron. 1868, p. 572; Bot. Mag. t. 6319; Veitchs' Man. Orch. Pl. pt. iii. p. 34.

Discovered on the Arracan Mountains near Tongu, in British Burmah, by Colonel Benson, through whom it was introduced, and flowered for the first time in Europe at Chelsea in the spring of 1868.

The specific name was given in allusion to the crystalline papillæ with which the anther case is covered.

DENDROBIUM CUMULATUM, *Lindl.*

Lindl. in Gard. Chron. 1855, p. 756; Rchb. *id.* 1868, p. 6; Bot. Mag. t. 5703; Veitchs' Man. Orch. Pl. pt. iii. p. 35.

Received at Kew and Chelsea from Moulmein through Colonel Benson in 1867.

It had previously appeared in the collection of Mr. F. Coventry, at Shirley, near Southampton, as early as 1855, but its native country was then unknown.

ORCHID SPECIES

DENDROBIUM GLOMERATUM, *Rolfe.*

Gard. Chron. 1894, vol. xv. p. 653, fig. 80; Orchid Review, 1894, vol. ii. p. 169.

A species imported from New Guinea, producing in small bunches from unpromising-looking pseudo-bulbs flowers of a warm rosy purple tint, with an orange-coloured labellum.

This species is without doubt the finest of the Pycnostachyate or "cluster-flowered" Dendrobes yet introduced.

DENDROBIUM GOULDII, *Rchb. f.*

Gard. Chron. 1867, p. 901.

One of the numerous Polynesian introductions of the late John Gould Veitch, in honour of whom it was named by Professor Reichenbach, but now unfortunately lost to cultivation.

DENDROBIUM HUTTONII, *Rchb. f.*

Rchb. in Gard. Chron. 1869, p. 686; Veitchs' Man. Orch. Pl. pt. iii. p. 49.

Discovered in Timor, one of the islands of the Malay Archipelago, by Henry Hutton in 1868, and an exceedingly rare, beautiful species with pure white sepals and petals, now seldom to be found.

DENDROBIUM INFUNDIBULUM, *Lindl.*, var. JAMESIANUM.

Syns. *D. Jamesianum*, Rchb. f.

Veitchs' Man. Orch. Pl. pt. iii. p. 50; Rchb. in Gard. Chron. 1869, p. 554; Fl. and Pom. 1869, p. 187.

Introduced through Colonel Benson, by whom it was discovered in British Burmah, and dedicated by Reichenbach as a distinct species to the late Mr. James Veitch junior.

The type species was collected by Thomas Lobb when travelling for us in British Burmah, but was not introduced on that occasion.

DENDROBIUM JOHANNIS, *Rchb. f.*

Gard. Chron. 1865, p. 890; Bot Mag. t. 5540.

A peculiar species having brown twisted sepals and petals, and a bright yellow lip, discovered in 1865 in North-East Australia by the late John Gould Veitch, who sent plants to Chelsea, where it flowered in August of that year.

DENDROBIUM JOHNSONIÆ, *F. Muell.*

Syns. *D. Macfarlanei*, Rchb. f.

The Garden, 1897, vol. li. p. 262, pl. 1113; Veitchs' Man. Orch. Pl. pt. iii. p. 59, fig.

Re-introduced in 1889 through the Rev. S. M. Macfarlane, who sent plants from New Guinea, in which country he laboured as a missionary.

DENDROBIUM KUHLII, *Lindl.*

Lindl. Bot. Reg. 1847, t. 47.

Introduced from Java by Thomas Lobb, but does not appear to have lived long in cultivation, or to have excited much interest.

DENDROBIUM LASIOGLOSSUM, *Rchb. f.*

Rchb. in Gard. Chron. 1868, p. 682; *id.* 1869, p. 277; Bot. Mag. t. 5825; Veitchs' Man. Orch. Pl. pt. iii. p. 52.

Discovered in the forests of Burmah by Colonel Benson, and sent to Chelsea and to the Royal Gardens, Kew, where it flowered for the first time in February 1868.

The specific name refers to the hairy lip or labellum.

DENDROBIUM LEUCOLOPHOTUM, *Rchb. f.*

Rchb. in Gard. Chron. 1883, vol. xviii. p. 552; Veitchs' Man. Orch. Pl. pt. iii. p. 53.

Introduced through Curtis from the Malay Archipelago, the precise locality being unknown.

Its chief value is the habit of flowering during November and December, when few other Dendrobes are in bloom.

The specific name is from the Greek, "a tuft of long white hair," and refers to the long one-sided racemes of white flowers, which bear a fancied resemblance to a horse's mane.

DENDROBIUM LINEALE, *Rolfe.*

Rolfe in Gard. Chron. 1889, vol. vi. p. 381.

A species introduced from New Guinea, and flowered for the first time in October 1889: the name lineale, "consisting of lines," was suggested by the numerous lines on the lips of the flower.

DENDROBIUM MACROPHYLLUM, *A. Rich.*, var. HUTTONI.

Veitchs' Catlg. of New Pl. for 1869, p. 24.

A white variety of the type, now rare in cultivation. The plants were sent from the Moluccas by Hutton, after whom it was named.

DENDROBIUM MACROPHYLLUM, *A. Rich.*, var. VEITCHIANUM.

Syns. *D. Veitchianum*, Lindl.

Bot. Mag. t. 5649; Lindl. in Bot. Reg. 1847, sub t. 25; Veitchs' Man. Orch. Pl. pt. iii. p. 60.

Sent to Exeter in 1846 by Thomas Lobb, who found it in the hottest jungles in the island of Java.

DENDROBIUM MESOCHLORUM, *Lindl.*

Lindl. in Bot. Reg. 1847, t. 36; Paxt. Fl. Gdn. vol. i. p. 63, fig. 43; Veitchs' Man. Orch. Pl. pt. iii. p. 61.

This very desirable plant bearing flowers with a violet-like perfume was introduced from India, through Thomas Lobb, who gave no locality.

DENDROBIUM MOOREI, *F. Muell.*

Rchb. in Gard. Chron. 1878, vol. x. p. 139; Veitchs' Man. Orch. Pl. pt. iii. p. 61.

A small white-flowered species sent to us in 1878 by Mr. Charles Moore, Director of the Botanic Gardens at Sydney, New South Wales, to whom it is dedicated. A native of Lord Howe's Island, it was discovered in 1869 by Mr. Fitzgerald, author of an illustrated work on the Australian Orchids.

DENDROBIUM PALPEBRÆ, *Lindl.*

Lindl. in Jour. Linn. Soc. x. p. 33 (1849); Paxt. Fl. Gdn. 1850-1851, i. p. 48; Veitchs' Man. Orch. Pl. pt. iii. p. 67.

Introduced in 1849 from Moulmein through Thomas Lobb.

The specific name Palpebræ, "eye-lids," refers to the fringe of long hairs like eye-lashes near the base of the lip. It is found sparingly in Burmah, varying in colour from white to dark rose.

DENDROBIUM PETRI, *Rchb. f.*

Rchb. in Gard. Chron. 1877, vol. vii. p. 107.

Introduced by Peter C. M. Veitch, who discovered it in Polynesia, in a locality not recorded, and after whom it was named by Dr. Reichenbach: probably now lost to cultivation.

DENDROBIUM POLYCARPUM, *Rchb. f.*

Rchb. f. in Gard. Chron. 1883, vol. xx. p. 492.

This species, introduced from the Sondaic area by Curtis, does not appear to be now in cultivation. The flowers are often self-fertilized, and an abundance of seed-capsules produced; hence the specific name.

DENDROBIUM PRÆCINCTUM, *Rchb. f.*

Rchb. in Gard. Chron. 1877, vol. iii. p. 750.

A small species of botanical interest only, introduced with an importation of Dendrobium Devonianum, and apparently not now in cultivation.

DENDROBIUM SUBCLAUSUM, *Rolfe.*

Rolfe in Kew Bulletin, October, 1894; Gard. Chron. 1895, vol. xviii. p. 655 (Report of R.H.S. Orchid Committee).

A brilliantly coloured and remarkable species introduced from the

Malay Archipelago, with bright cinnabar-orange-coloured flowers which opened for the first time in July 1894.

DENDROBIUM SUPERBIENS, *Rchb. f.*

Rchb. in Gard. Chron. 1876, vol. vi. p. 515; Veitchs' Man. Orch. Pl. pt. iii. p. 76.

One of the most striking of the Australian Dendrobes, a native of York Peninsula and some of the islands in Torres Strait, whence it was introduced by us in 1876, through the late Sir William MacArthur of Sydney, New South Wales.

DENDROBIUM SUPERBUM, var. ANOSUM, *Rchb. f.*

(a) *Hutton's var.*
(b) *Burke's var.*

Gard. Chron. 1884, vol. xxi. p. 306 (Burke's var.); Gard. Chron. 1869, p. 1206 (Hutton's var.); Veitchs' Man. Orch. Pl. pt. iii. p. 77.

The variety anosum was introduced to this country by the collector Cuming, and is remarkable for the almost entire absence of the rhubarb-like odour which characterizes the species.

Burke's variety is a very beautiful one, introduced in 1883 by the collector whose name it bears. The flowers are large, of a pure white colour, with the throat of the lip delicately pencilled with pale purple.

Hutton's variety was sent home in 1869 from one of the islands in the Malay Archipelago, and resembles Burke's variety, but the throat of the lip is deep purple.

DENDROBIUM TAURINUM, *Lindl.*, var. AMBOINENSE.

Orchid Review, 1897, vol. v. p. 304.

A form of the Philippine "Bull's Head" Dendrobe, introduced through David Burke from the island of Amboina, and first flowered at Chelsea in 1897.

The colour of the flowers differs from that of the type, the sepals being greenish-yellow suffused with bronzy brown; the petals are deep purple-brown, as are the side lobes of the lip, the front lobe more nearly resembling the sepals in colour.

DENDROBIUM TETRACHROMUM, *Rchb. f.*

Rchb. in Gard. Chron. 1880, vol. xiii. p. 712.

Introduced from Borneo through Curtis, and named tetrachromum by Professor Reichenbach from the "four colours" of its flowers.

DENDROBIUM TIPULIFERUM, *Rchb. f.*

Rchb. in Gard. Chron. 1877, vol. vii. p. 72.

A species, of botanical interest only, introduced from the Fiji Islands through Peter C. M. Veitch, and apparently lost to cultivation.

ORCHID SPECIES

DENDROBIUM TORTILE, *Lindl.*

Lindl. in Gard. Chron. 1847, p. 797, fig.; Bot. Mag. t. 4477; Veitchs' Man. Orch. Pl. pt. iii. p. 81; Orchid Review, 1900, vol. viii. p. 201, fig. 33.

Introduced through Thomas Lobb in 1847 from the Mergui district in Tenasserim, British Burmah.

The twisted sepals and petals of the flowers suggested the specific name. In colour they are rosy lilac, with a pale yellow lip blotched with purple at the base.

DENDROBIUM TRANSPARENS, *Wall.*

Paxt. Fl. Gdn. vol. i. t. 27; Bot. Mag. t. 4663; Veitchs' Man. Orch. Pl. pt. iii. p. 81.

Discovered by Dr. Wallich in the early part of the last century, but not introduced to European gardens till 1852, in which year Thomas Lobb sent plants to Exeter.

It was shortly afterwards sent to Kew, by Simons, from Assam.

DENDROBIUM XANTHOPHLEBIUM, *Lindl.*

Syns. *D. marginatum*, Batem.

Lindl. in Gard. Chron. 1857, p. 268; Bot. Mag. t. 5454; Veitchs' Man. Orch. Pl. pt. iii. p. 84.

Introduced from Moulmein through Thomas Lobb, and later by the Rev. C. Parish. It is now rarely seen in gardens.

EPIDENDRUM BICAMERATUM, *Rchb. f.*

Syns. *E. Karwinskyi*, Rchb. f.

Rchb. in Gard. Chron. 1869, p. 710; *id.* 1871, p. 1194; Veitchs' Man. Orch. Pl. pt. vi. p. 90.

A native of Mexico in the neighbourhood of Oaxaca, where it was first discovered by Karwinsky, and subsequently by Galcotti and others.

It was introduced in 1868, amongst an importation of Epidendrum vitellinum.

EPIDENDRUM CNEMIDOPHORUM, *Lindl.*

Gard. Chron. 1864, pp. 292 and 364; Bot. Mag. t. 5656; Veitchs' Man. Orch. Pl. pt. vi. p. 92.

Discovered in Guatemala by Mr. G. Ure-Skinner, who sent a few plants to some of his personal friends, and to Chelsea in 1864. The plant, rare in its native country, attains a height of 6 ft. or more, and the strongly-scented flowers are among the handsomest of the genus.

EPIDENDRUM CRINIFERUM, *Rchb. f.*

Rchb. in Gard. Chron. 1871, p. 1291; Bot. Mag. t. 6094; Veitchs' Man. Orch. Pl. pt. vi. p. 95.

Introduced through Endres, by whom it was discovered in Costa Rica in 1870. The specific name, from crinis, "a lock of hair," and ferre, "to bear," relates to the hair-like side lobes of the lip.

EPIDENDRUM LINDLEYANUM, *Rchb. f.*, var. CENTERÆ.

Syns. *Barkeria Lindleyana Centeræ*, Rchb.

Veitchs' Man. Orch. Pl. pt. vi. p. 105; Gard. Chron. 1873, p. 1597; The Garden, 1885, vol. xxvii. p. 396, pl. 490.

Introduced from Costa Rica through Endres in 1873, and dedicated to Mrs. Center, wife of the then superintendent of the Panama Railway.

The flowers are larger than those of the type, purplish-lilac in colour, with a number of large blotches about the column.

EPIDENDRUM PHYSODES, *Rchb. f.*

Gard. Chron. 1873, p. 289.

Sent from Costa Rica by Zahn.

A small-flowered species with whitish-brown flowers, of botanical interest only, and apparently not now cultivated.

EPIDENDRUM PSEUDEPIDENDRUM, *Rchb. f.*

Rchb. Xen. Orch. i. p. 160, t. 53; Rchb. in Gard. Chron. 1872, p. 763; Bot. Mag. t. 5929; Veitchs' Man. Orch. Pl. pt. vi. p. 113, figs.

Specimens of this species were orginally collected by Warscewicz, who presented them to Professor Reichenbach, and from this material the description and plate in the Xenia were prepared. For twenty years no more was heard of the plant till Zahn sent home specimens, which flowered for the first time at Chelsea, in July 1871.

The colouring of the flowers is peculiar, the sepals and petals being bright frog-green, and the labellum bright scarlet.

EPIDENDRUM SYRINGOTHYRSIS, *Rchb. f.*

Bot. Mag. t. 6145; Veitchs' Man. Orch. Pl. pt. vi. p. 121.

Introduced to Chelsea by Pearce in 1868, and flowered for the first time in May of the following year. It is a native of Bolivia, and had previously been gathered in the neighbourhood of Sorata, and also in the Andean valley of Challasuya by Mandon, who sent specimens to Professor Reichenbach.

The great size of the dense-flowered raceme, and its general resemblance in form and colour to the Lilac, suggested the specific name.

EPIDENDRUM THROMBODES, *Rchb. f.*

Linnea, xli. p. 79; Gard. Chron. 1883, vol. xx. p. 606 (W. B. Hemsley's List of Garden Orchids).

A species allied to Epidendrum aromaticum with yellow flowers blotched with purple-brown introduced from Peru in 1883. Of botanical interest only, it is now lost to cultivation.

EPIDENDRUM WALLISII, *Rchb. f.*

Rchb. in Gard. Chron. 1875, p. 66; *id.* 1878, p. 462; Veitchs' Man. Orch. Pl. pt. vi. p. 126.

Introduced through Gustav Wallis (after whom it is named) in 1874 from New Grenada, where it grows in light situations at an elevation of from 4,000-7,000 ft.

It is of value horticulturally on account of the flowers which are large for species of this genus, and continue to be produced almost throughout the year. They measure some 2 in. across, are of a yellow colour with spots of purple, and streaks of the last-named colour are prominent on the spreading whitish lip. Epidendrum Wallisii is also the parent of several very interesting hybrids, equally valuable for the long period over which they remain in bloom.

ERIA CURTISII, *Rchb. f.*

Rchb. in Gard. Chron. 1880, vol. xiv. p. 685.

Sent from Borneo by Curtis. The flowers are yellowish white, equal to those of Eria Ibera and similar species; the oblong bracts are unusually developed, in the way of those of E. bractescens, Lindl. It does not appear to be now in cultivation.

ERIA IGNEA, *Rchb. f.*

Rchb. in Gard. Chron. 1881, vol. xv. p. 782.

Imported from Borneo. The flowers are chrome-yellow, but the large bracts are almost vermilion in colour, from which feature the plant takes its specific name.

GALEANDRA BARBATA, *Lem.*

Lem. in l'Illus. Hort. iii. pp. 86, 89; *id.* vol. vii. t. 248; Rolfe in Gard. Chron. 1892, vol. xii. p. 431.

An Amazonian species introduced about the year 1856 and flowered shortly afterwards.

Its specific name is derived from the beard or tuft of hair on the disc of the lip.

GOODYERA (GEORCHIS) MACRANTHA, *Maxim.*

Fl. des Serres, 1867-68, tom. vii. p. 113; Gard. Chron. 1867, p. 1022, fig.

Brought to this country by the late John Gould Veitch on his return from Japan. The foliage is prettily marked as in certain Anœctochili, and renders the plant worthy of cultivation for that feature alone.

HOULLETIA BROCKLEHURSTIANA, *Lindl.*

Bot. Mag. t. 4072; Paxt. Mag. Bot. ix. p. 49; Veitchs' Man. Orch. Pl. pt. ix. p. 121.

This species first flowered in the collection of Mr. Brocklehurst of the Firs, near Macclesfield, in the year 1841, but remained scarce in gardens until William Lobb detected it on the Organ Mountains, and sent plants to Exeter in 1842. It is now found in many collections.

LÆLIA ANCEPS, *Lindl.*, var. VEITCHIANA.

Gard. Chron. 1883, vol. xix. p. 274; The Garden, 1884, vol. xxv. p. 534, pl. 446, fig. 7; Orchid Review, 1896, vol. iv. p. 53.

A light form found in an importation of the type. The sepals and petals are white with sometimes a faint tint of rose, and the largely developed lobes of the lip are of a soft lilac tint delicately pencilled with purple.

LIPARIS FORMOSANA, *Rchb. f.*

Rchb. in Gard. Chron. 1880, vol. xiii. p. 394.

Discovered in Formosa by Charles Maries, by whom seeds were sent to England, and flowered for the first time in March 1880.

The sepals and petals of the flowers are very light purple with green borders; the lip dark brown with a similar green border, and the peduncle a beautiful purple.

LYCASTE LASIOGLOSSA, *Rchb. f.*

Rchb. f. in Gard. Chron. 1872, p. 215; Bot. Mag. t. 6251; Veitchs' Man. Orch. Pl. pt. ix. p. 90.

Introduced from Guatemala in 1871, this most interesting member of the genus has flowers somewhat dull in colour, differing from all others in its shaggy lip, resembling the Paphinias, which suggested the specific name.

LYCASTE LINGUELLA, *Rchb. f.*

Rchb. in Gard. Chron. 1871, p. 738; Bot. Mag. t. 6303.

Introduced from Peru about the year 1870; a green-flowered species remarkable for the structure of the lip, which protrudes from the centre of the flower in the form of a little tongue.

LYCASTE LOCUSTA, *Rchb. f.*

Rchb. in Gard. Chron. 1879, vol. xi. p. 524; Orchid Review, 1898, p. 136; Bot. Mag. t. 8020.

An interesting species remarkable for its dull green flowers and the white fringe which extends all round the front of the lip.

Found by Walter Davis in Peru, it flowered for the first time at Chelsea in 1879. Subsequently apparently lost sight of, it has during recent years been re-introduced, and is now cultivated in Botanic Gardens on account of its very curious flowers.

LYCASTE SKINNERI, *Lindl.*, var. SUPERBA.

Fl. Mag. 1861, pl. 24.

A superb variety sent to us with several other distinct forms by Mr. G. Ure-Skinner from Guatemala, and flowered for the first time in April 1860.

MASDEVALLIA ATTENUATA, *Rchb. f.*

Rchb. f. in Gard. Chron. 1871, p. 834; Bot. Mag. t. 6273.

A species from Costa Rica of little horticultural value, but of interest botanically as a white-flowered species, a rather rare occurrence in this genus. The sepals are elongated into long yellow tails.

MASDEVALLIA BARLÆANA, *Rchb. f.*

Rchb. in Gard. Chron. 1876, vol. v. p. 170; Veitchs' Man. Orch. Pl. pt. v. p. 25.

Discovered by Walter Davis on the Andes of Peru, near Cuzco, and introduced in 1875.

It was dedicated to Senhor J. B. Barla, at that time Brazilian Consul at Nice, well known for his orchidologic works, as well as for his special knowledge of the Floras of Liguria and Sardinia.

MASDEVALLIA COCCINEA, *Lindl.*, var. HARRYANA.

Syns. *M. Harryana*, Rchb. f.; *M. Lindenii*, André, var. *Harryana*.

Veitchs' Man. Orch. Pl. pt. v. p. 34, fig. opposite; Rchb. in Gard. Chron. 1871, p. 1421; Fl. Mag. 1871, t. 555; Fl. and Pom. (1873) p. 169; La Belg. Hort. 1873, t. 21; Fl. des Serres, tom. xxi. t. 2250; l'Illus. Hort. 1873, p. 167, t. 142; Bot. Mag. t. 5990.

There are many varieties of the typical and very fine Masdevallia coccinea to which the one under notice is far superior from a horticultural point of view.

It was discovered by Chesterton in 1871 on the eastern side of the Cordillera near Sogamosa, where it has a vertical range of from 7,000-10,000 ft.

The flowers are extremely variable in colour, almost every shade, from deep crimson-purple, through magenta-crimson, crimson-scarlet, orange, yellow to cream-white being represented; the lighter shades of yellow are the rarest.

MASDEVALLIA DAVISII, *Rchb. f.*

Rchb. in Gard. Chron. vol. ii. pp. 710, 711; Bot. Mag. t. 6190; Veitchs' Man. Orch. Pl. pt. v. p. 38, figs.

This species, remarkable for its size and the colour of its flowers, was discovered by Walter Davis near the City of Cuzco on the eastern Cordillera of Peru, and flowered for the first time in this country in August 1874. It is found in the crevices of rocks on the slopes of the mountains at an immense elevation, probably not less than 10,500-12,000 ft., but within a restricted area, extending a few miles only along the flanks of the mountains, and within the vertical limits above mentioned.

MASDEVALLIA GARGANTUA, *Rchb. f.*

Rchb. in Gard. Chron. 1876, vol. vi. p. 516; Veitchs' Man. Orch. Pl. pt. v. p. 43.

Introduced in 1874 from the Frontino district in New Grenada through Gustav Wallis. When first expanded the flower emits a strong fetid odour. It is closely allied to Masdevallia elephanticeps, and by some authorities considered only a form of that species.

MASDEVALLIA IONOCHARIS, *Rchb. f.*

Rchb. in Gard. Chron. 1875, vol. iv. p. 388; Bot. Mag. t. 6262; Veitchs' Man. Orch. Pl. pt. v. p. 48.

A pretty free-flowering species introduced by us in 1874 from Peru through Walter Davis, who discovered it in the Andean valley of Sandia, in the province of Caravaya, at 9,000-10,000 ft. elevation. The flowers are white blotched with violet-purple, the "tails" yellow, spreading, and slender.

MASDEVALLIA LATA, *Rchb. f.*

Rchb. in Gard. Chron. 1877, vol. vii. p. 653.

A two-flowered species with dark reddish-brown sepals and yellowish tails, introduced from Central America through Zahn, but not now in cultivation in this country. The name lata was given by Professor Reichenbach on account of the broad basis of the sepals.

MASDEVALLIA MACRODACTYLA, *Rchb. f.*

Rchb. in Gard. Chron. 1872, p. 571.

Imported from New Grenada, and flowered in April 1872. The flowers are small, greenish-yellow with purple markings, and there are two brown nerves on the petals. The tails are long and the peduncles warty.

MASDEVALLIA PERISTERIA, *Rchb. f.*

Rchb. in Gard. Chron. 1874, p. 500; Bot. Mag. t. 6159; Fl. des Serres, tom. xxii. t. 2346; l'Illus. Hort. s. 3, t. 327; Veitchs' Man. Orch. Pl. pt. v. p. 57.

One of the handsomest coriaceous Masdevallias, introduced from New Grenada in 1873 through Gustav Wallis, who met with it in the province of Antioquia.

It derives its specific name Peristeria from the resemblance of its column and petals to the same organs in the Dove Plant—Peristeria elata.

The labellum is singularly coloured, and covered with numerous amethystine papillæ.

MASDEVALLIA POLYSTICTA, *Rchb. f.*

Rchb. in Gard. Chron. 1874, vol. i. p. 338, 290; *id.* 1875, vol. iii. p. 656, fig.; Bot. Mag. t. 6368; Veitchs' Man. Orch. Pl. pt. v. p. 58.

This species, flowered for the first time in England at Chelsea in the spring of 1875, is a native of Peru, and is said by Reichenbach to have been imported thence to the Botanic Gardens of Zurich by Mr. Ortiges.

The specific name, meaning "much spotted," refers to the spotted perianth.

MASDEVALLIA RADIOSA, *Rchb. f.*

Rchb. in Gard. Chron. 1877, vol. vii. p. 684; Veitchs' Man. Orch. Pl. pt. v. p. 59, fig. opposite.

Introduced from New Grenada in 1873-1874 through Gustav Wallis, by whom it was discovered near Frontino at an elevation of 8,000 ft. The flower is remarkable, tawny yellow in colour, densely spotted with blackish-purple, while the tails, 2 to 3 in. in length, are dull blackish-purple, paler towards the tips.

MASDEVALLIA REICHENBACHIANA, *Endres.*

Syns. *M. Normani*, Hort. Norman.

Rchb. in Gard. Chron. 1875, vol. iv. p. 257; *id.* 1881, vol. xvi. p. 230; Lindenia, vi. t. 250; Veitchs' Man. Orch. Pl. pt. v. p. 60.

Introduced from Costa Rica through Endres in 1873 and named at his request in compliment to Professor Reichenbach of Hamburg.

The nodding flowers are purple-brown above, yellowish-white beneath, and the three yellow tails reflex.

MASDEVALLIA SIMULA, *Rchb. f.*

Rchb. in Gard. Chron. 1875, p. 8; Veitchs' Man. Orch. Pl. pt. v. p. 63.

Introduced in 1874 from New Grenada through Chesterton.

It is a minute cæspitose plant, with gem-like flowers not more than ½ in. in diameter, but of surprising beauty when closely examined or seen through a magnifying glass.

MASDEVALLIA TRIARISTELLA, *Rchb. f.*

Rchb. in Gard. Chron. 1876, vol. vi. pp. 226, 559, fig.; Bot. Mag. t. 6268; Veitchs' Man. Orch. Pl. pt. v. p. 66.

Introduced through Endres, by whom it was discovered in Costa Rica in 1875.

It is the type species of Reichenbach's section of the genus, called Triaristellæ, which approach in their structure the genus Restrepia.

They are of value as botanical curiosities.

MASDEVALLIA VEITCHIANA, *Rchb. f.*

Rchb. in Gard. Chron. 1868, p. 814; Bot. Mag. t. 5739; Fl. des Serres, tom. xvii. t. 1803; Fl. Mag. t. 481; Fl. and Pom. 1873, p. 169, fig. 1; Veitchs' Man. Orch. Pl. pt. v. p. 67, fig.

Masdevallia Veitchiana was discovered in the lofty Andes of Peru by Pearce in 1866, and successfully introduced by him.

A few years later it was re-discovered in the same locality by Walter Davis, who states that it grows in the crevices and hollows of the rocks with but little soil, at an altitude of 11,000-13,000 ft.

It is a variable plant, the flowers differing in size, colour, and in the manner in which the papillæ is spread over the inner surface of the sepals. A large-flowered form, grandiflora, may be distinguished by having the upper sepal densely and uniformly covered with purple papillæ, while in the lateral two this covering is confined entirely to the outer half, the inner being of the purest orange-scarlet and destitute of papillæ.

MAXILLARIA CTENOSTACHYA, *Rchb. f.*

Rchb. in Gard. Chron. 1870, p. 39.

Imported from Costa Rica, but not now in cultivation. The tails of the yellow flowers are so long as to resemble the Brassias.

MILTONIA ENDRESII, *Nicholson.*

Syns. *Odontoglossum Warscewiczii*, Rchb. f.

Nich. Dict. Gard. vol. ii. p. 368; Rchb. in Gard. Chron. 1875, p. 270; Bot. Mag. t. 6163; Veitchs' Man. Orch. Pl. pt. viii. p. 101.

Originally discovered by Warscewicz in 1849, in restricted numbers, growing in only two localities on leguminous trees, it was twenty-two years later re-discovered by M. Linden's collector, Wallis, who tried unsuccessfully to introduce it. In 1873 it was found by Endres in Central America, and through him, after several attempts, we succeeded in introducing it. The first flowers were produced by a plant at Chelsea in 1875.

MILTONIA VEXILLARIA, *Benth.*

Syns. *Odontoglossum vexillarium*, Rchb.

Nich. Dict. Gard. vol. ii. p. 369; Rchb. in Gard. Chron. 1867, p. 901; *id.* 1872, p. 667; *id.* 1873, pp. 580, 644, figs.; Bot. Mag. t. 6037; Fl. Mag. n.s. t. 73; l'Illus. Hort. xx. t. 113; Rev. Hort. 1876, p. 390; La Belg. Hort. xxx. p. 257; Fl. des Serres, xx. t. 2058; Veitchs' Man. Orch. Pl. pt. viii. p. 110.

This well-known orchid was probably first discovered by the unfortunate Bowman, when collecting in New Grenada.

Subsequently found by Wallis, and again later by W. Roezl; both sent home plants which arrived dead or in a dying condition. With scanty information Henry Chesterton undertook, at our request, to

MASDEVALLIA VEITCHIANA

endeavour to bring a consignment home, in which he succeeded, and the first flowers opened at Chelsea in June 1873.

There are several natural varieties, and many in cultivation, that differ from the species in colour only.

MORMODES FRACTIFLEXUM, *Rchb. f.*

Rchb. f. in Gard. Chron. 1872, p. 141.

Imported from Costa Rica.

Professor Reichenbach says of it (*loc. cit.*), "It would be Mormodes Buccinator if only it had a strict and compact raceme." The sepals and petals are of a whitish-green with purplish streaks and dots, the lip white with radiating purple streaks.

MORMODES OCANÆ, *Rchb. f.*

Rchb. f. in Gard. Chron. 1879, vol. xii. pp. 582 and 178, figs. 133, 134; Bot. Mag. t. 6496; Veitchs' Man. Orch. Pl. pt. ix. p. 137, fig. p. 138.

Originally discovered by Schlim on the Eastern Cordillera of Colombia, near Ocaña. Professor Reichenbach described it in Walper's "Annales Botanices," from his specimens.

Subsequently gathered by Kalbreyer in the same region, it was successfully introduced, flowering for the first time in October 1879.

The flowers are of a peculiar shape, orange-yellow in colour, closely speckled with red-brown spots.

MORMODES SKINNERI, *Rchb. f.*

Gard. Chron. 1869, p. 50.

An interesting species obtained from Central America through the late Mr. G. Ure-Skinner, to whose memory it is dedicated.

The sepals and petals are honey-coloured with five longitudinal bars or lines of a dragon's blood colour. The lip is deep yellow with red spots and white hairs.

MORMODES WENDLANDI, *Rchb. f.*

Veitchs' Catlg. of Pl. 1881, p. 22.

A singular orchid, native of New Grenada, introduced through Kalbreyer.

The flowers are delightfully fragrant, and bright yellow in colour. The lip is peculiar and not inaptly described as resembling a cocked hat; the column has the characteristic twist of all species of Mormodes.

NASONIA PUNCTATA, *Hook. f.*

Bot. Mag. t. 5718.

A curious little orchid of botanical interest only, originally discovered by Hartweg in the mountains of El Sisme in Peru, and flowered for the first time at Chelsea in April 1868.

NOTYLIA ALBIDA, *Klotzsch.*

Rchb. in Gard. Chron. 1870, p. 987; Bot. Mag. t. 6311.

A native of Central America, first introduced to this country by Warscewicz, who sent plants to the gardens of the Horticultural Society.

Re-imported by us, it flowered for the first time in April 1872, dense racemes, producing white flowers some 6 in. or more in length.

ODONTOGLOSSUM BAPHICANTHUM, *Rchb. f.*

Syns. *O. odoratum*, var. *baphicanthum*, Veitch.

Gard. Chron. 1876, vol. vi. p. 260; *id.* 1883, vol. xix. p. 310; Veitchs' Man. Orch. Pl. pt. i. p. 55.

This originally appeared in one of our importations from New Grenada, and is probably of hybrid origin, and possibly a natural hybrid between Odontoglossum crispum and O. gloriosum.

ODONTOGLOSSUM BLANDUM, *Rchb. f.*

Rchb. in Gard. Chron. 1870, p. 1342; Veitchs' Man. Orch. Pl. pt. i. p. 14, fig. opposite.

Odontoglossum blandum was first discovered by Mr. Blunt in 1863-1865 in New Grenada, but the plants perished during transmission to Europe, and several subsequent consignments met with a similar fate. Its first appearance in a living state was at Stevens's Rooms, where plants were purchased by the Royal Horticultural Society, which flowered at Chiswick in 1871. It was still very scarce till Kalbreyer sent us a moderate importation of plants in 1879; it is still uncommon in collections.

ODONTOGLOSSUM BRACHYPTERUM, *Rchb. f.*

Rchb. f. in Gard. Chron. 1882, vol. xviii. p. 552; Veitchs' Man. Orch. Pl. pt. i. p. 71.

Believed to be a natural hybrid of which Odontoglossum nobile or O. Pescatorei is one, and O. luteo-purpureum the other parent, and sent by Kalbreyer from New Grenada.

ODONTOGLOSSUM CORONARIUM, *Lindl.*, var. DAYANUM.

Rchb. in Gard. Chron. 1876, vol. vi. p. 226.

Introduced from Peru. The sepals and petals are yellow, marbled with brown, and are distinct from the type not only in colour but in having three conical acute warts each side of the crest. Flowered for the first time in September 1875 with Mr. Day of Tottenham.

ORCHID SPECIES

ODONTOGLOSSUM DELTOGLOSSUM, *Rchb. f.*

Syns. *O. odoratum*, var. *deltoglossum*, Hort. Veitch.

Rchb. in Gard. Chron. 1881, vol. xv. p. 202; Veitchs' Man. Orch. Pl. pt. i. p. 56.

This supposed natural hybrid between Odontoglossom crispum and O. gloriosum appeared in an importation of these two species.

It is distinguished from O. Andersonianum, of similar origin, by a more deltoid lip, and the floral segments are sulphur-yellow blotched with brown.

ODONTOGLOSSUM DENISONÆ, *Hort.*, var. CHESTERTONII.

Syns. *O. crispum*, var. *Chestertonii*, Rchb. f.

Rolfe in Orchid Review, 1899, vol. vii. p. 361, fig. 18; Gard. Chron. 1876, p. 374; Veitchs' Man. Orch. Pl. pt. i. p. 26.

A beautiful Odontoglossum introduced through Chesterton, whose name it bears. It was formerly considered a variety of O. crispum, but is now thought to be a natural hybrid between O. luteopurpureum and O. crispum, the flowers being exactly intermediate in shape between those of the two species named.

ODONTOGLOSSUM KRAMERI, *Rchb. f.*

Rchb. in Gard. Chron. 1868, p. 98, fig.; Fl. Mag. t. 406; Bot. Mag. t. 5778; Fl. des Serres, t. 2469; l'Illus. Hort. t. 562.

A native of Costa Rica, where it was discovered by Carl Kramer, and introduced through him in 1868.

A rare plant, said to be restricted to a single locality on the mountain slopes near the Pacific Coast, it is now well nigh exterminated owing to the destruction of the forests for agricultural purposes.

ODONTOGLOSSUM LEEANUM, *Rchb. f.*

Syns. *O. odoratum*, var. *Leeanum*, Kent.

Rchb. in Gard. Chron. 1882, vol. xvii. p. 525.

This variety appeared in an importation from Columbia, and was named in compliment to Mr. W. Lee of Leatherhead, a leading amateur of orchids in his day.

It is supposed to be a natural hybrid between Odontoglossum gloriosum and O. crispum.

ODONTOGLOSSUM CORADINEI, *Rchb. f.*

Syns. *O. Lindleyanum*, var. *Coradinei*, Hort. Veitch.

Rchb. in Gard. Chron. 1872, p. 1068, fig.; Orchid Album, ii. t. 90; Gard. Chron. 1896, vol. xix. p. 233 (Rolfe on Natural Hybrids); Veitchs' Man. Orch. Pl. pt. i. p. 43.

Introduced in an importation of Odontoglossum crispum in 1872.

Reichenbach, in naming the plant, suggested it was probably a natural hybrid between O. triumphans and some species of the odoratum group. It is now recognized, however, that O. crispum and O. Lindleyanum are the two parents.

ODONTOGLOSSUM ŒRSTEDII, *Rchb. f.*

Rchb. in Gard. Chron. 1877, vol. vii. p. 302; The Garden, 1884, vol. xxvi. t. 454; Bot. Mag. t. 6820; Veitchs' Man. Orch. Pl. pt. i. p. 57, fig.

Professor Reichenbach believed that Warscewicz was the first to discover this plant, as a sketch in his possession made by Warscewicz seemed to be referable to this species.

It was afterwards found by Dr. Œrsted, and later by Kramer and Endres, the last-named sending plants to us in 1872.

ODONTOGLOSSUM PESCATOREI, *Lindl.*, var. VEITCHIANUM.

Rchb. in Gard. Chron. 1882, vol. xvii. p. 588; The Garden, 1884, vol. xxvi. t. 452; Veitchs' Man. Orch. Pl. pt. i. p. 59-60 (frontispiece).

A superb variety which appeared in one of our own importations of the type.

The flowers, which opened for the first time in this country in March 1882, are larger than the type and richly blotched with magenta-purple.

It is probable the whole stock of the plant is in the unrivalled collection belonging to Baron Sir Henry Schröder at The Dell, Egham.

ODONTOGLOSSUM PRÆNITENS, *Rchb. f.*

Rchb. in Gard. Chron. 1875, vol. iii. p. 524; Bot. Mag. t. 6229; Veitchs' Man. Orch. Pl. pt. i. p. 62.

A rare species introduced in 1874 through Gustav Wallis, by whom it was discovered on the eastern Cordillera of New Grenada in the province of Pamplona.

Only a few plants were received from the discoverer, and it has probably not since been re-imported.

ODONTOGLOSSUM RETUSUM, *Lindl.*

Bot. Mag. t. 7569.

A rare species discovered hy Hartweg in 1841 on rocks in the mountains of Saraguru, near Loxa, Ecuador, and first flowered at Chelsea in 1882. In habit and inflorescence it resembles Odontoglossum Edwardii, but the flowers are orange-red in colour with a green line at the base of the sepals and petals.

ODONTOGLOSSUM UROSKINNERI, *Lindl.*

Lindl. in Gard. Chron. 1859, pp. 708, 724; Veitchs' Man. Orch. Pl. pt. i. p. 69, fig.

Sent to Chelsea from Guatemala by Mr. Ure-Skinner, in 1854, but not

ODONTOGLOSSUM URO-SKINNERI

THE ROYAL GARDENS, KEW

flowered till 1859, a delay due to the defective treatment cool orchids then received.

The fairly large flowers, chestnut-brown mottled with green with a white lip, are produced usually during the months of July and August when but few species of the genus are in bloom.

ONCIDIUM ANTHOCRENE, *Rchb. f.*

Orchid Album, ix. t. 392; Veitchs' Man. Orch. Pl. pt. viii. p. 9.

Originally discovered by Gustav Wallis while collecting in New Grenada in 1872-1873, and subsequently by Chesterton through whom it was introduced.

The specific name, a "fountain of flowers," is a fanciful one.

ONCIDIUM BRYOLOPHOTUM, *Rchb. f.*

Rchb. f. in Gard. Chron. 1871, p. 738.

Introduced from Central America. A dimorphous species bearing large panicles of greenish flowers among which appear bright yellow ones with purplish streaks, the whole inflorescence compared by the author of the name to a German Christmas Tree. It does not appear to be in cultivation at the present time.

ONCIDIUM CHRYSODIPTERUM, *Veitch.*

Veitchs' Man. Orch. Pl. pt. viii. p. 23, fig.

A species acquired by us at an auction sale and flowered for the first time in the spring of 1891. The specific name, literally "golden wings," refers to the exceptionally bright and attractive yellow petals, contrasting strongly with the chestnut-brown of the remainder of the flower.

ONCIDIUM CURTUM, *Lindl.*

Lindl. in Bot. Reg. 1847, t. 68; Veitchs' Man. Orch. pl. viii. p. 33.

Introduced from the Organ Mountains, Brazil, through William Lobb in 1841-1842. The flowers are variable in colour and sometimes resemble those of Oncidium prætextum, but O. curtum may be easily distinguished by the very different crest on the lip.

ONCIDIUM EUXANTHINUM, *Rchb. f.*

Rchb. in Gard. Chron. 1869, p. 1158; Bot. Mag. t. 6322.

Imported from Brazil in 1869; the plate in the Botanical Magazine was prepared from a plant which flowered at Chelsea in 1871.

It has now become very scarce, if not quite lost to cultivation.

ONCIDIUM GLOSSOMYSTAX, *Rchb. f.*

Rchb. in Gard. Chron. 1879, vol. xii. p. 489.

A species of little interest horticulturally, introduced from New Grenada through Kalbreyer.

The flowers although small are interesting to the botanist, light yellow in colour with a few brown blotches, with the distinguishing feature of two pairs of keels on the disk of the lip, each keel being covered with very many white hairs.

ONCIDIUM METALLICUM, *Rchb. f.*

Rchb. in Gard. Chron. 1876, vol. v. p. 394.

Introduced from New Grenada, where it was discovered by Wallis.

The flowers are of a rich chestnut-brown colour with a distinct metallic hue, the borders of the superior sepal and smaller petals being blotched with rich yellow.

ONCIDIUM PRÆTEXTUM, *Rchb. f.*

Rchb. in Gard. Chron. 1873, p. 1206; *id.* 1881, vol. xv. p. 720; Bot. Mag. t. 6662; Rolfe in Orchid Review, 1904, vol. xii. p. 293.

This Brazilian Oncidium was first known in 1873 from specimens collected in the province of San Paulo by Mr. E. D. Jones, by whom they were sent to Mr. John H. Wilson of Liverpool.

Four years later we introduced plants from Rio de Janeiro, and exhibited them in flower before the Royal Horticultural Society in August 1878, when a First Class Certificate was awarded.

ONCIDIUM SUPERBIENS, *Rchb. f.*

Bot. Mag. t. 5980; Veitchs' Man. Orch. Pl. pt. viii. p. 81, fig.

A native of the forests of Venezuela and New Grenada, where it was discovered at about the same time by Fünck and Slim, in 1847, and by Purdie in the province of Ocaña. It was introduced to this country in 1871, and first flowered at Chelsea in the spring of 1872.

ONCIDIUM TECTUM, *Rchb. f.*

Rchb. in Gard. Chron. 1875, vol. iii. p. 780; Veitchs' Man. Orch. Pl. pt. viii. p. 82.

Introduced from New Grenada, through Gustav Wallis in 1874, this Oncidium is of little horticultural value, and seldom seen outside a Botanic garden.

The peculiar zigzag branching of the inflorescence, common to the Oncidia, is very strongly pronounced in this species.

ONCIDIUM WARSCEWICZII, *Rchb. f.*

Gard. Chron. 1871, p. 560, 1874, p. 48; Veitchs' Man. Orch. Pl. pt. viii. p. 91.

Originally discovered by Warscewicz on Chiriqui, in Veragua, in 1852, but lost sight of until re-introduced from Costa Rica in 1870.

It is one of the most distinct of the many species of Oncidium, and although somewhat resembling O. bracteatum is very different in the colour of its flowers.

PACHYSTOMA THOMSONIANUM, *Rchb. f.*

Syns. *Ancistrochilus Thomsonianus*, Rolfe.

Gard. Chron. 1879, vol. xii. pp. 582 and 625, fig.; Bot. Mag. t. 6471; Veitchs' Man. Orch. Pl. pt. vi. p. 4, fig.; The Garden, 1888, vol. xxxiii. p. 175, fig.; Rolfe in Orchid Review, 1904, vol. xii. p. 297, fig. 43.

Introduced through Kalbreyer, by whom it was discovered on the mountains of Old Calabar, West Tropical Africa, and dedicated at his request to the Rev. George Thomson, for many years a missionary in that part of the world.

It is remarkable for its beauty and that the nearest allies are Asiatic, connecting the floras of those widely sundered regions.

PHAIUS BLUMEI, *Lindl.*, var. BERNAYSII.

Syns. *P. grandifolius*, var. *Blumei*, sub-var. *Bernaysii*, Veitch; *P. Bernaysii*, Row. MSS.

Rchb. f. in Gard. Chron. 1873, p. 77; Bot. Mag. t. 6032; Veitchs' Man. Orch. Pl. pt. vi. p. 11.

Flowered in March 1873, and described by Professor Reichenbach (*loc. cit.*) from material supplied by us, and by Dr. Hooker in the Botanical Magazine.

It is possibly a form of the variety Blumei, differing only in the colour of the flowers, which are primrose-yellow, and of little value horticulturally on account of the blooms being often self-fertilized before they expand.

PHAIUS CALLOSUS, *Lindl.*

Syns. *Limodorum callosum*, Blume.

Gard. Chron. 1848, p. 287, with fig.; Rchb. Xen. Orch. t. 122.

A native of Java, first flowered in March 1848, the specific name, "thick-lipped," being derived from the prominent callus which passes from the lip down the tube. The flowers are reddish-brown in colour, tipped with dingy-white.

PHAIUS PHILIPPINENSIS, *N. E. Br.*

N. E. Brown in Gard. Chron. 1889, vol. vi. p. 239; Veitchs' Man. Orch. Pl. pt. vi. p. 13.

Discovered by David Burke on the slopes of the hills at 3,000-4,000 ft. elevation, in the Island of Mindanao, and interesting as being the first species of the genus Phaius to be discovered in the Philippines.

It flowered for the first time at Chelsea in 1889. As a species it is remarkably distinct, especially in the structure of its lip, which is truncate and slightly frilled.

PHALÆNOPSIS AMABILIS, *Blume.*

Syns. *P. grandiflora*, Lindl.

Lindl. in Gard. Chron. 1848, p. 39, with woodcut; Bot. Mag. t. 5184; Veitchs' Man. Orch. Pl. pt. vii. p. 23, figs.; Rolfe in Gard. Chron. 1886, vol. xxvi. p. 168.

Introduced into British gardens by Thomas Lobb, who sent plants from Java to Exeter in 1846, which flowered for the first time in this country in September of the following year. All collectors in that region since Lobb mention Phalænopsis amabilis, and agree in reporting it as growing near the sea-shore, sometimes high up on the trees and sometimes lower down. Burbidge found it in Labuan and North Borneo, Curtis detected it in North Celebes, and Burke met with a small-flowered variety in South-West New Guinea.

The species was known as early as 1750, in which year Rumphius figured it in his Herbarium Amboinense.

PHALÆNOPSIS CORNINGIANA, *Rchb. f.*

Rchb. in Gard. Chron. 1879, vol. xi. p. 620.

Described by Professor Reichenbach from materials supplied by us, and dedicated to Mr. Erastus Corning, the pioneer of orchid-growing in America, whose collection at Albany, New York, was so famous in its day.

PHALÆNOPSIS INTERMEDIA, *Lindl.*

Syns. *P. Lobbii*, Hort.

Gard. Chron. 1852, p. 230 (notice of exhibit); Lindl. in Paxt. Fl. Gdn. 1853, vol. iii. p. 163, fig. 310; Veitchs' Man. Orch. Pl. pt. vii. p. 44, fig.; Gard. Chron. 1886, vol. xxvi. p. 169; *id.* 1896, vol. xix. p. 106.

Introduced by Thomas Lobb in 1852 among an importation of Phalænopsis Aphrodite.

Later a French traveller, named M. Porte, brought two more plants from the Philippine Islands, after which thirteen years elapsed before a further addition was made by Messrs. Low & Co.

Lindley was the first to surmise the existence of natural hybrids on seeing a flower of Phalænopsis intermedia, which combined the characters of P. Aphrodite and those of P. rosea.

In 1886 Seden flowered a hybrid at Chelsea, which had as parents P. Aphrodite and P. rosea, which proved to be identical with the P. intermedia of Lindley, and confirmed the supposition.

PHALÆNOPSIS MACULATA, *Rchb. f.*

Rchb. in Gard. Chron. 1881, vol. xvi. p. 134; Veitchs' Man. Orch. Pl. pt. vii. p. 31.

Introduced from Sarawak in Borneo in 1880, through Curtis, by whom it was discovered growing on the limestone hills at an altitude of

1,000-1,500 ft., on damp almost bare rocks, under the shade of trees. It is one of the smallest of the genus, the flowers being only some ½ to ¾ in. in diameter.

PHALÆNOPSIS MARIÆ, *Burbidge.*

Burbidge in Orchid Album, ii. t. 80, *et* sub t. 87; Bot. Mag. t. 6964; Veitchs' Man. Orch. Pl. pt. vii. p. 32; Rolfe in Gard. Chron. 1886, vol. xxvi. p. 277.

Discovered by Burbidge when in the Sulu Archipelago in 1878, and dedicated by him to his wife. It was subsequently detected by David Burke on the hills near the south-east coast of the island of Mindanao, plentiful on the trunks and branches of trees in dense shade. It is a handsome species, allied to Phalænopsis Lueddemanniana and P. sumatrana.

PHALÆNOPSIS ROSEA, *Lindl.*

Lindl. in Gard. Chron. 1848, p. 671, fig.; Paxt. Fl. Gdn. 1852, t. 72; Bot. Mag. t. 5212; Fl. des Serres, tom. xvi. t. 1645; Veitchs' Man. Orch. Pl. pt. vii. p. 34.

Introduced from Manila through Thomas Lobb in 1848, it is one of the commonest of the Philippine Islands' Phalænopses, and is found in abundance in the hot valleys and along the streams in the neighbourhood of Manila.

Under cultivation it has helped in the production of many fine hybrids, of which Phalænopsis × Artemis (P. amabilis × P. rosea); P. × Cassandra (P. Stuartiana × P. rosea); P. × Hebe (P. Sanderiana × P. rosea); P. × Vesta (P. rosea leucaspis × P. Aphrodite), are the most noteworthy.

PHALÆNOPSIS SUMATRANA, *Rchb. f.*, var. PAUCIVITTATA.

Rchb. in Gard. Chron. 1882, vol. xvii. p. 628; Veitchs' Man. Orch. Pl. pt. vii. p. 40.

PHALÆNOPSIS SUMATRANA, *Rchb. f.*, var. SANGUINEA.

Rchb. in Gard. Chron. 1881, vol. xv. p. 782; Veitchs' Man. Orch. Pl. pt. vii. p. 40.

Both these varieties occurred in an importation of the species sent by Curtis from Borneo.

The former variety has fewer and paler markings on the sepals and petals, whilst in the variety sanguinea they are suffused with red-brown.

PHALÆNOPSIS × VEITCHIANA, *Rchb. f.*

Rchb. in Gard. Chron. 1872, p. 935; *id.* 1884, vol. xxi. p. 270; Fl. Mag. n.s. t. 213; Veitchs' Man. Orch. Pl. pt. vii. p. 47.

A supposed natural hybrid, one of the rarest and most distinct, between Phalænopsis Schilleriana and P. rosea, which appeared as a solitary specimen amongst our importation prior to 1872.

PHALÆNOPSIS VIOLACEA, *Teijsm.*

Rchb. in Gard. Chron. 1878, vol. x. p. 234; *id.* 1881, vol. xvi. p. 145, fig.; Fl. Mag. 1879, n.s. t. 342; Veitchs' Man. Orch. Pl. pt. vii. p. 41, figs.

Originally discovered by Teijsman near Pelambang, in Sumatra, in

1859, and by him sent to the Botanic Garden at Leyden, where it first flowered in Europe in 1861.

Nothing more was heard of the plant until Mr. Murton, of the Botanic Garden of Singapore, sent plants to Mr. M. H. Williams, of Tredrea, in Cornwall, and to Chelsea, in both of which establishments it flowered in 1878.

It remained rare in European collections until 1880, when Curtis sent a consignment from Sumatra, where it was discovered growing under the same conditions as Phalænopsis sumatrana.

PLEUROTHALLIS INSIGNIS, *Rolfe.*

Syns. *P. glossopogon*, Nicholson, non Rchb. f.

Rolfe in Gard. Chron. 1887, vol. i. p. 477; Bot. Mag. t. 6936.

A curious orchid of botanical interest only, the native country of which is not recorded, but is in all probability Venezuela.

In growth something like a Masdevallia, the flowers are remarkable for their acuminate sepals and long bristle-like petals.

POLYCYNIS GRATIOSA, *Endr. & Rchb. f.*

Rchb. in Gard. Chron. 1871, p. 1451.

Discovered in Costa Rica by Endres; closely allied to Polycynis lepida, but differing from that species in slight structural details of the lip.

RENANTHERA MATUTINA, *Lindl.*

Syns. *Aërides matutinum*, Blume.

Veitchs' Man. Orch. Pl. pt. vii. p. 85; Lindl. Bot. Reg. 1847, misc.

First discovered by Blume in 1824, growing on trees at the foot of Mount Salak, Java, from which locality it was introduced twenty years later through Thomas Lobb. For a long time subsequent to its introduction it remained very rare, but subsequent importations caused it to be more generally distributed. The flowers are some 2 in. in diameter, bright reddish-crimson toned with yellow, changing with age to orange-yellow.

RESTREPIA ELEGANS, *Karst.*

Bot. Mag. t. 5966.

A lovely little orchid, first cultivated in Europe by Messrs. Linden of Brussels, and first flowered in this country by us in February 1872.

It is a native of Caraccas, where it inhabits mossy tree trunks at elevations of 5,000-6,000 ft. The flowers are too small to be of any horticultural value, but are of great interest to the botanist.

RODRIGUEZIA LEOCHILINA, *Rchb. f.*

Gard. Chron. 1871, p. 970.

Introduced from Costa Rica and flowered at Chelsea in July 1871. It is closely allied to Rodriguezia maculata, from which species it differs in its even white lip.

SACCOLABIUM BIGIBBUM, *Rchb. f.*

Bot. Mag. t. 5767; Veitchs' Man. Orch. Pl. pt. vii. p. 113.

Discovered by Colonel Benson in Upper Burmah and sent to us in 1868.

It is still occasionally imported amongst Burmese orchids.

The plant is of dwarf habit, the flowers small, of a yellow colour, with a triangular whitish fringed lip.

SACCOLABIUM GIGANTEUM, *Lindl.*

Bot. Mag. t. 5635; Gard. Chron. 1862, p. 1194; Fl. des Serres, tom. xvii. t. 1765; Veitchs' Catlg. of Pl. 1868, p. 25; *id.* Man. Orch. Pl. pt. vii. p. 114, fig.

First discovered in the early part of the last century by one of Dr. Wallich's collectors near Prome in Lower Burmah. In 1859 it was next heard of, having been sent to Dr. Sumner, Bishop of Winchester, in whose garden at Farnham Castle it flowered in the autumn of 1862.

Plants continued to be extremely rare until re-introduced through Colonel Benson in 1866 from Prome and Thayetmayo.

SACCOLABIUM HUTTONI, *Hook.*

Syns. *Aërides Huttoni*, Hort. Veitch.

Veitchs' Catlg. of Pl. 1868, p. 23; Bot. Mag. t. 5681; Veitchs' Man. Orch. Pl. pt. vii. p. 70.

Introduced through Henry Hutton in 1866, and received only shortly after his early death in the Eastern Archipelago.

The exact locality whence Hutton introduced his plants is not known, and the plant remained scarce until again found by Curtis in 1882 in North Celebes, growing on mangrove trees near the sea-shore.

SACCOLABIUM MINIATUM, *Lindl.*

Lindl. in Bot. Reg. 1847, sub t. 26; *id.* t. 58; Bot. Mag. t. 5326; Veitchs' Man. Orch. Pl. pt. vii. p. 117; Orchid Review, 1896, vol. iv. p. 178.

Introduced from Java by Thomas Lobb in 1846, but now rarely seen in orchid collections in this country. The racemes of small but richly coloured flowers are produced in May and continue a long time in perfection. According to The Orchid Review, above quoted, there is probably an error in recording Java as its home, as it has not since been collected in that country and does not appear in the earlier herbaria. It has since been met with by Dr. Watt on the Naga Hills east of Khasia, and as Lobb visited that locality, there is a probability that the plants were collected there in the first instance.

SARCANTHUS CHRYSOMELAS, *Rchb. f.*

Rchb. f. in Gard. Chron. 1869, p. 662.

Introduced from Moulmein through Colonel Benson, by whom it was discovered. This species has handsome foliage and gold and purple flowers.

SARCANTHUS FLEXUS, *Rchb. f.*

Rchb. f. in Gard. Chron. 1881, vol. xvi. p. 492.

A botanical curiosity imported from Borneo : the flowers are yellowish-brown tipped with reddish-brown, and slightly larger than those of Sarcanthus paniculatus, Lindl.

SARCOCHILUS LUNIFERUS, *Benth.*

Syns. *Thrixspermum luniferum*, Rchb. f.

Bot. Mag. t. 7044; Gard. Chron. 1868, p. 786.

A curious plant of great botanical interest introduced in 1868 through the Rev. Mr. Parish, from near Moulmein in Tenasserim, British Burmah.

It is remarkable in that the leaves are seldom developed, and when produced rarely last any length of time.

The flowers are small, yellow spotted with red, with a white lip, which latter from its shape as seen in a front view suggested the specific name.

SCAPHOSEPALUM BREVE, *Rolfe.*

Syns. *Masdevallia brevis*, Rchb. f.

Rchb. f. in Gard. Chron. 1883, vol. xx. p. 588.

Imported from Demerara. The flowers are small, the upper sepals brownish with the expanded portions orange-yellow and purple : the lower sepal yellow spotted with purple and the tail dark purple. It is of great botanical interest.

SCAPHOSEPALUM GIBBEROSUM, *Rolfe.*

Syns. *Masdevallia gibberosa*, Rchb. f.

Rchb. in Gard. Chron. 1876, vol. v. p. 8; Bot. Mag. t. 6990.

A singular species, a native of New Grenada, whence it was introduced through Gustav Wallis, having highly curious and botanically interesting flowers, of little value horticulturally on account of their small size.

SPATHOGLOTTIS AUREA, *Lindl.*

Lindl. in Jour. Hort. Soc. London, 1850, p. 34; Rchb. in Gard. Chron. 1888, vol. iv. p. 92, with fig.; Veitchs' Man. Orch. Pl. pt. vi. p. 7, fig. reproduced.

Originally introduced in 1849 from Mount Ophir in Malacca through Thomas Lobb, who discovered it growing near Nepenthes sanguinea and Rhododendron jasminiflorum.

Only a few plants arrived, and these gradually died out after once flowering.

Nothing more was heard of it in a living state until 1886, when it appeared in an importation of orchids offered for sale at Stevens's rooms.

SPATHOGLOTTIS PETRI, *Rchb. f.*

Rchb. in Gard. Chron. 1877, vol. viii. p. 392; Bot. Mag. t. 6354; Veitchs' Man. Orch. Pl. pt. vi. p. 8.

Discovered in the Fiji Islands in 1876 by Peter C. M. Veitch, after whom it is named; sent by him to Chelsea, where it flowered for the first time in the following year.

The species is remarkable for its deciduous bracts, organs, in all other members of the genus persistent, remaining even long after the ripening of the fruit.

STANHOPEA GIBBOSA, *Rchb. f.*

Gard. Chron. 1869, p. 1254.

A species from South America closely allied to Stanhopea Wardii. The flowers, often 6 in. in diameter, are yellow, barred and spotted with crimson, darkest on the petals.

STANHOPEA XYTRIOPHORA, *Rchb. f.*

Rchb. in Gard. Chron. 1868, p. 842.

A pitcher-bearing species with yellow flowers introduced from Peru, distinct from other known species, approaching rather the Coryanthes. It does not appear to be in cultivation.

STAUROPSIS GIGANTEA, *Benth.*

Syns. *Vanda gigantea*, Lindl.

Benth. in Jour. Linn. Soc. vol. xviii. p. 331; Veitchs' Man. Orch. Pl. pt. vii. p. 2; Lindl. in Gard. Chron. 1858, p. 312; Bot. Mag. t. 5189.

First discovered by Wallich in Moulmein in 1826, and later in the same locality by Thomas Lobb, through whom it was introduced.

The plant flowered for the first time in this country in the then famous collection of the late Mr. Robert Warner at Broomfield, in April 1858.

The specific name must be assumed to refer to the large size of the flowers and leaves rather than to the habit of the plant, which under cultivation does not exceed moderate dimensions.

STELIS BRÜCKMÜLLERI, *Rchb. f.*

Bot. Mag. t. 6521.

A quaint little orchid of botanical interest, introduced from the Mexican Andes, with minute flowers of a purple colour, hairy inside.

STELIS GLOSSULA, *Rchb. f.*

Gard. Chron. 1870, p. 1373.

A curious little orchid of botanical interest only, imported from Costa Rica, with brownish flowers in two transverse rows, and bracts larger than the whole of the flower.

STELIS ZONATA, *Rchb. f.*

Rchb. f. in Gard. Chron. 1883, vol. xx. p. 556.

A botanical curiosity introduced from Demerara, allied to Stelis muscifera of Lindley, but smaller in all its parts, and interesting from its coloured zone.

STENIA GUTTATA, *Rchb. f.*

Rchb. in Gard. Chron. 1880, vol. xiv. p. 134.

A species closely allied to the rare Stenia pallida, but the sepals and petals are blunt and shorter, and spotted with Indian-purple on a straw-coloured ground.

It was found in Peru by Walter Davis, and flowered at Chelsea in Ju 1880.

THUNIA BENSONIÆ, *Hook. f.*

Syns. *Phaius Bensoniæ*, Hemsley.

Bot. Mag. t. 5694; Orchid Album, ii. t. 97.

Discovered by Colonel Benson in the neighbourhood of Rangoon in 1866, and flowered for the first time in this country in the Royal Gardens, Kew, and Chelsea in July 1867.

The flowers are amethyst purple in colour, the lip frilled at the edges and marked in the centre with numerous longitudinal frilled keels.

TRICHOCENTRUM PINELI, *Lindl.*

Gard. Chron. 1854, p. 772.

Discovered near Rio by Chevalier Pinel, who collected specimens, and after whom it was named. It flowered for the first time in this country at Chelsea in 1854.

TRICHOCENTRUM PURPUREUM, *Lindl.*

Lindl. in Gard. Chron. 1854, p. 772.

A plant of botanical interest only, described and named by Dr. Lindley from a plant which flowered in 1854, and which we believe came from Demerara.

TRICHOGLOTTIS COCHLEARIS, *Rchb. f.*

Rchb. in Gard. Chron. 1883, vol. xix. p. 142; Jour. of Hort. 1886, vol. xii. p. 194, fig. 34; *id.* 1893, vol. xxvi. p. 233, fig. 47.

A rare species, introduced in 1882 through Curtis, who met with it in the Island of Sumatra.

The flowers are white with purple bars inside and outside the sepals and petals; the lip is spoon-shaped, very thick with a few purple blotches.

TRICHOPILIA GRATA, *Rchb. f.*

Gard. Chron. 1868, p. 1338.

Introduced from Peru in 1868. The flowers are sweetly scented as are those of Trichopilia fragrans to which species grata is allied; they are yellowish-green with a white expanded lip, the free end being orange colour.

TRICHOPILIA MARGINATA, *Henf.*, var. LEPIDA.

Syns. *T. coccinea*, Warsc., var. *lepida*; *T. lepida*, Veitch.

Veitchs' Man. Orch. Pl. pt. ix. p. 183; Fl. Mag. n.s. t. 98.

This variety appeared amongst an importation of the species from Costa Rica in 1873. It is a very rare form with flowers larger than those of the type, and the margin of the lip more crisped.

VANDA BENSONI, *Batem.*

Bot. Mag. t. 5611; Gard. Chron. 1867, p. 180, fig.; Fl. des Serres, tom. xxii. t. 2392; Veitchs' Man. Orch. Pl. pt. vii. p. 89.

This Vanda was sent to us by Colonel Benson, who discovered it in Lower Burmah in 1866.

It flowered shortly after its arrival at Chelsea in the summer of the same year, and proved closely allied to Vanda Roxburghii and V. concolor, but the absence of all tessellation and the spotting and yellow colour of the inside of the flowers are distinguishing features.

VANDA CÆRULEA, *Griff.*

Lindl. in. Paxt. Fl. Gdn. vol. i. t. 36; Fl. des Serres, 1853, vol. vi. t. 609; l'Illus. Hort. 1860, t. 246; Veitchs' Man. Orch. Pl. pt. vii. p. 90, fig.; Orchid Album, vi. t. 282.

First discovered by William Griffith, the Indian botanist and explorer, in November 1837 on the Khasia Hills, in which locality it was later re-discovered by Sir J. D. Hooker and Dr. Thomson, but was not introduced to cultivation.

Thomas Lobb sent home plants from the Khasia Hills to Exeter, where one of them flowered for the first time in December 1850, and was exhibited at a meeting of the Horticultural Society of London, held in Regent Street, and received with marked favour.

The large flowers of soft light blue, tessellated with azure blue, are of great beauty.

VANDA CŒRULESCENS, *Griff.*

Gard. Chron. 1870, p. 529, fig. 97; Bot. Mag. t. 5834; Fl. Mag. n.s. t. 259; Veitchs' Man. Orch. Pl. pt. vii. p. 92.

Discovered near Bhamo in Burmah in 1837 by Griffiths, who collected specimens of the plant, but nothing more was heard of it until Colonel Benson re-discovered it in 1867, and sent plants the following year to Chelsea, where it flowered for the first time in February 1869.

Although by no means comparable either in size or colour with the beautiful Vanda cærulea, it is a very elegant plant, and the pale lilac-blue flowers cause it to be well worthy of cultivation.

VANDA DENISONIANA, *Benson & Rchb. f.*

Rchb. in Gard. Chron. 1869, p. 528; Bot. Mag. t. 5811; l'Illus. Hort. 1872, t. 105; Veitchs' Man. Orch. Pl. pt. vii. p. 94, fig.

Discovered by Colonel Benson on the Arracan Mountains and sent to Chelsea in 1868, where it flowered for the first time in this country in April 1869.

It is named in honour of Lady Londesborough in appreciation of Lord Londesborough's great love of orchids.

The flowers are ivory-white in colour with five longitudinal greenish-white lines on the lip.

VANDA HOOKERIANA, *Rchb. f.*

Rchb. in Gard. Chron. 1882, vol. xviii. p. 488; The Garden, 1883, vol. xxiii. t. 370; l'Illus. Hort. 1883, t. 484; Veitchs' Man. Orch. Pl. pt. vii. p. 96, fig. opposite.

This lovely Vanda was seen by several travellers, including Thomas Lobb, and was known in herbaria for some time previous to its introduction. In 1879 a correspondent in Labuan sent us living plants which were immediately acquired by Lord Rothschild.

One of these plants flowered for the first time at Tring Park in September 1882, and since that time the flowering of Vanda Hookeriana has been of frequent occurrence in that great garden.

VANDA INSIGNIS, *Blume.*

Lindl. in Paxt. Fl. Gdn. vol. ii. p. 19, fig.; Rchb. in Gard. Chron. 1868, p. 1259; Bot. Mag. t. 5759; Veitchs' Man. Orch. Pl. pt. vii. p. 97.

Introduced to Chelsea from the Moluccas by Hutton in 1866, and flowered for the first time in 1868.

It continued very rare in British collections until 1882, when it was re-imported through Curtis, at that time collecting in the Malay Archipelago. The flowers are tawny-yellow in colour with dark brown oblong spots, and the lip is bright rose-purple.

VANDA INSIGNIS, *Blume*, var. SCHRÖDERIANA, *Rchb. f.*

Rchb. in Gard. Chron. 1883, vol. xx. p. 392; The Garden, 1884, vol. xxv. t. 429.

This variety, which is remarkable for the colour deviation from the type, was introduced in Curtis's consignment.

VANDA SUAVIS, *Lindl.*

Syns. *V. tricolor*, var. *suavis*, Veitch.

Lindl. in Gard. Chron. 1848, p. 351, fig.; Bot. Mag. t. 5174; Paxt. Fl. Gdn. t. 42, fig. 3; Fl. des Serres, 1862, tt. 1604-1605; Veitchs' Man. Orch. Pl. pt. vii. p. 107, fig.

Introduced from Java through Thomas Lobb, and for many years one of the rarest Vandas in cultivation; it was first exhibited in flower on April 4th 1848.

Always associated with Vanda tricolor in its native home, it was imported mixed with the latter, although in small quantities.

The flowers are fragrant, produced in racemes in the axils of the leaves; they are white in colour spotted with red-purple, and the basal half of the lip is deep purple.

VANDA TRICOLOR, *Lindl.*

Bot. Reg. 1847, sub t. 59; Bot. Mag. t. 4432, Fl. des Serres, 1850, tom. vi. t. 641; Veitchs' Man. Orch. Pl. pt. vii. p. 106.

Vanda tricolor was introduced from Java in 1846 through Thomas Lobb, who discovered it in the western part of the island, at 1,500-2,500 ft. elevation, growing chiefly on large trees.

Resembling V. suavis in foliage, habit and inflorescence, it may be distinguished by the ground colour of its pale yellow flowers, which is white in V. suavis.

ZYGOPETALUM BURKEI, *Rchb. f.*

Rchb. f. in Gard. Chron. 1883, vol. xx. p. 684; Orchid Album, t. 142; Veitchs' Man. Orch. Pl. pt. ix. p. 44.

Introduced in 1881 through David Burke, by whom it was discovered on Roraima in British Guiana.

It inhabits rocks in the swamp, in which Cypripedium Lindleyanum and Heliamphora nutans have their home, at elevations of about 6,000 ft.

The colouring of the parts of the flower is very striking;—the sepals and petals are green with seven to nine longitudinal chocolate-brown stripes, which sometimes become broken up into dots; the lip is milk white, irregularly dentate along the margin, with about thirteen violet purple ribs, and the column is yellow streaked with purple.

ZYGOPETALUM BURTII, *Benth.*

Syns. *Batemannia Burtii*, Rchb.

Rchb. in Gard. Chron. 1872, p. 1099; Bot. Mag. t. 6003.

Originally discovered by Endres in 1867 in Costa Rica, and shortly afterwards imported from that country. A plant obtained from us flowered for the first time in Great Britain in the collection of the late Mr. Burnley Hume, at Winterton, Norfolk, in the summer of 1872.

The flowers are fleshy, 3 to 4 in. in diameter, white at the very base of the segments, then yellow, and the apical half red-brown with some yellow spots. Comparatively few plants have been introduced.

ZYGOPETALUM DAYANUM, *Benth.* var.

Syns. *Pescatorea Dayana*, Rchb., var. *rhodacea*.

Rchb. in Gard. Chron. 1873, p. 575; *id.* 1874, p. 226; Bot. Mag. t. 6214; Veitchs' Man. Orch. Pl. pt. ix. p. 50.

Discovered by Gustav Wallis in New Grenada, introduced in 1873, and named in honour of the late Mr. John Day of Tottenham.

The flowers, 3 in. in diameter, are coloured cream-white and green on the sepals and petals, and the lip white stained with crimson.

ZYGOPETALUM LAMELLOSUM, *Benth.*

Syns. *Pescatorea lamellosa*, Rchb. f.

Veitchs' Man. Orch. Pl. p. ix. p. 54; Bot. Mag. t. 6240; Rchb. in Gard. Chron. 1875, vol. iv. p. 225.

Introduced from New Grenada through Gustav Wallis, who gave no precise locality, and flowered for the first time in August 1875; it is long lost to cultivation.

The flowers, about $2\frac{1}{4}$ in. in diameter, are of a nearly uniform yellow colour with a yellowish-white lip marked by an orange and brown crest.

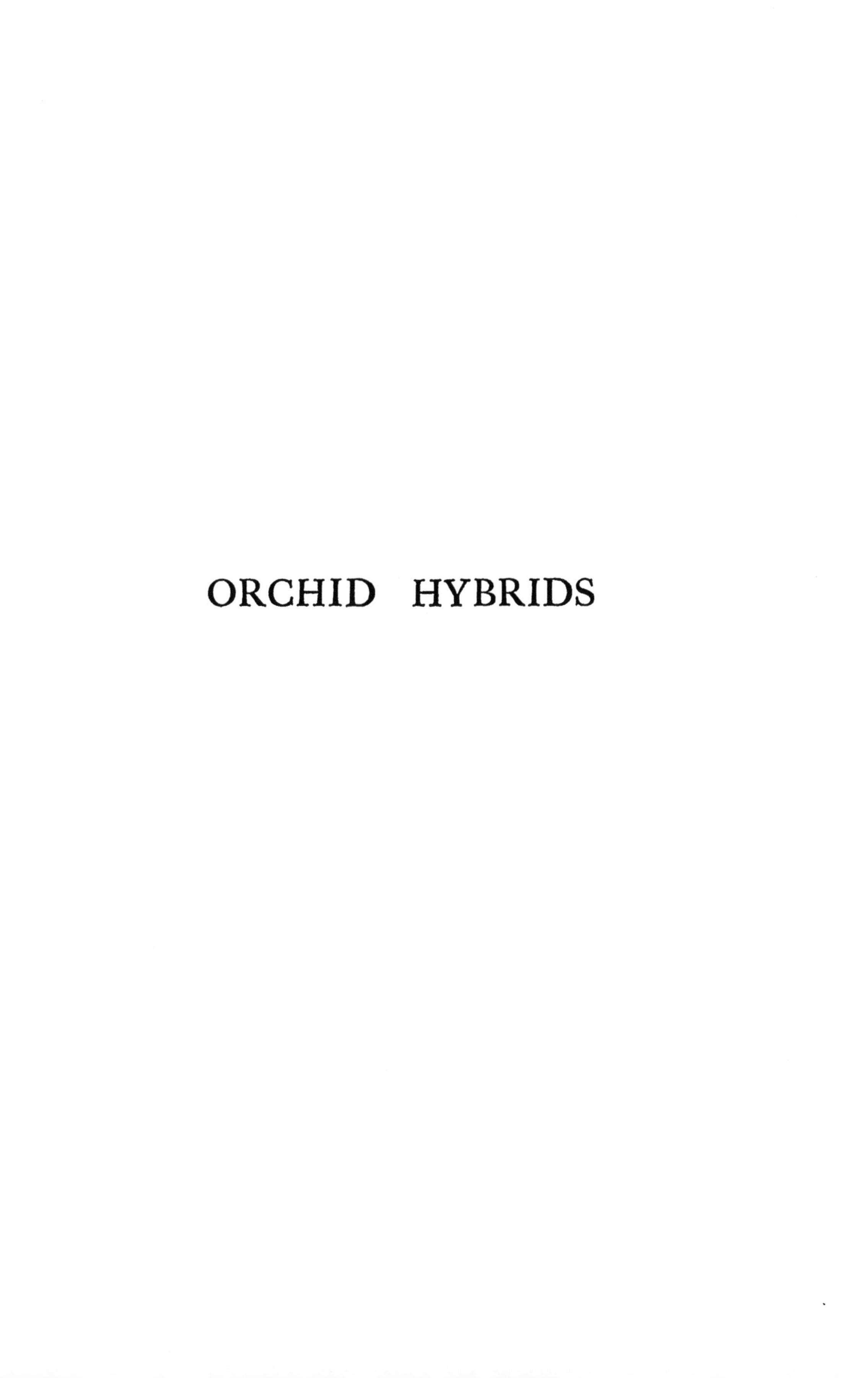

ORCHID HYBRIDS

ORCHID HYBRIDS

A LIST OF THE PRINCIPAL ORCHID HYBRIDS RAISED BY MESSRS. VEITCH SINCE 1853

ANGRÆCUM × VEITCHII.

Gard. Chron. 1899, vol. xxv. p. 31 (Report of R.H.S. Orchid Committee); *id.* p. 35, fig. 10; Orchid Review, 1901, vol. ix. p. 37.

Obtained by Seden from Angræcum sesquipedale and A. eburneum, and the first hybrid between two species of this remarkable genus.

A plant bearing three open flowers was exhibited for the first time on January 10th 1899. The Orchid Committee recommended the plant a First Class Certificate, and the raiser a Silver Flora Medal. The flowers are ivory-white with a long green spur.

ANGULOA × MEDIA, *Rchb. f.*

Syns. *A.* × *intermedia*, Rolfe.

Gard. Chron. 1881, vol. xvi. p. 38; Rolfe in Gard. Chron. 1888, vol. iii. p. 798; Veitchs' Man. Orch. Pl. pt. ix. p. 102; Orchid Review, 1893, vol. i. p. 40.

The offspring of Anguloa Clowesii and A. Ruckeri, the latter being the pollen parent; a similar hybrid had previously been raised by Mr. J. C. Bowring, of Forest Farm, Windsor, which died shortly after flowering, and it also occurs in a wild state. An imported plant flowered in 1893 in the collection of R. N. Measures, Esq., of Streatham, and proved identical with Seden's artificially raised plants.

ANŒCTOCHILUS (GOODYERA) × DOMINII.

Gard. Chron. 1861, p. 531; Williams' Orch. Man. 7th edt. p. 410.

Raised by John Dominy from Goodyera (Hæmaria) discolor, and Anœctochilus Lowii (Dossinia marmorata). The leaves are dark mottled olive green with five to nine flesh-coloured ribs, and the plant is probably not now in cultivation.

CALANTHE × BELLA.

Rchb. in Gard. Chron. 1881, vol. xv. p. 234; Veitchs' Man. Orch. Pl. pt. vi. p. 73; Reichenbachia, 1893, p. 31, t. 63, fig. 2.

Raised from Calanthe vestita Turneri and C. × Veitchii; the flowers, delicate light rose suffused with white, with a deep maroon blotch on the lip, are as large as the best forms of C. × Veitchii.

CALANTHE × DOMINII, *Lindl.*

Lindl. in Gard. Chron. 1858, p. 4; Bot. Mag. t. 5042; Veitchs' Man. Orch. Pl. pt. vi. p. 76; Dict. Ic. des Orchidées, Calanthe, hybr. pl. 2.

This cross, interesting as the first hybrid orchid to bloom, although not the first seedling to be raised by hand, flowered for the first time at Exeter in 1856, and was named by Dr. Lindley in honour of John Dominy, the foreman who effected the cross and raised the seedlings.

The parents used were Calanthe falcata and C. masuca.

CALANTHE × GIGAS.

Orchid Review, 1893, vol. i. pp. 61, 86; Jour. of Hort. 1893, Feb. 16th, p. 129, fig. 24; The Garden 1893, vol. xliv. p. 236, pl. 926.

A very beautiful hybrid raised by Seden from Calanthe vestita gigantea crossed with the pollen of C. Regnieri Sanderiana.

The flowers, borne on a strong spike, are nearly 3 in. across; the sepals milk-white, the petals faintly tinted with rose, and the lip a bright rose striated with white and a deep red-crimson blotch at the base.

CALANTHE × HARRISII.

Gard. Chron. 1895, vol. xviii. p. 721 (Report of R.H.S. Orchid Committee).

Raised by Seden from Calanthe Turneri and C. × Veitchii; pure white handsome flowers.

CALANTHE × LENTIGINOSA, *Rchb. f.*

Rchb. f. in Gard. Chron. 1883, vol. xix. p. 44; Veitchs' Man. Orch. Pl. pt. vi. p. 74.

A hybrid from Calanthe labrosa and C. × Veitchii, the latter being the seed parent.

There are two sub-varieties, *rosea* and *carminata*, the last-named one of the darkest of all hybrid Calanthes.

CALANTHE × MASUCO-TRICARINATA.

Gard. Chron. 1895, vol. xvii. p. 210 (Report of R.H.S. Orchid Committee).

This pretty hybrid raised from the two Indian varieties expressed by the name has white flowers suffused with rosy lilac, about 1 in. in diameter.

CALANTHE × SEDENII, *Hort. Veitch & Rchb. f.*

Rchb. f. in Gard. Chron. 1878, vol. ix. p. 168; Veitchs' Man. Orch. Pl. pt. vi. p. 75.

Raised by Seden from Calanthe × Veitchii crossed with C. vestita rubro-oculata, and flowered for the first time in 1878; one of the best rose-coloured of all Calanthes.

PRINCIPAL ORCHID HYBRIDS

CALANTHE × VEITCHII, *Lindl.*

Lindl. in Gard. Chron. 1859, p. 1016; Bot. Mag. t. 5375; Fl. Mag. t. 280; Veitchs' Man. Orch. Pl. pt. vi. p. 75, figs.; Dict. Ic. des Orchidées, Calanthe, hybr. pl. 1.

Raised by John Dominy at Exeter in 1856, and now the most popular and widely-grown representative of the genus. The bright and unusually attractive rose-coloured flowers on gracefully arching spikes are produced during the winter months, and the ease with which the plants can be grown account for its popularity. The parents are Calanthe (Limatodes) rosea and C. vestita.

CALANTHE × VEITCHII, var. ALBA.

Veitchs' Catlg. of Pl. 1897, p. 10.

Although originally raised by us at the same time and from the same cross as the typical C. × Veitchii, this variety has always been rare. It differs from the type in having pure white flowers.

CATTLEYA × ADELA.

Gard. Chron. 1898, vol. xxiv. p. 16 (Report of R.H.S. Orchid Committee).

Raised by Seden from Cattleya Trianæ and C. Percivaliana. The flower has lilac rose-tinted sepals and petals and a rich dark purple lip.

CATTLEYA × APOLLO.

Gard. Chron. 1896, vol. xx. p. 471 (Report of R.H.S. Orchid Committee).

Raised from Cattleya Mossiæ and C. Aclandiæ. The flowers in form and firm substance resemble C. Aclandiæ, but are almost as large as C. Mossiæ and of great brilliance.

CATTLEYA × ATALANTA.

Orchid Review, 1894, vol. ii. p. 275; Veitchs' Catlg. of Pl. 1897, p. 6.

Obtained from Cattleya Warscewiczii and C. guttata Leopoldii, the former being the pollen parent, from which the front lobe of the lip derived its brilliant colour.

CATTLEYA × BACTIA.

Gard. Chron. 1901, vol. xxx. p. 330 (Report of R.H.S. Orchid Committee); Orchid Review, 1901, vol. ix. p. 350.

Raised from Cattleya Bowringiana and C. guttata; the form of the flower is that of C. guttata, but the colour approaches more nearly that of C. Bowringiana.

CATTLEYA × BRABANTIÆ.

Fl. Mag. 1867, t. 360; Veitchs' Man. Orch. Pl. pt. ii. p. 89.

A hybrid raised by John Dominy and named in compliment to the Duchess of Brabant, afterwards Queen of the Belgians.

The species used were Cattleya Loddigesii and C. Aclandiæ.

CATTLEYA × BROWNIÆ, VEITCHS' var.

Gard. Chron. 1901, vol. xxx. p. 330 (Report of R.H.S. Orchid Committee).

A remarkable hybrid raised by Seden from Cattleya Harrisoniana and C. Bowringiana, in that the flower is very large, whilst those of both the parents are comparatively small.

CATTLEYA × CHAMBERLAINIANA, *Rchb. f.*

Rchb. f. in Gard. Chron. 1881, vol. xvi. p. 427; Veitchs' Man. Orch. Pl. pt. ii. p. 90; Gard. Chron. 1889, vol. v. p. 491 (Rolfe's List of Gdn. Orchids); Dict. Ic. des Orchidées, Cattleya hybr. pl. 17.

Raised by Seden from Cattleya guttata Leopoldii and C. Dowiana, and named in honour of the Right Hon. Joseph Chamberlain, a well-known admirer of orchidaceous plants.

The flowers of the hybrid are like those of the first-named parent in shape, but larger, the sepals and petals less spreading and of a remarkable warm brown tint which strikingly contrasts with the rich crimson purple lip.

CATTLEYA × CHLOE.

Orchid Review, 1900, vol. viii. p. 316 (Report of R.H.S. Orchid Committee).

A hybrid raised at Langley from Cattleya Bowringiana and C. bicolor.

CATTLEYA × CHLORIS.

Gard. Chron. 1893, vol. xiv. p. 525, fig. 88; Orchid Review, 1893, vol. i. pp. 339, 352.

Raised by Seden from Cattleya Bowringiana and C. maxima. The flowers are 5 in. across; the sepals and petals bright rose-purple and the lip deep purple-crimson with a lighter orange-barred throat.

CATTLEYA × CLYTIE.

Raised from Cattleya Bowringiana and C. velutina.

CATTLEYA × CYBELE.

Raised from Cattleya gaskelliana and C. Lueddemanniana.

CATTLEYA × DEVONIENSIS.

Syns. *C. × devoniana.*

Williams' Orch. Man. 7th edt. p. 160.

One of the early hybrids raised by Dominy, of which no record was kept, and which received a First Class certificate from the Royal Horticultural Society, October 11th 1864.

PRINCIPAL ORCHID HYBRIDS

CATTLEYA × DOMINIANA.

Lindl. in Gard. Chron. 1859, p. 948; Veitchs' Man. Orch. Pl. pt. ii. p. 90.

Raised by John Dominy at Exeter between Cattleya labiata and C. intermedia.

The several seedlings raised from this cross produce flowers alike in form but varying in colour; some have received varietal names, as *alba* (Fl. Mag. 1867, t. 367), with pale sepals and petals; and *lutea*, with a large yellow disc on the lip.

CATTLEYA × ELLA.

Gard. Chron. 1898, vol. xxiv. p. 202 (Report of R.H.S. Orchid Committee); Orchid Review, 1898, p. 317; Dict. Ic. des Orchidées, Cattleya hybr. pl. 13.

A distinct hybrid raised by Seden from Cattleya bicolor and C. Warscewiczii. The pale rosy lilac sepals have a white area at the base; the petals are broader than the sepals and of a darker tint; the lip resembles that of C. bicolor, having pinkish side-lobes and a front lobe of glowing purple.

CATTLEYA × ELVINA.

Gard. Chron. 1896, vol. xx. p. 534 (Report of R.H.S. Orchid Committee).

Raised by Seden from Cattleya Trianiæ and C. Schilleriana.

CATTLEYA × EMPRESS FREDERICK, var. LEONATA.

Gard. Chron. 1897, vol. xxii. pp. 428, 429, fig. 127.

Raised by Seden from Cattleya Mossiæ and C. Dowiana, the first plant flowering in the collection of Baron Schröder in June 1892. The variety *Leonata* differs from the original in having the sepals and petals of a bright rose colour instead of white.

CATTLEYA × ENID.

Gard. Chron. 1898, vol. xxiv. p. 92; Dict. Ic. des Orchidées, Cattleya hybr. pl. 23.

A hybrid raised by Seden from Cattleya Mossiæ and C. Warscewiczii. The sepals and petals are light rose-purple with a finely crisped lip of deep crimson-purple with a rich yellow throat.

CATTLEYA × EROS.

Gard. Chron. 1895, vol. xviii. p. 192 (Report of R.H.S. Orchid Committee); Orchid Review, 1895, vol. iii. p. 259.

Raised from Cattleya Mossiæ and C. Walkeriana. The flowers have the open shape of the last-named, but are larger, bright rose in colour, with a velvety crimson lip.

CATTLEYA × EUPHRASIA.

Gard. Chron. 1896, vol. xx. p. 310 (Report of R.H.S. Orchid Committee); Orchid Review, 1896, vol. iv. p. 297.

Raised from Cattleya Warscewiczii and C. superba, the latter being the pollen parent. The flower is of good form and substance, most like the seed parent in general character: the sepals and petals are bright rose-purple, the lip sub-entire, rich crimson-purple in front, and the throat nearly white with two yellow blotches at the sides.

CATTLEYA × EURYDICE.

Gard. Chron. 1895, vol. xviii. p. 527 (Report of R.H.S. Orchid Committee).

Raised by Seden from Cattleya labiata and C. Aclandiæ. The sepals and petals, of a pinkish lavender hue, bear a few purple spots; the rich crimson lip is yellow on the disc passing to cream-white at the base.

CATTLEYA × FABIA.

Orchid Review, 1894, vol. ii. p. 375.

Raised from Cattleya labiata and C. Dowiana. The light rosy-pink flowers have mottled segments and a lip approaching that of C. labiata, with some yellow in the throat.

CATTLEYA × FABIOLA.

Gard. Chron. 1896, vol. xx. p. 601.

Raised from Cattleya Bowringiana and the hybrid C. Harrisi.

CATTLEYA × INTERTEXTA.

Gard. Chron. 1897, vol. xxi. p. 177.

Raised from Cattleya Mossiæ and C. Warneri; the flowers adhere closely in form to those of the last-named parent.

CATTLEYA × MANGLESII, *Rchb. f.*

Rchb. f. in Gard. Chron. 1880, vol. xiv. p. 556; Veitchs' Man. Orch. Pl. pt. ii. p. 91.

Raised by Dominy from Cattleya Lueddemanniana crossed with C. Loddigesii, and flowered for the first time in August 1866.

CATTLEYA × MANTINII.

Veitchs' Catlg. of Pl. 1896, p. 4.

A charming hybrid raised by Mons. George Mantin, President of the Orchid Committee of the Societé Nationale d'Horticulture de France, and also by Seden from Cattleya Bowringiana and C. Dowiana, the latter being the pollen parent. Specimens from both progenies were exhibited simultaneously at the Royal Horticultural Society's Meeting on October 29th 1895.

PRINCIPAL ORCHID HYBRIDS

CATTLEYA × MARSTERSONIÆ, *Rchb. f.*

Rchb. f. in Gard. Chron. 1878, vol. x. p. 556; Veitchs' Man. Orch. Pl. pt. ii. p. 92.

Raised and named in honour of Mrs. Seden *née* Marsterson, and one of the first as also one of the most beautiful of the series of hybrids obtained by crossing one of the labiata forms with other species of Cattleya.

The parents were C. Loddigesii and C. labiata vera.

CATTLEYA × MELPOMENE.

Gard. Chron. 1897, vol. xxii. p. 315 (Report of R.H.S. Orchid Committee).

Raised by Seden from Cattleya Forbesii and C. Mendelii. The light rose-coloured flower has the lip white tinged with pink and a yellow throat.

CATTLEYA × MINUCIA.

Gard. Chron. 1892, vol. xii. p. 379 (Report of R.H.S. Orchid Committee); Orchid Review, 1893, vol. i. p. 357.

Raised by Seden from Cattleya Loddigesii and C. Warscewiczii. The flowers resemble somewhat a form of C. labiata, but with distinct traces of C. Loddigesii in the lip.

CATTLEYA × MIRANDA.

Gard. Chron. 1897, vol. xxi. p. 51 (Report of R.H.S. Orchid Committee); Orchid Review, 1897, vol. v. p. 80.

Raised by Seden from Cattleya amethystoglossa (guttata Prinzii) and C. Trianæ, the second hybrid in which C. amethystoglossa participates in the parentage.

CATTLEYA × NIOBE.

Gard. Chron. 1902, vol. xxxi. p. 280 (Report of R.H.S. Orchid Committee); Orchid Review, 1902, vol. x. p. 149.

Raised from a cross between Cattleya Aclandiæ and C. Mendelii. The seedling is singularly dwarf in habit, and has large wax-like flowers of a rose-colour sparsely spotted with purple.

CATTLEYA × OLIVIA.

Gard. Chron. 1897, vol. xxii. p. 315 (Report of R.H.S. Orchid Committee).

Raised by Seden from Cattleya intermedia and C. Trianæ: the flower is of a delicate peach-blossom colour.

CATTLEYA × PHEIDINÆ.

Gard. Chron. 1893, vol. xiv. p. 470 (Report of R.H.S. Orchid Committee); Orchid Review, 1893, vol. i. p. 363.

Raised by Seden from Cattleya intermedia and C. maxima; the reverse of the cross which produced, thirty-four years earlier, C. × Dominiana.

From this last-named it differs in having the lip closely veined all over, as in C. maxima.

CATTLEYA × PHILO.

Gard. Chron. 1892, vol. xi. p. 535 (Report of R.H.S. Orchid Committee).

Raised by Seden from Cattleya iricolor and C. Mossiæ, the first in which the rare C. iricolor participated.

CATTLEYA × PHILO, var. ALBIFLORA.

Orchid Review, 1893, vol. i. p. 357.

A light coloured form of the preceding derived from the same parents.

CATTLEYA × PICTURATA, *Rchb. f.*

Syns. *C.* × *hybrida*, var. *picta*.

Rchb. f. in Gard. Chron. 1877, vol. viii. p. 584; Fl. Mag. 1881, t. 473; Gard. Chron. 1889, vol. v. p. 746 (Rolfe's List of Garden Orchids); Veitchs' Man. Orch. Pl. pt. ii. p. 91.

A hybrid raised by Dominy at Exeter from Cattleya guttata and C. intermedia.

CATTLEYA × PORPHYROPHLEBIA, *Rchb. f.*

Rchb. f. in Gard. Chron. 1885, vol. xxiv. p. 552; Veitchs' Man. Orch. Pl. pt. ii. p. 92.

A hybrid raised by Seden from Cattleya intermedia and C. superba.

The rich purple veinings on the lip of the flower are remarkable, and suggested the name, "a purple vein."

CATTLEYA × PRINCESS.

Gard. Chron. 1899, vol. xxvi. p. 335 (Report of R.H.S. Orchid Committee).

Raised by Seden from Cattleya Trianæ and C. Lueddemanniana.

The sepals and petals are of a pale rose colour, the lip bright reddish purple with diverging orange-coloured lines, and all parts very broad.

CATTLEYA × QUINQUECOLOR.

Fl. Mag. t. 511; Williams' Orch. Man. 7th edt. p. 183.

A hybrid raised by Dominy, which received the name quinquecolor, or five-coloured, on account of the many colours the different parts of the flower assume. The parents were Cattleya Forbesii and C. Aclandiæ.

CATTLEYA × SUAVIOR.

Veitchs' Man. Orch. Pl. pt. ii. p. 92; Gard. Chron. 1889, vol. v. p. 802 (Rolfe's List of Garden Orchids).

Raised by Seden from Cattleya intermedia and C. Mendelii.

PRINCIPAL ORCHID HYBRIDS

CATTLEYA × VESTALIS.

Gard. Chron. 1899, vol. xxvi. p. 402 (Report of R.H.S. Orchid Committee).

Raised by Seden from Cattleya Dowiana aurea and C. maxima, the latter being the seed-bearer. The flowers of the hybrid resemble those of C. maxima, but are larger and of a blush-white or pale pink hue, with a purple lip passing to rich orange colour at the base.

CATTLEYA × WENDLANDIANA.

Gard. Chron. 1894, vol. xvi. p. 447 (Report of R.H.S. Orchid Committee); Orchid Review, 1894, vol. ii. p. 144; Dict. Ic. des Orchidées, Cattleya hybr. pl. 12.

Raised from Cattleya Warscewiczii and C. Bowringiana, two of the most distinct of all Cattleyas: the best characters of both are blended in the progeny.

The brilliant rose-purple is reproduced, but the throat is more expanded and has two bright yellow blotches.

The plant was named in compliment to the late Herr Wendland, Superintendent of the Berggarten, Herrenhausen, Hanover.

CHYSIS × CHELSONI, *Rchb. f.*

Rchb. f. in Gard. Chron. 1874, p. 535; *id.* 1880, vol. xiii. p. 717, fig.; Fl. Mag. 1878; n.s. t. 297; l'Orchidophile, 1883, p. 497; Veitchs' Man. Orch. Pl. pt. vi. p. 27, fig.; Williams' Orch. Man. edt. 7, p. 195, fig.

This, the first hybrid raised in this genus, was flowered at Chelsea, Chysis Limminghei being the seed and C. bractescens the pollen parent.

CHYSIS × LANGLEYENSIS.

Gard. Chron. 1896, vol. xix. p. 593 (Report of R.H.S. Orchid Committee).

Raised by Seden from Chysis bractescens and the hybrid C. × Chelsoni.

CYMBIDIUM × EBURNEO-LOWIANUM.

Gard. Chron. 1889, vol. v. p. 363; Veitchs' Man. Orch. Pl. pt. ix. p. 23; Dict. Ic. des Orchidées, Cymbidium, hybr. pl. 1.

The first hybrid in the genus produced artificially: as the name implies, the parents were Cymbidium eburneum and C. Lowianum.

The seedling plants took nine years to flower. Since the appearance of this first cross another hybrid has been produced by using the species reversed, and is known as C. × Lowio-eburneum.

CYMBIDIUM × LOWIO-GRANDIFLORUM.

Gard. Chron. 1902, vol. xxxi. p. 116; Orchid Review, 1902, vol. x. p. 100.

Raised by Seden from Cymbidium Lowianum and C. grandiflorum, the former being the seed-bearer. The flowers resemble a good form of

C. Lowianum with the addition of some red-brown spots at the base of the front lobe of the lip, which show the influence of C. grandiflorum.

CYPRIPEDIUM × ADRASTUS.

Syns. C. × *Euryades.*
C. × *Hera.*

Gard. Chron. 1892, vol. xi. pp. 214, 243 (Report of R.H.S. Orchid Committee); Dict. Ic. des Orchidées, Cypripedium, hybr. pl. 44; Jour. of Hort. 1894, p. 103, fig. 16; Orchid Review, 1899, vol. vii. p. 61.

Raised by Seden from Cypripedium × Leeanum and C. villosum Boxalli, and exhibited in February 1892 under the name of C. × Hera, but owing to another variety already bearing that name it was changed to that which it now bears.

CYPRIPEDIUM × ÆSON.

Gard. Chron. 1892, vol. xiii. p. 744 (Report of R.H.S. Orchid Committee); Orchid Review, 1893, vol. i. p. 61.

Raised from Cypripedium insigne and C. Druryi, with much resemblance to the first-named parent. A great improvement on the original hybrid is known as *Æson giganteum.*

CYPRIPEDIUM (SELENIPEDIUM) × AINSWORTHII, var. CALURUM.

Syns. *C.* × *calurum*, Rchb.

Nich. Dict. Gard. vol. iii. p. 413; Rchb. in Gard. Chron. 1881, vol. xv. p. 41; Veitchs' Catlg. of Pl. 1882, p. 22; Fl. and Pom. 1884, p. 145, pl. 619; Orchid Album, iii. t. 136.

Raised by Seden from Cypripedium (Selenipedium) longifolium and C. × Sedenii. The very curious and beautiful petals, resembling C. longifolium, but longer and twisted, suggested the name given by Professor Reichenbach.

The type was raised by Mitchell, gardener to Dr. Ainsworth, of Manchester.

CYPRIPEDIUM (SELENIPEDIUM) × ALBO-PURPUREUM, *Rchb. f.*

Rchb. f. in Gard. Chron. 1877, vol. viii. p. 38; l'Orchidophile, 1883, pp. 508, 509, fig.; Veitchs' Man. Orch. Pl. pt. iv. p. 102.

Raised by Seden from Cypripedium (Selenipedium) Schlimii and the first hybrid Cypripede (C. × Dominianum) of the Selenipedia group.

The influence of the pollen parent preponderates in the form of the floral segment, and that of the seed parent in the colour.

PRINCIPAL ORCHID HYBRIDS

CYPRIPEDIUM × ANTIGONE.

Rolfe in Gard. Chron. 1890, vol. viii. p. 716; Jour. of Hort. 1891, vol. xxiii. p. 262, fig. 49.

A hybrid raised by Seden from Cypripedium Lawrenceanum and C. niveum, the reverse cross of that which produced C. × Aphrodite, altogether a more robust plant.

CYPRIPEDIUM × APHRODITE.

Veitchs' Man. Orch. Pl. pt. iv. p. 77; Gard. Chron. 1893, vol. xiv. p. 342 (Report of R.H.S. Orchid Committee); Gard. Mag. 1894, Feb. 10, p. 76, fig.

Raised from Cypripedium niveum and C. Lawrenceanum; and a very handsome hybrid, with foliage scarcely less attractive than the flowers, which are white with rosy veins and dots.

CYPRIPEDIUM × ARETE.

Gard. Chron. 1892, vol. xii. p. 744 (Report of R.H.S. Orchid Committee); *id.* 1893, vol. xiii. p. 8; Orchid Review, 1903, vol. i. p. 32.

Raised by Seden from Cypripedium concolor and C. Spicerianum, the latter being the pollen parent.

The creamy white flowers are all covered with a profusion of rose dots, and the upper sepal, base of petals and edge of lip are tinged with green.

CYPRIPEDIUM × ARTEMIS.

Gard. Chron. 1895, vol. xvii. p. 199 (Chapman's List of Hybrid Orchids).

Raised from Cypripedium Dayanum and C. Swanianum.

CYPRIPEDIUM × ARTHURIANUM, *Rchb. f.*

Rchb. f. in Gard. Chron. 1874, p. 676; Veitchs' Man. Orch. Pl. pt. iv. p. 77, fig.; l'Orchidophile, 1887, p. 209, col. pl.; Orchid Review, 1893, vol. i. p. 305, fig.; *id.* 1905, vol. xiii. p. 40; Dict. Ic. des Orchidées, Cypripedium hybr. pl. 12.

Raised at Chelsea by Seden from Cypripedium insigne and the then rare and beautiful C. Fairieanum.

Only a single seedling was raised, and this flowered for the first time in 1874, when Professor Reichenbach dedicated it to the late Arthur Veitch. It was the second of the Fairieanum hybrids to be raised, and is one of the most robust of the group. The influence of C. Fairieanum is obvious in the undulate petals and in the veining of the dorsal sepal.

CYPRIPEDIUM × ARTHURIANUM, var. PULCHELLUM.

Jour. of Hort. 1892 (Feb.), fig. 66; Orchid Review, 1893, vol. i. p. 305; *id.* 1894, vol. ii. p. 44; *id.* 1905, vol. xiii. p. 40, fig.; Veitchs' Catlg. of Pl. 1894, p. 10.

A variety obtained from a cross in which Cypripedium insigne Chantini was used with C. Fairieanum instead of the typical C. insigne.

The dorsal sepal in this variety is broader than in the type, the spots fewer in number and twice as large; the petals shorter and the nerves darker.

CYPRIPEDIUM × ASTREA.

Gard. Chron. 1892, vol. xii. p. 191 (Report of R.H.S. Orchid Committee).

Raised by Seden from Cypripedium Spicerianum and C. Philippinense.

The upper sepal is pure white with a green tinge at the base and a purple line in the centre; the twisted petals are tinted with rose and the lip with lilac.

CYPRIPEDIUM × BARON SCHRÖDER.

Gard. Chron. 1896, vol. xx. p. 667 (Report of R.H.S. Orchid Committee); Jour. of Hort. 1896, p. 533, fig.; Gard. Mag. 1896, p. 890, fig.; Orchid Review, 1901, vol. ix. p. 81, fig. 16.

Raised by Seden from Cypripedium × œnanthum superbum and the once rare C. Fairieanum, and one of the most perfect of the hybrid Cypripedes in which C. Fairieanum participates as a parent. The dorsal sepal is heavily spotted with reddish purple on a lighter ground, and the petals veined and suffused with similar colours and spotted on the lower half. It is a hybrid of the third generation, four species being concerned in its ancestry; C. Fairieanum, C. insigne, C. villosum, and C. barbatum.

CYPRIPEDIUM (SELENIPEDIUM) × BRYSA.

Gard. Chron. 1892, vol. xi. p. 343 (Report of R.H.S. Orchid Committee); Orchid Review, 1893, vol. i. p. 358.

Raised by Seden from Cypripedium (Selenipedium) × Sedeni candidulum and C. (S.) Boissierianum.

The flowers resemble a large form of the first-named parent with a greenish tinge in the colouring.

CYPRIPEDIUM × CALANTHUM, *Rchb. f.*

Rchb. f. in Gard. Chron. 1880, vol. xiv. p. 652; Veitchs' Man. Orch. Pl. pt. iv. p. 79.

A hybrid from Cypripedium barbatum Crossii and C. Lowii, flowered for the first time in September 1878.

CYPRIPEDIUM × CALOPHYLLUM.

Rchb. in Gard. Chron. 1881, vol. xv. p. 169; Veitchs' Man. Orch. Pl. pt. iv. p. 80.

One of the earliest hybrids to be raised artificially by us, and never described until it was again flowered by Mr. B. S. Williams. The parents are Cypripedium barbatum and C. venustum.

CYPRIPEDIUM × CALYPSO.

Veitchs' Catlg. of Pl. 1890, p. 14; Gard. Chron. 1891, vol. ix. p. 86 (Report of R.H.S. Orchid Committee); Dict. Ic. des Orchidées; Cypripedium hybr. pl. 26.

Raised from Cypripedium villosum Boxalli and C. Spicerianum, and distinguished among the C. Spicerianum hybrids by most bright and varied colours.

The white purple-banded dorsal sepal is inherited from C. Spicerianum as is also the red-brown lip; the brilliant colour so remarkable in this hybrid is due to the influence of C. villosum Boxalli.

There are now in cultivation many forms raised since the type appeared, distinguished by varietal names.

CYPRIPEDIUM × CAPTAIN HOLFORD.

Gard. Chron. 1899, vol. xxvi. p. 198 (Report of R.H.S. Orchid Committee).

Raised by Seden from Cypripedium hirsutissimum and C. superbiens. The large handsome flowers have a decided resemblance to C. hirsutissimum in the dorsal sepal; the broad sepals are white, tinged with green at the base, densely spotted with dark purple; the lip is large, of a dull rose colour.

CYPRIPEDIUM (SELENIPEDIUM) × CARDINALE, *Rchb. f.*

Syns. *Phragmopedilum* × *cardinale*, Rolfe.

Rchb. f. in Gard. Chron. 1882, vol. xviii. p. 488; The Garden, 1885, vol. xxvii. p. 520, pl. 495; Veitchs' Man. Orch. Pl. pt. iv. p. 102; Orchid Review, 1893, vol. i. p. 81, fig. 5; *id.* 1903, p. 248, fig. 5; Orchid Album, vii. t. 370.

Raised by Seden from Cypripedium × Sedenii and C. (Selenipedium) Schlimii albiflorum.

The richly coloured lip in contrast to the almost pure white sepals and petals, and the elegant form of the flower render this hybrid one of the most attractive of the Selenipedia group.

CYPRIPEDIUM (SELENIPEDIUM) × CLEOLA.

Gard. Chron. 1892, vol. xii. p. 744 (Report of R.H.S. Orchid Committee); *id.* 1893, vol. xiii. p. 8; Orchid Review, 1893, vol. i. p. 326; Dict. Ic. des Orchidées, Selenipedium hybr. pl. 2.

Raised from Cypripedium (Selenipedium) Schlimii album and C. reticulatum (Boissierianum). The plant has the characteristic habit of the Selenipedium section to which both the parents belong.

CYPRIPEDIUM (SELENIPEDIUM) × CLONIUS.

Gard. Chron. 1893, vol. xiv. p. 536 (Report of R.H.S. Orchid Committee); Jour. of Hort. 1893, pp. 394, 395, fig. 59.

Raised by Seden from Cypripedium (Selenipedium) × conchiferum and C. (S.) caudatum Wallisii.

The flowers somewhat resemble those of C. × conchiferum, but have more attenuated petals; the pouch is clear waxy white with the infolded lobes spotted with purple; the petals white with green lines and rose-tinted drooping tail-like tips.

CYPRIPEDIUM × CREON.

Jour. of Hort. 1892, vol. xxiv. p. 205, fig. 31; Gard. Chron. 1891, vol. ix. p. 214 (Report of R.H.S. Orchid Committee).

Raised from Cypripedium × œnanthum superbum and C. Harrisianum, having fine hybrids for parents; the result is disappointing.

CYPRIPEDIUM × CRETHUS.

Gard. Chron. 1892, vol. xii. p. 622 (Report of R.H.S. Orchid Committee); Orchid Review, 1893, vol. i. p. 359.

Raised from Cypripedium Spicerianun crossed with the pollen of C. Argus.

The upper sepal is white-bordered, with black dots about the surface, the petals yellowish-green, also black-spotted, and the lip coppery green with a rosy suffusion round the orifice.

CYPRIPEDIUM (SELENIPEDIUM) × DOMINIANUM, *Rchb. f.*

Rchb. f. in Gard. Chron. 1870, p. 1181; Fl. Mag. 1870, t. 499; Veitchs' Man. Orch. Pl. pt. iv. p. 103, fig. opposite; The Garden, 1891, vol. xxxix. p. 412, pl. 803; l'Orchidophile, 1882, pp. 452, 453, fig.

Raised by John Dominy at Chelsea from Cypripedium (Selenipedium) caricinum and C. caudatum, and named by Professor Reichenbach in compliment to him. This, the first hybrid raised among the Selenipedia, is still one of the most admired of the group.

CYPRIPEDIUM (SELENIPEDIUM) × DOMINIANUM, var. CLYMENE.

Orchid Review, 1894, vol. ii. p. 160; Gard. Chron. 1893, vol. xiii. p. 456 (Report of R.H.S. Orchid Committee).

A variety of the preceding having pale coloured flowers. It was raised by Seden from C. (Selenipedium) caricinum and C. caudatum Wallisii.

CYPRIPEDIUM × DRURYO-HOOKERÆ.

Gard. Chron. 1896, vol. xix. p. 530 (Report of R.H.S. Orchid Committee); Dict. Ic. des Orchidées, Cypripedium hybr. pl. 39.

The parentage of this seedling is expressed by the name.

The flowers are of wax-like substance and singular in colour; the dorsal sepal green passing to white at the border; the petals clear green strongly tinted with rose-lilac passing to white at the tips with a strong median line of purple; the lip is yellowish-green, tinted and veined with rose.

PRINCIPAL ORCHID HYBRIDS

CYPRIPEDIUM × ELECTRA, *Rolfe.*

Rolfe in Gard. Chron. 1888, vol. iii. p. 297; Veitchs' Man. Orch. Pl. pt. iv. p. 83.

A hybrid raised from unknown parents, but probably of the same origin as Cypripedium × Galatea.

CYPRIPEDIUM × EURYALE.

Veitchs' Man. Orch. Pl. pt. iv. p. 83.

Raised by Seden from Cypripedium Lawrenceanum and C. superbiens.

CYPRIPEDIUM × EURYANDRUM, *Rchb. f.*

Rchb. f. in Gard. Chron. 1875, vol. iv. p. 772; Fl. Mag. n.s. t. 187; Fl. des Serres, xxii. t. 2278; Veitchs' Man. Orch. Pl. pt. iv. p. 83, fig.

Raised by Seden from Cypripedium barbatum and C. Stonei, and well known as one of the most distinct of its race.

CYPRIPEDIUM × EURYLOCHUS.

Gard. Chron. 1892, vol. xi. p. 664 (Report of R.H.S. Orchid Committee); Orchid Review, 1893, vol. i. p. 359.

Raised by Seden from Cypripedium ciliolare and C. hirsutissimum. The flowers, attractive and prettily spotted, have petals curiously elongated and deflexed.

CYPRIPEDIUM × EVENOR.

Gard. Chron. 1892, vol. xi. p. 664 (Report of R.H.S. Orchid Committee); Orchid Review, 1893, vol. i. p. 359.

A very interesting hybrid with purple-spotted flowers on a cream-white ground, raised from Cypripedium Argus and C. bellatulum.

CYPRIPEDIUM × GALATEA, *Rolfe.*

Rolfe in Gard. Chron. 1888, vol. iii. p. 168; Veitchs' Man. Orch. Pl. pt. iv. p. 84.

A hybrid of unknown parentage, probably from Cypripedium Harrisianum and C. insigne Maulei, and to this cross probably also belong C. × Acis, C. × Orestes, and C. × Electra.

CYPRIPEDIUM × GERMINYANUM, *Rchb. f.*

Rchb. f. in Gard. Chron. 1886, vol. xxv. p. 200; Veitchs' Man. Orch. Pl. pt. iv. p. 85; Jour. of Hort. 1893, pp. 67, 74, fig. 10; Dict. Ic. des Orchidées, Cypripedium hybr. pl. 52.

Raised by Seden from Cypripedium villosum and C. hirsutissimum, and dedicated to Count Adrien de Germiny, of Gonville, near Rouen, who was the owner of one of the finest collections of orchids in France.

CYPRIPEDIUM (SELENIPEDIUM) × GIGANTEUM.

Syns. *C.* × *macrochilum*, var. *giganteum*.

Orchid Review, 1894, vol. ii. p. 186; Gard. Mag. 1894, p. 256, fig.; Jour. of Hort. 1894, pp. 386, 387, fig. 62; Gard. Chron. 1894, vol. xv. p. 602 (Report of R.H.S. Orchid Committee).

A hybrid, raised by Seden, resembling a very large form of Selenipedium × macrochilum, and derived from Cypripedium caudatum Lindeni (Phragmopedilum caudatum, var. Uropedium) and the hybrid Selenipedium × grande. The lip is broad and well rounded in front with white side lobes beautifully spotted and tinged with purple; the petals, white, striped with green, with claret coloured tips, attain a length of over 18 in.

CYPRIPEDIUM (SELENIPEDIUM) × GRANDE.

Rchb. in Gard. Chron. 1881, vol. xv. p. 462; *id.* 1882, vol. xviii. p. 488; Lindenia, vi. p. 7, t. 242; Veitchs' Man. Orch. Pl. pt. iv. p. 104, fig.

One of the finest of the Selenipedium hybrids, raised from Cypripedium (Selenipedium) longifolium Roezlii (Hartwegii) crossed with the pollen of C. (Selenipedium) caudatum.

It is the most robust in the group to which it belongs, the sword-shaped leaves being from 20 to 30 in. long, the flower-scape frequently more than a yard in height, and the flowers 7 to 8 in. across the sepals from tip to tip.

CYPRIPEDIUM (SELENIPEDIUM) × GRANDE, var. MACROCHILUM.

Syns. *C.* × *macrochilum*.

Orchid Review, 1893, vol. i. p. 326; O'Brien in Gard. Chron. 1891, vol x. pp. 199, 343, fig.

Raised from Cypripedium (Selenipedium) longifolium and C. (S.) caudatum Lindeni (Phragmopedilum caudatum, var. Uropedium), and the most extraordinary hybrid in the group to which it belongs, being the product of the practically pouchless Phragmopedilum crossed with the normal pouched Selenipedium, and possessing a remarkably long and elongated pouch.

CYPRIPEDIUM × HARRISIANUM, *Rchb. f.*

Rchb. f. in Gard. Chron. 1869, p. 108; Fl. Mag. 1869, pl. 431. Fl. des Serres, tom. xxii. tt. 2289-2290; Veitchs' Man. Orch. Pl. pt. iv. p. 86, fig.; Dict. Ic. des Orchidées, Cypripedium hybr. pl. 11.

Raised by John Dominy about the year 1864 from Cypripedium villosum and C. barbatum at Chelsea, where it flowered for the first time in 1869, and the first Cypripedium artificially raised: it has since been obtained by several operators from the same cross.

The name Harrisianum was given in compliment to Dr. Harris, of Exeter, who gave Dominy the idea of hybridizing orchids.

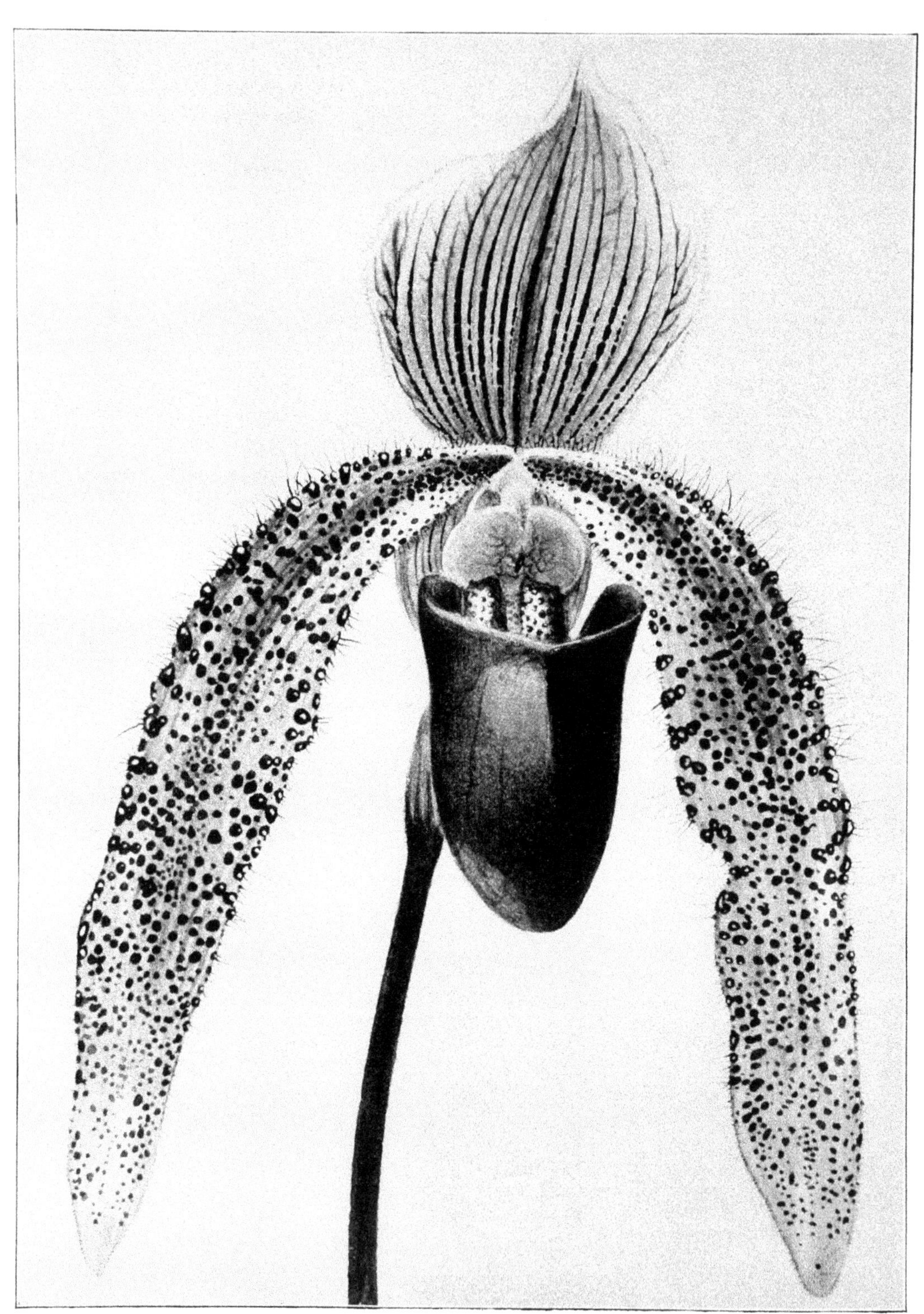

CYPRIPEDIUM ×"JAMES H. VEITCH"

There are several forms, of which the one named *superbum* is probably the most distinct.

CYPRIPEDIUM × H. BALLANTINE.

Syns. *C.* × *Ballantinei.*

Rolfe in Gard. Chron. 1890, vol. viii. p. 408; Veitchs' Catlg. of Pl. 1893, p. 13, fig.; Orchid Review, 1905, vol. xiii. p. 18.

Raised by Seden from Cypripedium purpuratum and C. Fairieanum, named after Mr. H. Ballantine, the grower of Baron Schröder's fine collection of orchids, and flowered for the first time in August 1890.

It is undoubtedly one of the finest of the hybrids from C. purpuratum, and clearly shows the influence of both parents.

CYPRIPEDIUM × IANTHE.

Gard. Chron. 1892, vol. xi. p. 343 (Report of R.H.S. Orchid Committee); Orchid Review, 1893, vol. i. p. 358; Veitchs' Catlg. of Pl. 1894, p. 13.

Probably from Cypripedium Harrisianum and C. venustum: the flowers much resemble C. Harrisianum in shape, but are distinct in colour.

CYPRIPEDIUM × JAMES H. VEITCH.

Gard. Chron. 1894, vol. xvi. p. 258 (Report of R.H.S. Orchid Committee); *id.* p. 287, fig. 40; Orchid Review, 1894, vol. ii. p. 309; Jour. of Hort. 1894, Sept. 6th, p. 227, fig. 33.

Raised by Seden from Cypripedium Stonei platytænium and C. Curtisii, and one of the finest of all hybrids. The great breadth of petals peculiar to the celebrated variety of Cypripedium Stonei is transmitted to the offspring; they are broadly ribbon-like, pendulous, 4½ in. in length, about 1 in. broad, and light yellow evenly spotted with dark red-purple warts except near the apices. The lip is helmet-shaped and almost uniform red-crimson with the infolded side lobes paler but dotted with darker warts.

CYPRIPEDIUM × LEEANUM, var. SUPERBUM.

Gard. Chron. 1885, vol. xxiii. p. 277; Rchb. in Gard. Chron. 1886, vol. xxv. p. 168; l'Orchidophile, 1885, p. 240; Veitchs' Man. Orch. Pl. pt. iv. p. 88, with fig.

This variety was raised by Seden from Cypripedium insigne Maulei and C. Spicerianum. The hybrid known as C. Leeanum was first raised at Burford Lodge, Dorking, and flowered sometime after that of Seden's production, which, being the finer variety, was called *superbum.*

CYPRIPEDIUM (SELENIPEDIUM) × LEUCORRHODUM, *Rchb.*

Syns. *Phragmopedilum* × *leucorrhodum*, Rolfe.

Rchb. in Gard. Chron. 1885, vol. xxiii. p. 270; Veitchs' Man. Orch. Pl. pt. iv. p. 104; Orchid Review, 1893, vol. i. p. 169, fig. 11.

Raised from Cypripedium (Selenipedium) Roezlii and C. (S.) Schlimii albiflorum, the influence of the pollen parent shows strongly in this hybrid. The general colour is blush white with a suffusion of delicate pink on the upper sepal and a lip light rose pink suffused with white.

CYPRIPEDIUM × LITTLE GEM.

Orchid Review, 1903, vol. xi. p. 341; *id.* 1905, vol. xiii. p. 108, fig. 25.

A seedling raised by Seden from Cypripedium × Baron Schröder and C. × Harrisianum, which flowered for the first time in April 1903.

Several seedlings were raised from the same cross, but they differ widely in size as in form.

In the typical form, now known as the Westfield variety, the flower is small, the dorsal sepal white, flushed with dark rose, and the markings are of chocolate-purple; the lip is purple-brown; other varieties show more of the influence of C. × Harrisianum, and have flowers above the medium size and of a rich deep colour.

CYPRIPEDIUM × LUCIDUM, *Rchb. f.*

Rchb. f. in Gard. Chron. 1877, vol. viii. p. 521.

Raised by Seden by crossing Cypripedium Lowii with the pollen of C. villosum. The upper sepal is greenish with much brown at the base, and in the centre there are many spots; the petals, narrow at the base, are dilated at the apex and shining; the lip is like that of C. villosum, and chiefly brownish-violet in colour.

CYPRIPEDIUM × MACROPTERUM, *Rchb. f.*

Rchb. f. in Gard. Chron. 1882, vol. xviii. p. 552; *id.* 1883, vol. xx. p. 294; Veitchs' Man. Orch. Pl. pt. iv. p. 90.

Raised from Cypripedium Lowii and C. superbiens, and remarkable as being one of the comparatively few hybrids in which the influence of the seed parent is shown in the flower and that of the pollen parent in the foliage; the influence of C. superbiens is seen only in the spotting on the basal half of the petals and in the large helmet-shaped lip and in the staminode.

CYPRIPEDIUM × MARMOROPHYLLUM, *Rchb. f.*

Rchb. f. in Gard. Chron. 1876, vol. vi. p. 130; Veitchs' Man. Orch. Pl. pt. iv. p. 90.

Raised by Seden from Cypripedium Hookeræ crossed with C. barbatum. The influence of the pollen parent preponderates in the flower; that of the seed parent is most decided in the foliage.

CYPRIPEDIUM × MARSHALLIANUM, *Rchb.*

Rchb. in Gard. Chron. 1875, vol. iv. p. 804; Veitchs' Man. Orch. Pl. pt. iv. p. 91.

Raised from Cypripedium venustum pardinum crossed with C. concolor.

PRINCIPAL ORCHID HYBRIDS

CYPRIPEDIUM × MEDEIA.

Gard. Chron. 1895, vol. xvii. p. 199 (Chapman's List of Hybrid Orchids).

Raised from Cypripedium Spicerianum and C. hirsutissimum.

CYPRIPEDIUM × MELANCHUS.

Gard. Chron. 1893, vol. xiii. p. 456 (Report of R.H.S. Orchid Committee); Orchid Review, 1893, vol. i. p. 147.

Raised by Seden from Cypripedium Hookeræ and C. Stonei, the latter being the pollen parent. The dorsal sepal is almost of the same shape as that of C. Stonei, but is suffused with olive-green and the stripes are less distinct. The petals are broader than in C. Stonei, and bright purple-brown with numerous small spots. The lip is veined with light brown on a paler ground. The flower resembles the pollen parent, and the influence of the seed parent is chiefly seen in the foliage.

CYPRIPEDIUM × MELIS.

Orchid Review, 1895, vol. iii. p. 320.

Raised from Cypripedium Boxallii and C. philippinense.

CYPRIPEDIUM × MEROPS.

Orchid Review, 1894, vol ii. p. 159.

Raised by Seden from Cypripedium ciliolare and C. Druryi. The flowers are pale buff yellow with red-brown blotches and stains.

CYPRIPEDIUM × MICROCHILUM, *Rchb. f.*

Rchb. f. in Gard. Chron. 1882, vol. xvii. p. 77; Veitchs' Man. Orch. Pl. pt. iv. p. 92, with fig.; Dict. Ic. des Orchidées, Cypripedium, pl. 17.

Raised from Cypripedium niveum crossed with the pollen of C. Druryi, and first flowered in 1882.

One of the most distinct of hybrids well-nigh intermediate between the two parents, the lip is nearly as in C. Druryi, white, veined with pale green, but very small.

CYPRIPEDIUM × MILO.

Gard. Chron. 1894, vol. xvi. p. 670 (Report of R.H.S. Orchid Committee); Orchid Review, 1895, vol. iii. p. 30.

Raised by Seden from Cypripedium insigne Chantinii and C. œnanthum superbum; there are many varieties, but none superior to that known as *grandis*.

CYPRIPEDIUM × MINOS.

Gard. Chron. 1893, vol. xiv. p. 692 (Report of R.H.S. Orchid Committee); Orchid Review, 1895, vol. iii. p. 23; Dict. Ic. des Orchidées, sub Cypripedium hybr. pl. 47.

Raised from Cypripedium Spicerianum crossed with pollen from C. × Arthurianum. The pure white upper sepal is tinted with purple at the base; the lip and petals are yellowish-brown, the latter retaining the downward curve so conspicuous in C. × Arthurianum.

Two of the most distinct varieties are distinguished by the names *magnificum* and *superbum.*

CYPRIPEDIUM × MORGANIÆ.

Rchb. in Gard. Chron. 1880, vol. xiv. p. 134; Gard. Chron. 1886, vol. xxvi. p. 241, fig. 49; *id.* 1890, vol. vii. pl. p. 544; Veitchs' Man. Orch. Pl. pt. iv. p. 92, fig.; The Garden, 1883, vol. xxiii. p. 58, pl. 372; Orchid Review, 1904, vol. xii. p. 41, fig.; Dict. Ic. des Orchidées, Cypripedium hybr. pl. 27.

Raised from Cypripedium superbiens crossed with C. Stonei, and named in compliment to Mrs. Morgan of New York, in her day an ardent amateur of orchids. At the time of its first appearance the finest hybrid of its class, the large pouch and long, drooping, spotted petals rendering it peculiarly attractive.

CYPRIPEDIUM × MORGANIÆ, var. LANGLEYENSE.

Orchid Review, 1894, vol. ii. p. 79; Veitchs' Catlg. of Pl. 1895, p. 6.

This is the first hybrid to flower in which the remarkable Cypripedium Stonei platytænium participated in the parentage.

It differs from the original C. × Morganiæ in the flowers being larger, the spots on the petals more isolated and brighter, and the lip more highly coloured.

CYPRIPEDIUM × NIOBE.

Gard. Chron. 1890, vol. vii. p. 9; Orchid Album, vol. x. t. 438; W. J. Bean in The Garden, 1891, vol. xxxix. p. 483, pl. 806; l'Orchidophile, 1892, p. 81, fig.; Orchid Review, 1893, vol. i. p. 7, fig. 1; *id.* 1905, vol. xiii. p. 41, fig.; Dict. Ic. des Orchidées, Cypripedium hybr. pl. 13.

Raised from Cypripedium Spicerianum with pollen from the once rare C. Fairieanum, from seed sown in 1884, the first flower opening in 1889. They show a combination of the characters of the two parents, and are remarkable in that they possess the broadest dorsal sepal in the group to which they belong.

C. × Niobe forms the subject of the first figure in the first volume of the Orchid Review.

CYPRIPEDIUM × NITENS, *Rchb. f.*

Rchb. f. in Gard. Chron. 1878, vol. ix. p. 398; Veitchs' Man. Orch. Pl. pt. iv. p. 93; Dict. Ic. des Orchidées, Cypripedium hybr. pl. 22.

Raised from Cypripedium villosum and C. insigne Maulei, and first flowered in 1878. The name nitens was suggested by the highly lustrous appearance of the flowers.

CYPRIPEDIUM × NORMA.

Gard. Chron. 1895, vol. xvii. p. 82; Orchid Review, 1905, vol. xiii. p. 104, figs. 22 and 23.

Raised from Cypripedium Spicerianum and C. × Niobe, the first of the hybrids in which the beautiful C. × Niobe took part; it first flowered in January 1895.

A finer form, known as the Westfield variety, has a broader dorsal sepal and is of a darker shade.

CYPRIPEDIUM × NUMA, *Rolfe.*

Rolfe in Gard. Chron. 1890, vol. vii. p. 608.

An uninteresting hybrid from Cypripedium Lawrenceanum and C. Stonei.

CYPRIPEDIUM × ŒNANTHUM, *Rchb. f.*

Rchb. f. in Gard. Chron. 1876, vol. v. p. 297; Veitchs' Man. Orch. Pl. pt. iv. p. 93.

Raised by Seden at Chelsea from Cypripedium × Harrisianum and C. insigne Maulei, the first hybrid Cypripedium to flower, of which one of the parents is itself a hybrid.

CYPRIPEDIUM × ŒNANTHUM, var. SUPERBUM.

Veitchs' Catlg. of Pl. 1885, p. 11; Veitchs' Man. Orch. Pl. pt. iv. p. 94, fig.; Lindenia, t. 33.

An improved form of the type and a very magnificent slipper.

CYPRIPEDIUM × ŒNONE.

Gard. Chron. 1892, vol. xii. p. 744 (Report of R.H.S. Orchid Committee); Veitchs' Catlg. of Pl. 1894, p. 13.

Raised from Cypripedium superbiens and C. Hookeræ, the former being the pollen parent.

The flowers are fairly intermediate between those of the two species, though the lip is nearer that of C. superbiens, brownish in front, pale green beneath.

CYPRIPEDIUM × ORESTES.

Veitchs' Man. Orch. Pl. pt. iv. p. 94.

Raised by Seden from Cypripedium × Harrisianum and probably C. insigne Maulei, but there is much doubt on the point.

CYPRIPEDIUM × ORION.

Gard. Chron. 1893, vol. xiii. p. 80 (Report of the R.H.S. Orchid Committee); Orchid Review, 1893, vol. i. p. 61; Dict. Ic. des Orchidées, Cypripedium hybr. pl. 62.

Raised from Cypripedium insigne crossed with the pollen of C. concolor.

The upper sepal is large, whitish washed with yellow, with spots and tints of purple; the lip and petals are creamy yellow, also washed and spotted with purple.

CYPRIPEDIUM × ORPHANUM, *Rchb. f.*

Rchb. f. in Gard. Chron. 1886, vol. xxvi. p. 166; Veitchs' Man. Orch. Pl. pt. iv. p. 94.

Probably obtained from Cypripedium barbatum and C. Druryi, but by an oversight the parentage was not recorded, and refuge was taken in this name.

CYPRIPEDIUM × PATENS, *Rchb. f.*

Rchb. f. in Gard. Chron. 1877, vol. viii. p. 456.

Raised at Chelsea from Cypripedium Hookeræ crossed with C. barbatum; of no great interest.

CYPRIPEDIUM (SELENIPEDIUM) × PENELAUS.

Orchid Review, 1893, vol. i. p. 61; Jour. of Hort. 1893, March 16th, p. 219, fig. 45.

Raised from the pouchless Cypripedium caudatum (Phragmopedilum caudatum Uropedium) and the hybrid C. × Ainsworthii calurum.

CYPRIPEDIUM (SELENIPEDIUM) × PERSEUS.

Gard. Chron. 1892, vol. xii. p. 622 (Report of R.H.S. Orchid Committee); Orchid Review, 1893, vol. i. p. 358.

Raised by Seden from Cypripedium (Selenipedium) × Sedenii, porphyreum and C. Lindleyanum, and one of the first hybrids to flower, of which this distinct species was a parent.

This occurred during the year 1892.

CYPRIPEDIUM (SELENIPEDIUM) × PERSEUS, var. PHÆDRA.

Orchid Review, 1893, vol. i. p. 52; Gard. Chron. 1893, vol. xiii. p. 80; Veitchs' Catlg. of Pl. 1894, p. 13.

A hybrid of the Selenipedium group, from Cypripedium Lindleyanum of Mount Roraima, and C. (Selenipedium) × Sedenii candidulum, three distinct species being concerned in the parentage.

The dorsal sepal is cream-white mottled with light rose for two-thirds of its length; the petals are about 3 in. long, light rose pink with a white medium line, and the helmet-shaped lip is rose-pink with yellowish-white lobes dotted with crimson.

CYPRIPEDIUM × PHERES.

Gard. Chron. 1892, vol. xii. p. 745 (Report of R.H.S. Orchid Committee); Orchid Review, 1893, vol. i. p. 32.

Raised from Cypripedium hirsutissimum and C. insigne, the latter being the seed-bearer; a not very successful experiment.

CYPRIPEDIUM × PORPHYROCHLAMYS.

Rchb. f. in Gard. Chron. 1884, vol. xxi. p. 476; Veitchs' Man. Orch. Pl. p. iv. p. 96; Orchid Album, ix. t. 426; Orchid Review, 1893, vol. i. p. 105, fig. 7.

Raised from Cypripedium barbatum Crossii and C. hirsutissimum, the first hybrid in which the last-named species participated. The prevailing colour of the flowers is deep crimson-purple; the upper sepal has a white margin and dark veins; the petals are yellowish-green at the base, and the lip brownish-purple.

CYPRIPEDIUM × PRIAM.

Orchid Review, 1900, vol. viii. p. 373; *id.* 1905, vol. xiii. p. 105.

Raised from Cypripedium × Niobe and C. insigne Chantinii, and first flowered in November 1900; a light-coloured form of the type is known as *Leucas*.

CYPRIPEDIUM × PRIAPUS.

Syns. *C.* × *Œolus.*

Gard. Chron. 1890, vol. vii. p. 526 (Report of R.H.S. Orchid Committee).

Raised by Seden from Cypripedium philippinense and C. villosum; the latter it much resembles.

CYPRIPEDIUM × PYCNOPTERUM, *Rchb. f.*

Rchb. in Gard. Chron. 1876, vol. v. p. 622; Veitchs' Man. Orch. Pl. pt. iv. p. 96.

CYPRIPEDIUM × PYCNOPTERUM, *Rchb. f.*, var. PORPHYROSPILUM.

Rchb. in Gard. Chron. 1879, vol. xii. p. 489; Veitchs' Man. Orch. Pl. pt. iv. p. 97.

Both these plants were raised from seed from the same capsule; the second can but be regarded as a variety of the former, which it surpasses in size of flower and depth of colour. The parents are Cypripedium venustum and C. Lowii.

CYPRIPEDIUM × RADIOSUM, *Rchb. f.*

Rchb. in Gard. Chron. 1885, vol. xxiv. p. 424; Veitchs' Man. Orch. Pl. pt. iv. p. 97.

Raised by Seden from Cypripedium Lawrenceanum and C. Spicerianum, and of little value.

CYPRIPEDIUM × REGINÆ.

Gard. Chron. 1896, vol. xx. p. 534 (Report of R.H.S. Orchid Committee).

Raised from Cypripedium Leeanum and C. Fairieanum.

The flowers suggest those of C. Arthurianum, but the upper two-thirds of the dorsal sepal is pure white with a few lines of purple.

CYPRIPEDIUM × SAPPHO.

Gard. Chron. 1895, vol. xvii. p. 199 (Chapman's List of Hybrid Orchids).

Raised by Seden from Cypripedium Lowii and C. barbatum.

CYPRIPEDIUM (SELENIPEDIUM) × SCHRÖDERÆ, *Rchb. f.*

Rchb. f. in Gard. Chron. 1883, vol. xix. p. 432; Orchid Album, v. t. 196; Veitchs' Man. Orch. Pl. pt. iv. p. 105, fig.; Orchid Review, 1898, vol. vi. p. 361, fig. 17.

Raised from Cypripedium (Selenipedium) caudatum and C. (S.) × Sedenii. It is one of the finest of the Selenipedia hybrids, and was named as a compliment to the late Baroness Schröder of the Dell, Egham.

CYPRIPEDIUM (SELENIPEDIUM) × SCHRÖDERÆ CANDIDULUM.

Gard. Chron. 1896, vol. xix. p. 88 (Report of R.H.S. Orchid Committee).

From Cypripedium (Selenipedium) caudatum Wallisii and C. (S.) × Sedenii candidulum.

CYPRIPEDIUM (SELENIPEDIUM) × SEDENII, *Rchb. f.*

Rchb. in Gard. Chron. 1873, p. 1431; id. 1886, p. 596, 597, fig. (peloria); Fl. Mag. n.s. 1876, t. 206; *id.* 1878, t. 302; Veitchs' Man. Orch. Pl. pt. iv. p. 105, fig. p. 106; l'Orchidophile, 1882, p. 179, fig.; Rev. Hort. 1879, p. 470, col. pl.; Dict. Ic. des Orchidées, Selenipedium, hybr. pl. 1.

A hybrid raised by Seden, after whom it is named, by cross-fertilizing Cypripedium (Selenipedium) Schlimii with the pollen of C. (S.) longifolium. The plant is very floriferous, of easy culture, and one of the most popular hybrids in this special group of Slipper Orchids.

CYPRIPEDIUM (SELENIPEDIUM) × SEDENII, var. PORPHYREUM.

Syns. *C.* × *porphyreum*, Rchb.

Gard. Chron. 1878, vol. ix. p. 366; Veitchs' Man. Orch. Pl. pt. iv. p. 106.

Raised by Seden from Cypripedium (Selenipedium) longifolium Hartwegii (Roezlii), crossed with C. (S.) Schlimii, a graceful and delicate object.

CYPRIPEDIUM (SELENIPEDIUM) × SEDENII, var. CANDIDULUM.

Lindenia, t. 245; Orchid Album, t. 481; Dict. Ic. des Orchidées, Selenipedium hybr. pl. 2.

Raised from Cypripedium (Selenipedium) longifolium crossed with C. (S.) Schlimii albiflorum; the flowers are more highly coloured than those of the type.

CYPRIPEDIUM × SELLIGERUM.

Veitchs' Catlg. of Pl. 1878, p. 13, with fig.; Gard. Chron. 1880, vol. xiii. p. 776, with fig. 133; Veitchs' Man. Orch. Pl. pt. iv. p. 97; Dict. Ic. des Orchidées, Cypripedium hybr. pl. 37.

Raised from Cypripedium barbatum crossed with C. philippinense, and flowered for the first time during the summer of 1878. The dorsal sepal is finely lined with purple, and the drooping petals are narrow and most delicately twisted.

CYPRIPEDIUM × SIRIUS.

Orchid Review, 1895, vol. iii. p. 32.

Obtained by Seden from Cypripedium barbatum Warneri and C. Godefroyæ.

CYPRIPEDIUM (SELENIPEDIUM) × SUAVE.

Orchid Review, 1901, vol. ix. p. 93.

Raised from Cypripedium (Selenipedium) × Sedenii candidulum and C. (S.) Klotzschianum.

In the details of the flower this hybrid approaches the first-named plant, which was the seed-bearer, but it also shows intermediate characters between the two species.

CYPRIPEDIUM × SUPERCILIARE, *Rchb. f.*

Rchb. in Gard. Chron. 1876, vol. v. p. 795; Veitchs' Man. Orch. Pl. pt. iv. p. 98; Dict. Ic. des Orchidées, Cypripedium hybr. pl. 38.

A hybrid from Cypripedium barbatum and C. superbiens. The dorsal sepal is very broad, marked with purple and green lines on a white ground; the petals are strongly ciliated with long black hairs and marked with large points of the same colour.

CYPRIPEDIUM × TAUTZIANUM, *Rchb. f.*

Rchb. in Gard. Chron. 1886, vol. xxvi. p. 681; Veitchs' Man. Orch. Pl. pt. iv. p. 98.

Raised from Cypripedium niveum and C. barbatum, and dedicated to the late F. G. Tautz, Esq., of Studley House, Hammersmith, S.W., at that time possessor of one of the best collections of Cypripedes in the country, and a most ardent orchidist.

CYPRIPEDIUM × T. B. HAYWOOD.

Rolfe in Gard. Chron. 1889, vol. v. p. 428.

A hybrid from Cypripedium superbiens and C. Curtisii, named in compliment to the late T. B. Haywood, Esq., of Woodhatch, Reigate, a great amateur.

CYPRIPEDIUM × TELEMACHUS.

Gard. Chron. 1892, vol. xi. p. 816 (Report of R.H.S. Orchid Committee); Orchid Review, 1893, vol. i. p. 359.

Raised by Seden from Cypripedium Lawrenceanum and C. niveum, a similar parentage to C. × Aphrodite, of which it is merely a variety, differing in having the greater part of the surface of its flowers suffused rosy crimson.

CYPRIPEDIUM × TESSELATUM, *Rchb. f.*

Rchb. in Gard. Chron. 1875, vol. iv. p. 614; Dict. Ic. des Orchidées, Cypripedium hybr. pl. 14.

Raised at Chelsea from a cross between Cypripedium concolor and C. barbatum. There is a form of the original hybrid named *porphyreum* more vigorous than the type and more brilliantly coloured.

CYPRIPEDIUM × TESSELATUM, var. PORPHYREUM, *Rchb. f.*

Rchb. in Gard. Chron. 1881, vol. xv. p. 41; *id.* 1885, vol. xx. p. 492; Veitchs' Man. Orch. Pl. iv. p. 99.

Raised from Cypripedium concolor and C. barbatum. At the time of its introduction the peculiar and distinct shade of rose-purple seen in the flowers of this hybrid was unique amongst Cypripedes.

CYPRIPEDIUM × THALIA.

Orchid Review, 1905, vol. xiii. p. 107, fig. 24.

Obtained from Cypripedium × Baron Schröder crossed with C. insigne Chantinii, and a beautiful flower with an orbicular white dorsal line spotted with dark purple. The petals and lip are yellow marked with purple-brown; a variety named *punctatum* is more spotted than the type.

CYPRIPEDIUM × THIBAUTIANUM, *Rchb. f.*

Rchb. f. in Gard. Chron. 1886, vol. xxv. p. 104; Veitchs' Man. Orch. Pl. pt. iv. p. 99.

Raised from Cypripedium Harrisianum crossed with the pollen of C. insigne Maulei; the same parents as produced C. × œnanthum, and named in compliment to the late M. Thibaut of Sceaux, near Paris, formerly a well-known nurseryman and lover of Cypripedia.

CYPRIPEDIUM × TITYUS.

Gard. Chron. 1892, vol. xii. p. 622 (Report of R.H.S. Orchid Committee); Orchid Review, 1893, vol. i. p. 358.

An interesting hybrid obtained from Cypripedium Spicerianum crossed with the pollen of C. × œnanthum superbum, and flowered for the first time during 1892.

CYPRIPEDIUM × VERNIXIUM, *Rchb. f.*

Rchb. f. in Gard. Chron. 1879, vol. xi. p. 398; Veitchs' Man. Orch. Pl. pt. iv. p. 99.

Raised by Seden from Cypripedium Argus crossed with C. villosum.

PRINCIPAL ORCHID HYBRIDS

CYPRIPEDIUM × VEXILLARIUM, *Rchb.*

Gard. Chron. 1870, p. 1373; *id.* 1880, vol. xiii. p. 780, fig. 135; Orchid Album, t. 447; The Garden, 1874, vol. v. p. 102, fig.; Veitchs' Man. Orch. Pl. pt. iv. p. 100, fig.; Orchid Review, 1893, vol. i. p. 297, fig. 15; *id.* 1905, vol. xiii. p. 16; Dict. Ic. des Orchidées, Cypripedium hybr. pl. 2.

Raised by Dominy at Chelsea from Cypripedium barbatum and C. Fairieanum, this hybrid has the peculiar curved and drooping petals of the latter, and the dorsal sepal resembles, more or less, that of C. barbatum. Flowering for the first time in 1870, it was one of the earliest hybrids of the Fairieanum series to bloom.

CYPRIPEDIUM × WINNIANUM, *Rchb. f.*

Rchb. f. in Gard. Chron. 1886, vol. xxv. p. 362; Veitchs' Man. Orch. Pl. pt. iv. p. 100; Dict. Ic. des Orchidées, Cypripedium hybr. pl. 18.

Raised from Cypripedium villosum and C. Druryi, and dedicated to Charles Winn, Esq., of Selly Hill, Birmingham, a noted amateur. The flowers are intermediate between those of the two parents, the prevailing ground colour a soft light brown.

CYPRIPEDIUM × ZENO.

Orchid Review, 1895, vol. iii. p. 32.

Raised from Cypripedium × nitens and C. insigne Chantinii.

DENDROBIUM × ADRASTA.

Gard. Chron. 1892, vol. xi. p. 503 (Report of R.H.S. Orchid Committee); Orchid Review, 1893, vol. i. p. 358.

Raised by Seden from Dendrobium Pierardii and D. superbum, and the first artificial hybrid in which D. Pierardii participated.

The flowers are very pretty, having pale pink sepals and petals, and have a decided primrose-yellow lip.

DENDROBIUM × ÆNEAS.

Gard. Chron. 1893, vol. xiii. p. 366 (Report of R.H.S. Orchid Committee); Orchid Review, 1893, vol. i. p. 128.

Raised from Dendrobium moniliforme crossed with D. crystallinum. The sepals and petals are creamy white, with tips of pale rose; the cream-coloured lip tipped with rose has a curious fulvous bronze disc.

DENDROBIUM × AINSWORTHII, var. INTERTEXTUM.

Orchid Review, 1895, vol. iii. p. 103.

Raised by Seden from Dendrobium nobile and Lee's variety of D. aureum. The waxy-white flowers are large and handsome, the lip primrose-yellow with a maroon disc; a handsome hybrid.

DENDROBIUM × AINSWORTHII, var. SPLENDIDISSIMUM.

See Dendrobium × splendidissimum.

DENDROBRIUM × ALCIPPE.

Syns. *D.* × *Aureum.*

Gard. Chron. 1894, vol. xv. p. 475 (Report of R.H.S. Orchid Committee); Orchid Review, 1894, vol. ii. p. 173.

A hybrid raised by Seden from Dendrobium lituiflorum Freemannii and D. Wardianum.

The sepals and petals are bright rose-purple, paler at the base; the lip is chiefly very white, with an Indian purple disc and an apical border of rose-purple.

DENDROBIUM × ASPASIA.

Syns. *D.* × *Wardiano-aureum.*

Gard. Chron. 1890, vol. vii. p. 336 (Report of R.H.S. Orchid Committee); Orchid Review, 1893, vol. i. p. 137, fig. 9; Veitchs' Catlg. of Pl. 1896.

A curious and charming hybrid from Dendrobium aureum and D. Wardianum.

The plant has a habit nearly intermediate between that of the two parents; the flowers have the general outline of those of D. Wardianum, the sepals and petals are white tipped with rose-purple, and the lip is cream-white with an orange-coloured disc and an irregularly pencilled crimson blotch.

DENDROBIUM × CORDELIA.

Orchid Review, 1894, vol. ii. p. 172.

Raised from Dendrobium aureum and D. euosmum leucopterum; the flowers, 3 in. in diameter, resemble those of D. aureum in shape. The sepals and petals are ivory-white, the former narrowly margined with light pink. The lip is light yellow, the front lobe and apex white.

DENDROBIUM × CYBELE, *Rolfe.*

Rolfe in Gard. Chron. 1878, vol. ix. p. 202; Veitchs' Man. Orch. Pl. pt. iii. p. 87.

A hybrid from Dendrobium Findlayanum and D. nobile, in which the pollen parent has exerted the most influence; the conspicuous yellow lip of the seed parent, D. Findlayanum, is here totally lost and has had no perceptible influence.

DENDROBIUM × DOMINIANUM, *Rchb. f.*

Rchb. in Gard. Chron. 1878, vol. ix. p. 202; Veitchs' Man. Orch. Pl. pt. iii. p. 88.

Raised by John Dominy at Exeter from Dendrobium nobile and D. Linawianum, and named in compliment to him.

PRINCIPAL ORCHID HYBRIDS

DENDROBIUM × DULCE.

Gard. Chron. 1892, vol. xi. p. 214 (Report of R.H.S. Orchid Committee); Orchid Review, 1893, vol. i. p. 358.

Raised from Dendrobium Linawianum and D. aureum, the former being the male parent, a pretty hybrid with rose-coloured flowers.

DENDROBIUM × EDITHÆ.

Gard. Chron. 1895, vol. xvii. p. 337 (Report of R.H.S. Orchid Committee).

Raised by Seden from Dendrobium aureum and D. nobile nobilius.

DENDROBIUM × ENDOCHARIS, *Rchb.*

Rchb. in Gard. Chron. 1876, vol. v. p. 298; Veitchs' Man. Orch. Pl. pt. iii. p. 88, fig.; Veitchs' Catlg. 1880, p. 12, fig.

The chaste white flowers, with their delightful perfume, render this one of the most admired of all hybrid Dendrobes; it was raised from Dendrobium japonicum crossed with D. aureum.

DENDROBIUM × EUOSMUM, *Rchb.*

Rchb. in Gard. Chron. 1885, vol. xxiii. p. 174; Veitchs' Man. Orch. Pl. pt. iii. p. 89; Orchid Review, 1894, vol. ii. p. 112.

Raised by Seden from Dendrobium × endocharis and D. nobile.

The influence of the pollen parent is conspicuous in the form and colour of the flower, while that of the seed parent—itself a hybrid—is chiefly noticeable from a delightful fragrance.

DENDROBIUM × EUOSMUM, var. LEUCOPTERUM.

Gard. Chron. 1886, vol. xxv. p. 488.

One of the most beautiful hybrid Dendrobes yet raised. The flowers are white, the disc of the lip Indian-purple.

DENDROBIUM × EUOSMUM, var. VIRGINALE.

Gard. Chron. 1895, vol. xvii. p. 337 (Report of R.H.S. Orchid Committee).

The flowers of this form are pure white with a purple blotch at the base of the lip: one of the finest of all Dendrobes.

DENDROBIUM × EURYALUS.

Gard. Chron. 1894, vol. xv. p. 409 (Report of R.H.S. Orchid Committee); Jour. of Hort. April 26th, fig.; Orchid Review, 1894, vol. ii. p. 142.

Raised by Seden from Dendrobium nobile and D. × Ainsworthii, the former the pollen parent.

The flowers have the general shape of D. nobile; the sepals and petals are light rosy purple, the lip of the same colour with a large-feathered maroon blotch in the centre.

DENDROBIUM × EURYCLÆA.

Gard. Chron. 1892, vol. xi. p. 503 (Report of R.H.S. Orchid Committee); Orchid Review, 1893, vol. i. p. 358.

Raised from Dendrobium lituiflorum and D. Wardianum, the reverse of the cross that produced D. × micans, and flowered for the first time in 1892, but now probably lost to cultivation.

DENDROBIUM × ILLUSTRE.

Gard. Chron. 1895, vol. xviii. p. 15, fig. 4; Orchid Review, 1895, vol. iii. p. 243; Jour. of Hort. 1895, June 27th, pp. 561, 566, fig. 99; Gard. Mag. 1895, June 29th, p. 385, fig.

The two species crossed are unusually dissimilar and have little in common, though they both inhabit Burmese territory. Dendrobium Dalhousieanum, introduced in 1837, has tall terete stems, leafy when young, and D. chrysotoxum clavate pseudo-bulbs with a few leaves at the top. In the hybrid the amalgamation of the two species is most distinctly traceable.

DENDROBIUM × MENTOR.

Gard. Chron. 1893, vol. xiii. p. 580 (Report of R.H.S. Orchid Committee); Orchid Review, 1893, vol. i. p. 189.

Raised by Seden from Dendrobium primulinum and D. superbum; a pretty hybrid with light rose-coloured flowers.

DENDROBIUM × MICANS, *Rchb. f.*

Rchb. f. in Gard. Chron. 1879, vol. xi. p. 332; Veitchs' Man. Orch. Pl. pt. iii. p. 89.

Obtained from Dendrobium Wardianum crossed with D. lituiflorum, and a very interesting hybrid and free grower.

DENDROBIUM × NIOBE.

Gard. Chron. 1893, vol. xiii. p. 456 (Report of R.H.S. Orchid Committee); Orchid Review, 1893, vol. i. p. 146.

From Dendrobium tortile and D. nobile, the latter, the pollen parent, very much influenced the colour of the flowers; these are bright rose-purple and the lip has a deep maroon disc as in the pollen parent.

DENDROBIUM × OPHIR.

Orchid Review, 1902, vol. x. p. 100.

Raised from Dendrobium aureum and D. signatum, the former being the seed-bearer; the flowers are of a charming tone of yellow, the lip a deeper shade of colour and pubescent.

DENDROBIUM × ILLUSTRE

THE DELL, EGHAM

DENDROBIUM × PORPHYROGASTRUM.

Gard. Chron. 1895, vol. xviii. p. 102 (Report of R.H.S. Orchid Committee).

Raised from Dendrobium Dalhousieanum and D. Huttonii. The large flowers are a curious rosy lilac in colour.

DENDROBIUM × RHODOSTOMA, *Rchb.*

Rchb. in Gard. Chron. 1876, vol. v. p. 795; Veitchs' Man. Orch. Pl. pt. iii. p. 90.

Exhibited for the first time at the Brussels Centennial flower show in 1876; the parents are Dendrobium Huttonii and D. sanguinolentum.

DENDROBIUM × SPLENDIDISSIMUM, *Rchb.*

Rchb. in Gard. Chron. 1879, vol. xi. p. 298; Veitchs' Man. Orch. Pl. pt. iii. p. 91.

Raised by crossing Dendrobium aureum with D. nobile. The flowers are of firm texture and glisten as if varnished. The hybrid first flowered in 1879, received the above name from Reichenbach, though really a variety of D. Ainsworthii, a much finer form is known as *grandiflora*.

DENDROBIUM × STRIATUM.

Gard. Chron. 1892, vol. xii. p. 566 (Report of R.H.S. Orchid Committee); Orchid Review, 1893, vol. i. p. 358.

A curious hybrid, raised by Seden from Dendrobium moniliforme crossed with D. Dalhousieanum, two species belonging to entirely different groups.

The shield-like lip is a feature of the flowers.

DENDROBIUM × THWAITESIÆ, VEITCHS' var.

Gard. Chron. 1904, vol. xxxv. p. 274, with fig.

A handsome yellow-flowered Dendrobe, one of the finest in cultivation, raised at Langley from Dendrobium splendidissimum grandiflorum and D. × Wiganæ, and flowered for the first time in April 1904.

DENDROBIUM × VIRGINIA.

Gard. Chron. 1894, vol. xv. p. 343 (Report of R.H.S. Orchid Committee); Orchid Review, 1894, vol. ii. p. 171.

The first hybrid from Dendrobium Bensoniæ and D. moniliforme (japonicum). The flowers are a lovely clear white with a small maroon spot near the base of the lip. On account of the shortness of the internodes the pseudo-bulbs are quite hidden on that part of the bulbs where the flowers are produced.

DENDROBIUM × WARDIANO-JAPONICUM.

Gard. Chron. 1892, vol. xi. p. 343 (Report of R.H.S. Orchid Committee).

Raised from the two species expressed by the compound name; a very pretty and delicately coloured hybrid.

DIALÆLIA VEITCHII.

Orchid Review, 1905, vol. xiii. p. 115.

A bigeneric hybrid raised by Seden from Diacrium bicornutum and Lælia cinnabarina as seed parent.

The flowers, borne in the same manner as those of Diacrium, measure some 2 in. across, have narrow segments, white tinged with lilac, the younger showing a slight bronzy tint; this remarkable success flowered for the first time in March 1905.

DISA × DIORES.

Gard. Chron. 1894, vol. xv. p. 49 (Report of R.H.S. Orchid Committee); Orchid Review, 1894, vol. ii. p. 239.

Raised by Seden from Disa × Veitchii and D. grandiflora, the latter one of the parents of D. × Veitchii.

The flowers closely approach those of D. grandiflora, but are paler in colour, the dorsal sepal being nearly white.

DISA × LANGLEYENSIS, *Hort. Veitch.*

Gard. Chron. 1894, vol. xvi. p. 35, fig. 5; Orchid Review, 1894, vol. ii. p. 202.

A hybrid between Disa tripetaloides and D. racemosa, flowered at Langley, and subsequently at the Royal Gardens, Kew.

The flowers, of a beautiful rose-pink shade, are borne in racemes of ten to twelve and more.

DISA × LUNA.

Gard. Chron. 1902, vol. xxxi. p. 314 (Report of R.H.S. Orchid Committee).

The parents of this hybrid are Disa racemosa and D. × Veitchii, the first named being one of the parents of the latter. In size the flowers exceed those of D. racemosa, and are bright rose-purple in colour; the interior of the flower is whitish with a slight tint of rose and a network of purple.

DISA × VEITCHII.

Gard. Chron. 1894, vol. xvi. p. 93, fig. 14; The Garden, 1892, vol. xlii. p. 408, pl. 882; Dict. Ic. des Orchidées, Disa hybr. pl. 1.

One of the first and one of the best hybrids in the genus from Disa grandiflora and D. racemosa.

The flowers, rose-purple with dark crimson and yellow markings in the centre, last long in perfection. Shown for the first time in flower in London on June 9th 1891.

PRINCIPAL ORCHID HYBRIDS

EPICATTLEYA GUATEMALENSIS, *Rolfe.*

Syns. *Cattleya* × *Guatemalensis*, Veitch.

Rolfe in Gard. Chron. 1889, vol. v. p. 491, in List of Garden Orchids; Fl. Mag. 1861, t. 61; Veitchs' Man. Orch. Pl. pt. ii. p. 86.

A remarkable hybrid sent from Guatemala by Mr. G. Ure-Skinner, who found it with Cattleya Skinneri and Epidendrum aurantiacum on the stem of the same tree; and it may be assumed that this plant is a natural bigeneric hybrid of the two species with which it was found.

EPICATTLEYA MATUTINA.

Gard. Chron. 1897, vol. xxi. p. 210 (Report of R.H.S. Orchid Committee); Gard. Chron. 1897, vol. xxi. p. 232, fig. 77, p. 233; Orchid Review, 1897, vol. v. p. 110.

Raised from Epidendrum radicans and Cattleya Bowringiana. The plant has much the same habit as E. radicans, even to the air-root-bearing characteristic, and the base of the stem shows a tendency to thicken, but there is little evidence of the Cattleya.

The flowers are some 2 in. in diameter, yellow in colour, tinged with vermilion.

EPICATTLEYA MRS JAMES O'BRIEN, *Hort.*

Gard. Chron. 1899, vol. xxv. p. 31 (Report of R.H.S. Orchid Committee), fig. 11, p. 37.

Raised by Seden from Epidendrum × O'Brienianum and Cattleya Bowringiana; a brilliant coloured plant.

EPICATTLEYA RADIO-BOWRINGIANA.

Gard. Chron. 1898, vol. xxiii. p. 385 (Report of R.H.S. Orchid Committee); *id.* p. 391, fig.; Orchid Review, 1898, vol. vi. p. 198.

A bigeneric hybrid from Epidendrum radiatum and Cattleya Bowringiana with the habit of Epidendrum radiatum, with ovoid, flattish pseudo-bulbs; rosy purple flowers, and a lip of a deeper tint.

EPIDENDRUM × CLARISSA.

Gard. Chron. 1900, vol. xxvii. p. 239 (Report of R.H.S. Orchid Committee); *id.* 1901, vol. xxix. p. 242.

Raised from Epidendrum × elegantulum and E. Wallisii. The flowers, sepals and petals white with reddish markings; the lip violet with white at the base and along the margin.

The variety *superba*, with blooms of a darker shade and bolder spots, is a superior form of the type.

EPIDENDRUM × ELEGANTULUM.

Gard. Chron. 1886, vol. xix. p. 361, fig. 49; Jour. of Hort. 1896, March 16th, p. 251, fig. 46; Orchid Review, 1896, vol. iv. pp. 108 and 124; Dict. Ic. des Orchidées, Epidendrum hybr. pl. 1.

A hybrid raised from Epidendrum × Endresio-Wallisii and E. Wallisii.

EPIDENDRUM × ELEGANTULUM, var. AUREUM.

Orchid Review, 1896, vol. iv. p. 108.

A paler coloured form of the type in which the sepals and petals are entirely light yellow and the disc of the lip nearly half red-purple in radiating lines.

EPIDENDRUM × ELEGANTULUM, var. LEUCOCHILUM.

Gard. Chron. 1898, vol. xxiii. p. 238 (Report of R.H.S. Orchid Committee); Dict. Ic. des Orchidées, Epidendrum hybr. pl. 1a.

A variety of the type with yellow sepals and petals and a pronounced pure white lip.

EPIDENDRUM × ENDRESIO-WALLISII, *Hort. Rolfe.*

Orchid Review, 1893, vol. i. p. 104; Gard. Chron. 1892, vol. xi. p. 88 (Report of R.H.S. Orchid Committee).

A hybrid obtained by crossing two species widely different in size and appearance.

The pollen parent is the rare and lovely Epidendrum Endresii, a native of Costa Rica, few plants of which have been imported alive; the seed parent is E. Wallisii, from the Frontino district of New Grenada.

The flowers from different plants vary considerably in colour, and are about 1 in. in diameter and most freely produced.

EPIDENDRUM × LANGLEYENSE.

Gard. Chron. 1899, vol. xxv. p. 402 (Report of R.H.S. Orchid Committee).

Raised at Langley from Epidendrum Pseudepidendrum and E. Wallisii, the latter being the seed parent.

EPIDENDRUM × O'BRIENIANUM, *Rolfe.*

Rolfe in Gard. Chron. 1888, vol. iii. p. 771, with figs.; Veitchs' Man. Orch. Pl. pt. vi. p. 128, with figs. reproduced; Dict. Ic. des Orchidées, Epidendrum hybr. pl. 2.

Raised and first flowered in the spring of the year 1888, the first artificially produced hybrid in the genus to be raised in this country.

The species used as parents were Epidendrum erectum and E. radicans, and the hybrid shows fairly intermediate characters.

It was named in compliment to Mr. James O'Brien, a well-known authority on Orchidaceous Plants.

EPIDENDRUM × RADICO-VITELLINUM, *O'Brien.*

J. O'Brien in Gard. Chron. 1897, vol. xxii. p. 16; Orchid Review, 1897, vol. v. p. 314.

Raised from the two species expressed by the name, and of slender growth, the root-bearing stems, furnished with alternate leaves. The evidence of the mother is seen in the soft green tint and slightly glaucous hue of the stems and leaves, the former having a slight tendency to enlarge at the nodes.

The flowers, on erect stems some 6 in. in length, are yellow, tinged with scarlet on the margin of the trilobed lip.

EPIDENDRUM × RADICO-STAMFORDIANUM.

Orchid Review, 1898, vol. vi. p. 198.

Raised at Langley from the two species indicated by the name, and the first artificially raised hybrid in which Epidendrum Stamfordianum participated.

In habit and inflorescence it resembles the pollen parent, E. radicans, and the influence of E. Stamfordianum is clearly seen in the shape of the flowers.

EPIDENDRUM × WALLISIO-CILIARE.

Gard. Chron. 1894, vol. xvi. p. 730; Orchid Review, 1894, vol. ii. p. 14.

Raised by Seden from the two species expressed by the compound name, Epidendrum ciliare being the seed-bearer. This singular-looking hybrid has yellow sepals and petals, and a white lip delicately fringed along the margin, with four or five short radiating maroon-purple lines on either side of the deep yellow crest. The influence of the seed parent is seen in the shape of the leaves, segments and the fimbriation of the lip.

EPILÆLIA RADICO-PURPURATA, *O'Brien.*

O'Brien in Gard. Chron. 1897, vol. xxii. p. 61, fig. p. 83; Orchid Review, 1897, vol. v. pp. 232, 273, fig. 12.

A bigeneric hybrid from Epidendrum radicans and Lælia purpurata, the former the pollen plant.

The hybrid resembles the male plant in habit, having erect, reed-like stems, which produce aerial roots. The flowers are 2 in. in diameter, rich orange-scarlet, with a lemon-yellow lip broadly margined with light reddish-purple.

The influence of the female parent is seen in the dwarf habit, the nearly entire lip, and in the modified colour.

EPIPHRONITIS VEITCHII.

Gard. Chron. 1890, vol. vii. p. 799 (Report of R.H.S. Orchid Committee); Watson's Orchids, new edit. 1903, p. 348, col. pl.; Orchid Review, 1893, vol. i. p. 116; Dict. Ic. des Orchidées, Epiphronitis, hybr. pl. 1.

A bigeneric hybrid obtained by crossing Epidendrum radicans with Sophronitis grandiflora, the latter being the seed parent. The two plants used as parents are totally distinct in habit, the one being but a few inches high and the other as many feet. In the hybrid the pollen parent greatly preponderates, but the stems are only about 1 ft. or $1\frac{1}{2}$ ft. high.

The flowers are increased in size and have a dash of crimson added in the sepals and petals; the disc of the lip is also more yellow, clear cut and spathulate in form.

GOODYERA × VEITCHII.

Williams' Orch. Man. ed. 7, p. 412.

Raised by John Dominy from Goodyera discolor and Anœctochilus Veitchii; a bigeneric hybrid of great botanical interest.

LÆLIA × CLIO.

Syns. *Brassolælia Clio.*

Orchid Review, 1902, vol. x. p. 86.

A hybrid between Lælia (Brassavola) glauca and L. cinnabarina, raised by Seden at Langley and first flowered in January 1902.

The flowers are light reddish-buff in colour, and in form fairly intermediate between those of the two parents.

LÆLIA × DIGBYANO-PURPURATA.

Syns. *Brassolælia Veitchii.*

Orchid Review, 1899, vol. vii. p. 31; *id.* 1902, vol. x. p. 85.

An interesting and striking hybrid raised from Lælia (Brassavola) Digbyana and L. purpurata.

The flowers resemble those of the last-named parent in shape, the sepals and petals are white, and the front of the lip is purple.

LÆLIA × DIGBYANO-PURPURATA, var. KING EDWARD VII.

Gard. Chron. 1902, vol. xxxi. pp. 206, 207, fig. 68, also suppl. col. pl. p. 413.

A very fine form with an enormous lip, sepals and petals white in colour, slightly tinged with rose, the lip primrose-yellow, veined with bright rose-purple markings and delicately fringed.

LÆLIA × EDISSA.

Gard. Chron. 1900, vol. xxvii. p. 143.

Raised by Seden from Lælia anceps and L. purpurata.

The flowers resemble those of the first-named species, but are larger; the lip is rich purple.

LÆLIA × EUTERPE, *Rolfe.*

Rolfe in Gard. Chron. 1888, vol. iv. s. 3, p. 533.

An interesting little hybrid raised by Seden from Lælia pumila Dayana and L. crispa.

PRINCIPAL ORCHID HYBRIDS

LÆLIA × FLAMMEA, *Rchb. f.*

Rchb. in Gard. Chron. 1874, p. 599; *id.* 1876, p. 394; Fl. and Pom. 1874, p. 133; Veitchs' Man. Orch. Pl. pt. ii. p. 96.

This, one of the most beautiful of all hybrid Lælias, is, as regards colour, unique even among orchids, and was raised from Lælia cinnabarina crossed with L. × Pilcheriana.

LÆLIA × FLAVINA.

Orchid Review, 1902, vol. x. p. 147.

A pretty hybrid raised by Seden from Lælia pumila and L. flava; the flowers, of good size, are primrose-yellow in colour, with an orange-coloured disc to the lip.

LÆLIA × LATONA.

Jour. of Hort. 1892, vol. xxiv. p. 353, fig.; Gard. Chron. 1892, vol. xi. p. 598 (Report of R.H.S. Orchid Committee); Veitchs' Catlg. of Pl. 1894, p. 14; Dict. Ic. des Orchidées, Lælia hybr. pl. 3.

A very beautiful and much appreciated hybrid raised from Lælia cinnabarina and L. purpurata.

The colour of the flowers, distinct and unusual, is a uniform orange-yellow; the lip, scarcely lobed, being deep red-purple with an orange-coloured border.

LÆLIA × MRS. M. GRATRIX.

Syns. *Brassolælia Gratrixiæ*, Rolfe.

Gard. Chron. 1900, vol. xxix. p. 17, fig. 5.; Orchid Review, 1899, vol. vii. pp. 349, 351; 1902, vol. x. p. 85; Dict. Ic. des Orchidées, Lælia hybr. pl. 10.

A hybrid between Lælia cinnabarina and L. Digbyana, first flowered in October 1899: the flowers are a beautiful orange-yellow, and the lip has the fringe peculiar to L. Digbyana.

LÆLIA × OMEN.

Gard. Chron. 1896, vol. xx. p. 667 (Report of R.H.S. Orchid Committee).

Raised by Seden from Lælia purpurata and L. autumnalis, with flowers of singular appearance and of a light rose colour, close in form to those of L. autumnalis.

LÆLIA × PILCHERIANA, *Rchb.*

Rchb. in Gard. Chron. 1868, p. 815; Fl. Mag. t. 340; Veitchs' Man. Orch. Pl. pt. ii. p. 96.

A hybrid raised by John Dominy from Lælia crispa and L. Perrinii, flowered for the first time in March 1867, and named in honour of Mr. Pilcher, formerly gardener to Sigismund Rucker, Esq., a successful cultivator of orchids.

LÆLIOCATTLEYA AMESIANA.

Syns. *Lælia* × *Amesiana*, Rchb. f.

Rchb. f. in Gard. Chron. 1884, vol. xxi. p. 109; Orchid Album, vi. t. 253; Veitchs' Man. Orch. Pl. pt. ii. p. 94.

Raised at Chelsea, and dedicated to the late Hon. F. L. Ames, of North Easton, Massachusetts, formerly one of the most liberal patrons of horticulture in America, and the owner of an unusually choice collection of plants.

LÆLIOCATTLEYA ASCANIA.

Gard. Chron. 1893, vol. xiii. p. 518 (Report of R.H.S. Orchid Committee); Jour. of Hort. May 4th, 1893, p. 351, fig. 65; Orchid Review, 1893, vol. i. p. 167.

Raised by Seden from Lælia xanthina and Cattleya Trianæ. The flowers are much like C. labiata in shape, but are smaller. The sepals are sulphur-yellow in colour, the petals much broader, and white with a tinge of sulphur-yellow, the lips similar in colour but that the front lobe is rich crimson.

LÆLIOCATTLEYA AURORA, *Rolfe.*

Rolfe in Gard. Chron. 1889, vol. vi. p. 380.

A hybrid from Lælia pumila Dayana and Cattleya Loddigesii, raised from seed sown in 1882, and first flowered in October 1889.

LÆLIOCATTLEYA BELLA.

Syns. *Lælia* × *bella*, Rchb. f.

Rolfe in Gard. Chron. 1889, vol. vi. p. 78 (List of Garden Orchids); Rchb. in Gard. Chron. 1884, vol. xxi. p. 174; Veitchs' Man. Orch. Pl. pt. ii. p. 94.

The fortunate result of crossing Lælia purpurata and Cattleya labiata vera, the reverse of the cross which produced Læliocattleya Antigone.

At the time of flowering, this beautiful hybrid was unsurpassed by any of its class.

LÆLIOCATTLEYA CALLISTOGLOSSA, *Rolfe.*

Syns. *Lælia* × *callistoglossa*, Rchb. f.

Rchb. f. in Gard. Chron. 1882, vol. xvii. p. 76; Orchid Album, t. 235; Veitchs' Man. Orch. Pl. pt. ii. p. 94, fig.; Dict. Ic. des Orchidées, Læliocattleya, hybr. pl. 8.

Raised by Seden from Lælia purpurata and Cattleya Warscewiczii.

The gorgeous lip of this hybrid is scarcely equalled in colour by any of the species belonging to the grand race of orchids from which it is derived.

LÆLIOCATTLEYA CALLISTOGLOSSA, var. IGNESCENS.

Gard. Chron. 1895, vol. xviii. p. 588 (Report of R.H.S. Orchid Committee); Veitchs' Catlg. of Pl. 1899, fig. p. 9.

This variety is the result of another cross from the same species that produced the type, but finer varieties were used, and the progeny improved in all its parts.

LÆLIOCATTLEYA CALOGLOSSA.

Syns. *Lælia* × *caloglossa*, Rchb. f.

Rchb. in Gard. Chron. 1877, vol. vii. p. 139; Veitchs' Man. Orch. Pl. pt. ii. p. 94.

Raised by John Dominy from Cattleya labiata vera crossed with either Lælia crispa or L. Boothiana (lobata).

LÆLIOCATTLEYA CANHAMIANA.

Syns. *Lælia* × *Canhamiana*, Rchb. f.

Rchb. in Gard. Chron. 1885, vol. xxiv. p. 6; Veitchs' Man. Orch. Pl. pt. ii. p. 95; Flora and Sylva, vol. ii. p. 376.

Raised at Chelsea from Lælia purpurata and Cattleya Mossiæ, and named after Charles Canham, a well-known grower of a former generation.

LÆLIOCATTLEYA CASSANDRA.

Syns. *Cattleya* × *Cassandra*, Rolfe.

Rolfe in Gard. Chron. 1888, vol. iv. p. 596; *id.* 1890, vol. viii. p. 529.

A hybrid, raised by Seden by crossing Cattleya Loddigesii with the pollen of one of the forms of Læliocattleya elegans, the latter itself a natural hybrid between Lælia purpurata and Cattleya guttata Leopoldii.

LÆLIOCATTLEYA CASSIOPE, *Rolfe.*

Rolfe in Gard. Chron. 1889, vol. vi. p. 620.

Raised from Lælia pumila and Læliocattleya exoniensis, the fourth of a series of bigeneric hybrids with Lælia pumila as one parent. The seed was sown in 1881, and the first flowers produced in November 1889, proving of little interest.

LÆLIOCATTLEYA CLONIA.

Gard. Chron. 1894, vol. xvi. p. 511 (Report of R.H.S. Orchid Committee); Orchid Review, 1894, vol. ii. p. 373; Dict. Ic. des Orchidées, Læliocattleya hybr. pl. 12.

Raised by Seden from Læliocattleya elegans Turneri and Cattleya Warscewiczii. The flowers resemble those of the last-named, and the front lobe of the lip is broad, circular and undulated.

LÆLIOCATTLEYA CLONIA, var. SUPERBA.

Gard. Chron. 1895, vol. xviii. p. 421, fig.

An improved form of the type, with a very much richer and broader lip.

LÆLIOCATTLEYA CORNELIA.

Gard. Chron. 1893, vol. xiv. p. 692 (Report of R.H.S. Orchid Committee).

Raised from Lælia pumila crossed with the pollen of Cattleya labiata.

LÆLIOCATTLEYA CORONIS.

Orchid Review, 1901, vol. ix. p. 23 (Report of R.H.S. Orchid Committee).

Raised from Cattleya labiata and Lælia cinnabarina. The flowers are copper-yellow in colour with a purple lip.

LÆLIOCATTLEYA CYBELE.

Orchid Review, 1901, vol. ix. p. 156.

A handsome hybrid from Læliocattleya Schilleriana and Cattleya Trianæ, with lavender-tinted sepals and petals, a bright crimson-purple lip, primrose-yellow at the disc passing to white at the base.

LÆLIOCATTLEYA DECIA.

Gard. Chron. 1894, vol. xvi. p. 699, fig. 89; Jour. of Hort. 1895, Feb. 14th, pp. 130, 131, fig. 21.

Raised by Seden from Cattleya Dowiana aurea and Lælia Perrinii.

LÆLIOCATTLEYA DECIA, var. ALBA.

Gard. Chron. 1896, vol. xx. p. 667 (Report of R.H.S. Orchid Committee); *id.* 1897, vol. xxi. p. 120, fig. 34.

The sepals and petals of this variety are pure white, and the lip has an indescribable silvery white tracery on a delicate rose-pink ground. It was adjudged the best hybrid orchid of the year in 1896.

LÆLIOCATTLEYA DIGBYANO-MENDELII, VEITCHS' var.

Gard. Chron. 1901, vol. xxx. p. 204, fig. 63, p. 207; Jour. of Hort. 1901, vol. xliii. pp. 237, 239, fig.; Gard. Mag. 1901, pp. 594, 597, fig.

The parentage of this fine hybrid is expressed by the compound name. It differs from other varieties of the same class in having larger flowers of uniform purplish-rose, and not the usual pallid whitish colour; the lip, 3 in. wide, is of a reddish tinge, the disc yellow.

LÆLIOCATTLEYA DIGBYANO-MOSSIÆ.

Syns. *Brassocattleya Veitchii*, Rolfe.

Rolfe in Gard. Chron. 1889, vol. v. p. 742, fig. 111; *id.* 1895, vol. xviii. p. 161, fig. reproduced; Orchid Review, 1902, vol. x. p. 83; Flora and Sylva, vol. ii. p. 264.

The product of a cross, as its name implies, between Lælia (Brassavola) Digbyana and Cattleya Mossiæ, raised by Seden, and flowered for the first time in May 1889, when the plant was seven years old. It is still one of the very choice and rare of this group.

LÆLIOCATTLEYA DOMINIANA LANGLEYENSIS

LÆLIOCATTLEYA DIGBYANO-TRIANÆ.

Syns. *Brassocattleya Sedenii*, Rolfe.

Gard. Chron. 1898, vol. xxiii. p. 9, fig. 3; Orchid Review, 1897, vol. v. p. 132; *id.* 1902, vol. x. p. 84.

A hybrid from the two fine species expressed by the name. In colour a bright rose-pink with the throat of the frayed lip buff-yellow passing into light yellow inside the rose-pink margin, the basal half of the disc veined with reddish-purple.

LÆLIOCATTLEYA DOMINIANA.

Syns. *Lælia* × *Dominiana*, Rchb. f.

Rchb. in Gard. Chron. 1878, vol. x. p. 332; Fl. Mag. n.s. t. 325; Veitchs' Man. Orch. Pl. pt. ii. p. 95.

Raised by Dominy from Lælia purpurata and Cattleya Dowiana.

LÆLIOCATTLEYA DOMINIANA, var. LANGLEYENSIS.

The Garden, 1899, vol. lvi. p. 142, pl. 1236.

Raised at Langley from Lælia purpurata and Cattleya Dowiana, and a beautiful interesting hybrid, as it clears up the uncertainty which previously existed as to the origin of Læliocattleya Dominiana, one of the early hybrids raised without a record by John Dominy.

LÆLIOCATTLEYA EPICASTA.

Gard. Chron. 1893, vol. xiv. p. 342 (Report of R.H.S. Orchid Committee); Jour. of Hort. 1895, April 25th, pp. 358, 364, fig. 60; Gard. Mag. 1895, May 4th, p. 261, fig.

Obtained by Seden from Lælia pumila and Cattleya Warscewiczii.

LÆLIOCATTLEYA EUMŒA.

Orchid Review, 1894, vol. ii. p. 21.

The first result of many attempts to cross the Mexican Lælias with the South American Cattleyas. Raised by Seden from Cattleya Trianæ and Lælia majalis, the flowers smaller than those of the Cattleya parent, have much the same shape, and the influence of the Lælia is seen in the colour and in certain slight modifications in the form.

LÆLIOCATTLEYA EUDORA.

Veitchs' Catlg. of Pl. 1899, p. 18.

A superb hybrid, the offspring of Cattleya Mendelli and Lælia purpurata, with flowers between 7 in. to 8 in. in diameter: one of the finest of all the Læliocattleyas.

LÆLIOCATTLEYA EUNOMIA.

Gard. Chron. 1895, vol. xviii. p. 307 (Report of R.H.S. Orchid Committee); Orchid Review, 1895, vol. iii. p. 292.

Raised by Seden from Lælia pumila Dayana and Cattleya Gaskelliana, and fairly intermediate in shape, but on the whole most resembling the pollen parent, C. Gaskelliana.

The sepals and petals are bright lilac-rose, the lip amethyst-purple with two small yellow blotches on the disc.

LÆLIOCATTLEYA EUPHROSYNE.

Gard. Chron. 1895, vol. xviii. p. 527 (Report of R.H.S. Orchid Committee).

Raised from Lælia pumila Dayana and Cattleya Warscewiczii; the flowers large and well formed, light rose-colour, with a bright purple lip with a central primrose-yellow disc.

LÆLIOCATTLEYA EXIMIA.

Gard. Chron. 1890, vol. vii. p. 800 (Report of R.H.S. Orchid Committee); Lindenia, t. 386.

Noteworthy as the first of a series of Cattleya Warneri crossed with Lælia purpurata. The flowers are highly coloured and the lip extremely handsome.

LÆLIOCATTLEYA EXONIENSIS.

Syns. *Cattleya* × *exoniensis*.

Rolfe in Jour. Linn. Soc. vol. xxiv. p. 169; Fl. Mag. 1866, t. 269; Gard. Chron. 1867, p. 1144; Veitchs' Man. Orch. Pl. pt. ii. p. 95.

One of Dominy's earliest and most successful efforts, and, at the time of its introduction the most beautiful of the noble race to which it belongs; the parents were probably Lælia crispa and Cattleya Mossiæ.

LÆLIOCATTLEYA FAUSTA, *Rolfe*.

Syns. *Cattleya* × *Fausta*, Rchb.

Rolfe in Gard. Chron. 1889, vol. v. p. 619; Rchb. in Gard. Chron. 1873, p. 289, figs. p. 290; The Garden, 1873, vol. iv. p. 435, fig.; Fl. Mag. t. 189; Veitchs' Man. Orch. Pl. pt. ii. p. 90.

Raised by Seden from a cross between Cattleya Loddigesii and Læliocattleya exoniensis, but no great improvement on the parents.

There are several varieties, which differ somewhat in colour, distinguished by the names *aurea*, *bella*, *crispa*, and *delicata*.

LÆLIOCATTLEYA FELIX.

Rchb. in Gard. Chron. 1876, vol. vi. p. 68; Veitchs' Man. Orch. Pl. pt. ii. p. 96.

Raised by John Dominy, probably from Cattleya Schilleriana crossed with Lælia crispa: a poor thing.

PRINCIPAL ORCHID HYBRIDS

LÆLIOCATTLEYA HIPPOLYTA.

Gard. Chron. 1890, vol. vii. p. 398 (Report of R.H.S. Orchid Committee); Dict. Ic. des Orchidées Læliocattleya, hybr. pl. 12.

Raised by Seden from Lælia cinnabarina and Cattleya Mossiæ, the latter the seed-bearer; the hybrid combines in a happy manner the characters of the two species.

LÆLIOCATTLEYA ILLIONE.

Orchid Review, 1901, vol. ix. p. 367; Gard. Chron. 1901, vol. xxx. p. 401 (Report of R.H.S. Orchid Committee).

Raised by Seden from Cattleya Bowringiana and Læliocattleya Dominiana Langleyensis, one of the best of the Cattleya Bowringiana crosses, the flowers being large and finely formed, of a bright, dark rosy purple, with a rosy red labellum.

LÆLIOCATTLEYA ISIS.

Gard. Chron. 1895, vol. xviii. p. 467.

Raised from Cattleya × Marstersoniæ and Lælia pumila, the former being the pollen parent.

LÆLIOCATTLEYA KING OF SPAIN.

Orchid Review, 1905, vol. xiii. p. 211.

A fine hybrid of unrecorded parentage, but probably derived from Læliocattleya Digbyano-Mossiæ and some Cattleya of the labiata group. The lip, the most attractive feature, is of large size, beautifully crisped, rich purple-crimson in colour with a yellow throat.

LÆLIOCATTLEYA LACUSTA.

Orchid Review, 1898, vol. vi. p. 41.

Raised from Cattleya bicolor and Lælia harpophylla. The flowers are intermediate in shape and colour; the sepals and petals orange-yellow, the side lobes of the lip white, and the front lobe scarlet-crimson extending down the disc.

LÆLIOCATTLEYA LADY ROTHSCHILD.

Veitchs' Catlg. of Pl. 1896, p. 8; Orchid Review, 1895, p. 353.

A lovely hybrid raised by crossing Cattleya Warscewiczii and Lælia Perrinii, the latter being the seed parent.

LÆLIOCATTLEYA LEUCOGLOSSA.

Syns. *Cattleya × leucoglossa.*

Orchid Review, 1893, vol. i. p. 357.

Raised from Cattleya Loddigesii and Læliocattleya Fausta.

LÆLIOCATTLEYA LEUCOGLOSSA, var. BELLA.

Orchid Review, 1894, vol. ii. p. 43.

The typical form has rose-pink sepals and petals, and a white lip with some yellow in the throat. The variety *bella* differs in having the front lobe of the lip dull violet-purple.

LÆLIOCATTLEYA LUCILIA.

Gard. Chron. 1899, vol. xxv. p. 403 (Report of R.H.S. Orchid Committee).

Raised by Seden from Cattleya Dowiana and Læliocattleya Schilleriana. The flowers are cream-white tipped with purple; the lip yellow veined in front with purple.

LÆLIOCATTLEYA MARDELLI.

Syns. *Cattleya* × *Mardelli*, Rchb.

Rchb. in Gard. Chron. 1879, vol. xi. p. 234; Fl. Mag. 1881, pl. 437; Veitchs' Man. Orch. Pl. pt. ii. p. 91.

Raised by Seden from Cattleya Lueddemanniana and Læliocattleya elegans, and of no great interest.

LÆLIOCATTLEYA MONA.

Raised from Cattleya Schröderæ and Lælia flava. The flowers are self-coloured throughout, pure yellow, of a similar shade to that of Cattleya flava, but larger.

LÆLIOCATTLEYA MYRA.

Gard. Chron. 1895, vol. xvii. p. 337 (Report of R.H.S. Orchid Committee); Orchid Review, 1895, vol. iii. p. 103; Dict. Ic. des Orchidées, Læliocattleya, hybr. pl. 10.

Raised by Seden from Cattleya Trianæ crossed with the pollen of Lælia flava: the flowers are creamy yellow with a sulphur-yellow disc to the lip, marked with reddish veins.

LÆLIOCATTLEYA NOVELTY.

Syns. *Lælia* × *Novelty*.

Orchid Review, 1893, vol. i. p. 288; Gard. Mag. Aug. 26th, 1893, pp. 530, 531, fig.

Raised from Lælia pumila Dayana and Læliocattleya elegans, but a similar hybrid had previously been raised by Dr. Harris of Lamberhurst, and neither had any great beauty.

LÆLIOCATTLEYA NYSA.

Gard. Chron. 1893, vol. xiv. p. 342 (Report of R.H.S. Orchid Committee); Orchid Review, 1893, vol. *i.* p. 307; *id.* 1894, vol. ii. p. 349; Jour. of Hort. 1894, Jan. 10th, fig.; *id.* 1900, vol. xli. p. 329, fig. 91; Dict. Ic. des Orchidées, Læliocattleya hybr. pl. 1.

Raised from Lælia crispa and Cattleya Warscewiczii, this hybrid varies considerably, and several seedlings have been named, such as *superba*, *picta*, and *purpurea* (as their names indicate), differing from the type in size, marking and colour of the flowers. The typical form is a light rosy-mauve with the front lobe of the lip deep purple-crimson, but the variety *superba* is one of the finest products of orchid hybridizing.

LÆLIOCATTLEYA OPHIR.

Orchid Review, 1901, vol. ix. p. 350.

Raised by Seden from Lælia xanthina and Cattleya Dowiana aurea, a striking plant with flowers with yellow sepals and petals and a tawny purple lip.

LÆLIOCATTLEYA ORPHEUS.

Syns. *Brassocattleya Orpheus.*

Gard. Chron. 1902, vol. xxxi. p. 50 (Report of R.H.S. Orchid Committee); Orchid Review, 1902, vol. x. p. 85.

This most interesting cross, obtained from Lælia (Brassavola) glauca and Cattleya Trianæ alba, is fairly intermediate in character, having white sepals slightly tinged with pink and white petals and a lip blotched with a yellow disc.

LÆLIOCATTLEYA PALLAS.

Gard. Chron. 1889, vol. vi. pp. 620, 701 (Report of R.H.S. Orchid Committee); Orchid Review, 1894, vol. ii. p. 21; Veitchs' Catlg. of Pl. 1895, p. 18, fig.; Rev. de l'Hort. Belge, 1897, p. 130, fig. 22; Dict. Ic. des Orchidées, Læliocattleya hybr. pl. 9.

Raised from Lælia crispa and Cattleya Dowiana; an exceptionally fine hybrid, and none are finer than the variety *superba*, exhibited for the first time in December 1889.

The sepals and petals are very pale yellow suffused with light blush pink; the lip deep rose-purple, undulated and crisped, with a rich golden yellow throat.

LÆLIOCATTLEYA PARYSATIS.

Orchid Review, 1894, vol. ii. p. 310.

Raised by Seden from Cattleya Bowringiana crossed with the pollen of Lælia pumila, from seed sown in 1888; the plants first flowered in 1893.

LÆLIOCATTLEYA PHILBRICKIANA.

Syns. *Lælia* × *Philbrickiana*, Rchb. f.

Rchb. f. in Gard. Chron. 1879, vol. xii. p. 102; Veitchs' Man. Orch. Pl. pt. ii. p. 96.

Raised from Cattleya Aclandiæ and Lælia elegans and dedicated to the late Judge Philbrick, of Oldfield, Bickley.

LÆLIOCATTLEYA PISANDRA.

Gard. Chron. 1893, vol. xiv. p. 536 (Report of R.H.S. Orchid Committee); Jour. of Hort. Nov. 2nd, 1893, pp. 394, 407, fig. 60; Orchid Review, 1893, vol. i. p. 374.

Raised from Cattleya Eldorado and Lælia crispa. The flowers are of a beautiful rose-purple with a deep velvety purple lip.

LÆLIOCATTLEYA PROSERPINE, *Rolfe.*

Rolfe in Gard. Chron. 1890, vol. viii. p. 352; Dict. Ic. des Orchidées, Læliocattleya hybr. pl. 2.

A hybrid raised by Seden from Lælia pumila Dayana and Cattleya velutina, from seed sown in 1883 and flowered in August 1890.

LÆLIOCATTLEYA QUEEN ALEXANDRA.

Gard. Chron. 1902, vol. xxxi. p. 116; Supp. col. fig. p. 413.

A very beautiful hybrid Læliocattleya raised by Seden from Læliocattleya bella and Cattleya Trianæ, with sepals and petals of an intense rosy lilac, and a lip deep ruby-purple with an orange-coloured disc. The petals are very broad and well displayed.

LÆLIOCATTLEYA REMULA.

Orchid Review, 1900, vol. viii. p. 228.

Raised from Cattleya Aclandiæ and Lælia tenebrosa, the latter the pollen parent.

The flowers favour those of the Cattleya parent in shape; the sepals and petals of a coppery yellow tint, unspotted, the lip a light purple with the side-lobes and disc nearly white.

LÆLIOCATTLEYA ROSALIND.

Gard. Chron. 1897, vol. xxi. p. 2, fig. 1; Jour. of Hort. January 21st, fig. 16.

Raised by Seden from Cattleya Trianæ and Læliocattleya Dominiana.

The sepals are French-white; the broad petals silvery white with a delicate tinge of rose-pink; the lip is rich yellow at the base with clear, white veining, the side lobes rosy purple, the front with a large rich purple blotch.

LÆLIOCATTLEYA SEDENII, *Rchb.*

Rchb. in Gard. Chron. 1877, vol. viii. p. 424; Veitchs' Man. Orch. Pl. pt. ii. p. 97.

Raised from Cattleya superba and Læliocattleya elegans, and named by Professor Reichenbach in compliment to the raiser; it is doubtful if this plant still exists.

LÆLIOCATTLEYA SEMIRAMIS.

Gard. Chron. 1895, vol. xviii. p. 588 (Report of R.H.S. Orchid Committee); *id.* 1901, vol. xxx. p. 401.

Raised from Lælia Perrinii and Cattleya Gaskelliana. The variety *superba* is the showiest of all the Perrinii hybrids.

LÆLIOCATTLEYA STATTERIANA.

Orchid Review, 1894, vol. ii. p. 21; Veitchs' Catlg. of Pl. for 1895, p. 7.

Raised from the autumn-flowering Lælia Perrinii crossed with the pollen of Cattleya labiata vera.

The purple front lobe of the lip with its milk-white disc offers a remarkable contrast, and is a marked characteristic of the flower.

LÆLIOCATTLEYA STELLA, *Rolfe.*

Rolfe in Gard. Chron. 1889, vol. vi. p. 322.

Raised by Seden from Lælia crispa and Læliocattleya elegans Wolstenholmiæ from seed sown in 1881; the plant flowered for the first time in July 1889.

LÆLIOCATTLEYA TIMORÆ, *N. E. Brown.*

N. E. Brown, in Gard. Chron. 1887, vol. ii. p. 428; Orchid Review, 1894, vol. ii. p. 255.

Raised by Seden from Lælia pumila Dayana, and Cattleya Lueddemanniana, and of no great interest.

LÆLIOCATTLEYA TIRESIAS.

Orchid Review, 1896, vol. iv. p. 15.

Raised from Cattleya Bowringiana crossed with the pollen of Læliocattleya elegans.

The flowers resemble those of the seed parent, the sepals and petals being rose-purple with a rich crimson-purple lip white at the base.

LÆLIOCATTLEYA TRIOPHTHALMA.

Syns. *Lælia* × *triophthalma, Cattleya* × *triophthalma*, Rchb.

Rolfe in Gard. Chron. 1891, vol. ix. p. 802; vol. x. p. 155; Orchid Review, 1893, vol. i. p. 101; *id.* 1894, vol. ii. p. 40; Rchb. in Gard. Chron. 1883, vol. xx. p. 526; Veitchs' Man. Orch. Pl. pt. ii. p. 97.

Raised by Seden from Cattleya superba and the beautiful hybrid Lælio cattleya exoniensis. The colouring of the disc gives a tripartite appearance suggesting the name.

The sepals and petals are blush-pink, the side-lobes of the lip the same colour as the sepals, but inside suffused and veined with rose.

LÆLIOCATTLEYA TYDEA.

Gard. Chron. 1894, vol. xv. p. 210 (Report of R.H.S. Orchid Committee); Orchid Review, 1894, vol. ii. p. 95; Jour. of Hort. 1894, March 8th, p. 181, fig. 30.

Raised from Lælia pumila and Cattleya Trianæ. The flowers are of a deep tint of rosy purple, and bear a resemblance to the first-named parent.

LÆLIOCATTLEYA VACUNA.

Orchid Review, 1901, vol. ix. p. 119.

A hybrid from Cattleya guttata and Lælia cinnabarina.

The flowers are pale yellow with the front lobe of the lip deep-crimson-purple.

LÆLIOCATTLEYA VEITCHIANA, *Rolfe.*

Syns. *Lælia* × *Veitchiana*, Rchb. *Cattleya* × *Veitchiana*, Hort.

Rolfe in Jour. Linn. Soc. vol. xxiv. p. 169; Rchb. in Gard. Chron. 1874, p. 566; Fl. Mag. n.s. t. 305; Veitchs' Man. Orch. Pl. pt. ii. p. 97.

A hybrid raised by Dominy from Cattleya labiata vera with Lælia crispa, and flowered for the first time in March 1874, when sixteen years old; exhibited before the Floral Committee of the Royal Horticultural Society, it was awarded a First Class Certificate.

LÆLIOCATTLEYA VICTORIA, *Rolfe.*

Syns. *Lælia* × *Victoria*, Hort.

Gard. Chron. 1889, vol. vi. p. 156 (Rolfe's List of Garden Orchids); Gard. Chron. 1888, vol. iv. p. 578 (Report of R.H.S. Orchid Committee); Gard. and For. 1888, p. 495.

Raised by Seden from Lælia crispa and the beautiful Læliocattleya Dominiana, a hybrid with the richest colouring possible; it passed to the collection of Baron Schröder, who first exhibited it in flower before the Royal Horticultural Society.

LÆLIOCATTLEYA VIOLETTA.

Gard. Chron. 1897, vol. xxi. p. 115 (Report of R.H.S. Orchid Committee).

Raised from Cattleya Gaskelliana and Lælia purpurata. The flowers have the Cattleya form with the rich colouring of the Lælia parent.

LÆLIOCATTLEYA ZENOBIA, *Rolfe.*

Rolfe in Gard. Chron. 1887, vol. i. p. 552.

Raised by Seden from Cattleya Loddigesii and Læliocattleya elegans Turneri.

LÆLIOCATTLEYA ZEPHYRA.

Gard. Chron. 1894, vol. xvi. p. 103 (Report of R.H.S. Orchid Committee); Orchid Review, 1894, vol. ii. p. 285.

Raised from Lælia xanthina and Cattleya Mendelii, the former the pollen parent.

The hybrid resembles the first-named in its flowers, the sepals and petals are light buff, the front lobe of the lip purple-crimson, and the remainder yellow, deeper at the throat.

LEPTOLÆLIA VEITCHII.

Gard. Chron. 1903, vol. xxxiii. p. 50, fig. 23; Orchid Review, 1902, vol. x. p. 157.

A bigeneric hybrid from Leptotes bicolor and Lælia cinnabarina, the former the seed parent.

The cream-coloured flowers flushed with pink are about 1½ in. in diameter; the leaves, intermediate in form between those of the two progenitors, are neither cylindrical as in Leptotes nor flattened as in the Lælia.

The first recorded hybrid between the two genera Leptotes and Lælia.

MASDEVALLIA × AJAX.

Gard. Chron. 1896, vol. xx. p. 137 (Report of R.H.S. Orchid Committee); Orchid Review, 1896, vol. iv. p. 228.

Raised by Seden from Masdevallia × Chelsoni and M. Peristeria, the latter being the pollen parent.

The general effect is that of M. × Chelsoni modified in shape and darker in colour on account of the numerous little dots derived from M. Peristeria.

MASDEVALLIA × ALCESTE.

Orchid Review, 1901, vol. ix. p. 154.

Raised from Masdevallia × Asmodia and M. Veitchiana, with large dark red flowers, an orange-yellow ground colour breaking through in places.

MASDEVALLIA × ASMODIA.

Gard. Chron. 1894, vol. xv. p. 762 (Report of R.H.S. Orchid Committee); Orchid Review, 1894, vol. ii. p. 202.

Three species are involved in the parentage of this hybrid raised from Masdevallia × Chelsoni and M. Reichenbachiana.

In shape it resembles the pollen parent, M. Reichenbachiana, is of about the same size, and in colour a peculiar reddish-purple with slightly darker veins and some dull yellow in the throat.

MASDEVALLIA × CAUDATO-ESTRADÆ, *Rolfe.*

Rolfe in Gard. Chron. 1889, vol. v. p. 714; Veitchs' Man. Orch. Pl. pt. v. p. 74.

A hybrid, as its name indicates, between Masdevallia Estradæ and M. caudata Shuttleworthii, the seed parent.

MASDEVALLIA × CHELSONI, *Rchb.*

Rchb. in Gard. Chron. 1880, vol. xiii. p. 554; Veitchs' Man. Orch. Pl. pt. v. p. 74.

MASDEVALLIA × CHELSONI, var. SPLENDENS, *Veitch, Rolfe.*

Rolfe in Gard. Chron. 1889, vol. v. p. 619; Veitchs' Man. Orch. Pl. pt. v. p. 74.

This, the first Masdevallia raised from seed in Europe, and the first hybrid to flower, is of unusual interest.

It was raised by Seden from Masdevallia amabilis and M. Veitchiana, as was the variety *splendens*, by reversing the cross, the hybridizer employing finer varieties of the species than those that produced the type.

MASDEVALLIA × ELLISIANA, *Rolfe.*

Rolfe in Gard. Chron. 1889, vol. vi. p. 74; Veitchs' Man. Orch. Pl. pt. v. p. 75.

A hybrid from Masdevallia coccinea Harryana and M. ignea, dedicated to the late Lady Howard de Walden, Ellis being the family name.

MASDEVALLIA × GAIRIANA, *Rchb. f.*

Rchb. f. in Gard. Chron. 1884, vol. xxii. p. 38; Veitchs' Man. Orch. Pl. pt. v. p. 75.

Raised from Masdevallia Veitchiana and M. Davisii and dedicated to John Gair, Esq., of The Kilns, Falkirk, at that time the possessor of the best collection of orchids in Scotland.

MASDEVALLIA × GLAPHYRANTHA, *Rchb.*

Rchb. in Gard. Chron. 1886, vol. xxvi. p. 648; Veitchs' Man. Orch. Pl. pt. v. p. 76.

A hybrid of no great interest raised from Masdevallia infracta crossed with M. Barlæana.

MASDEVALLIA × IMOGEN.

Veitchs' Catlg. of Pl. 1899, p. 21.

A distinct, interesting hybrid from Masdevallia Veitchiana and M. Schlimii, the latter the seed-bearer. The sepals have the rich scarlet of M. Veitchiana, shaded with the chestnut-brown of M. Schlimii.

MASDEVALLIA × SPLENDIDA, *Rchb.*

Gard. Chron. 1878, vol. ix. p. 493; Veitchs' Man. Orch. Pl. pt. v. p. 73.

A natural hybrid gathered on the Eastern Cordillera of Peru, near Cuzco, by Walter Davis, who sent it with a consignment of Masdevallia Veitchiana and M. Barlæana. Seden has since produced this artificially by fertilizing M. Veitchiana with M. Barlæana.

MASDEVALLIA × SPLENDIDA, var. PARLATOREANA, *Veitch.*

Syns. *M. Parlatoreana*, Rchb.

Rchb. in Gard. Chron. 1879, vol. xi. p. 172; Veitchs' Man. Orch. Pl. pt. v. p. 73.

A more attractive form than the above-named, dedicated by Professor Reichenbach to Professor Parlatore of Florence, the most distinguished Italian botanist of his time.

PRINCIPAL ORCHID HYBRIDS

MILTONIA × BLEUANA.

Gard. Chron. 1889, vol. v. p. 203; Veitchs' Man. Orch. Pl. pt. viii. p. 118.

Raised in this country for the first time the cross was effected in 1884, and the first flower from the progeny opened in 1891.

The same cross had previously been obtained by M. Bleu of Paris, whose hybrids flowered in 1889.

The parents were Miltonia vexillaria and M. Roezlii, the latter being the pollen parent, and the offspring may be said briefly to possess the vegetative characters of the mother plant and the flowers of the pollen parent. This success created quite unusual interest.

ODONTOGLOSSUM × EXCELLENS, *Rchb.*, Nat. hyb.

Rchb. in Gard. Chron. 1881, vol. xvi. p. 426; Gard. Chron. 1891, vol. ix. p. 754; Dict. Ic. des Orchidées, Odontoglossum hybr. pl. 1.

This beautiful Odontoglot first appeared in the collection of Sir Trevor Lawrence at Burford Lodge, Dorking. In his description Professor Reichenbach suggested that it might possibly be a hybrid between Odontoglossum Pescatorei (nobile) and O. triumphans or O. tripudians.

Seden proved the correctness of the Professor's hypothesis by raising O. × excellens from O. Pescatorei crossed with O. triumphans, further remarkable as the first Odontoglossum hybrid to be raised in this country and successfully brought to the flowering stage.

PHAIOCALANTHE INSPERATA.

Gard. Chron. 1897, vol. xxii. p. 315 (Report of R.H.S. Orchid Committee).

A bigeneric hybrid raised by Seden from Phaius grandifolius and Calanthe Masuca, the latter the pollen parent.

PHAIOCALANTHE IRRORATA, *Rolfe.*

Syns. *Phaius × irroratus*, Rchb. f.

Rolfe in Jour. Linn. Soc. 1887, xxiv. p. 168; Rchb. in Gard. Chron. 1867, p. 264, fig.; Fl. Mag. 1869, t. 426; Veitchs' Man. Orch. Pl. pt. vi. p. 17.

A hybrid raised by Dominy between Phaius grandifolius and Calanthe vestita Turneri nivalis, which flowered for the first time in 1867.

The creamy white flowers are intermediate between the two parents as regards their expansion, both have a rosy hue over the limb, and a pale yellow hue over the disc of the lip.

PHAIOCALANTHE IRRORATA, var. PURPUREA.

Veitchs' Man. Orch. Pl. pt. vi. p. 17.

Raised by Seden from the same cross. As distinguished from Phaiocalanthe irrorata the sepals and petals are of a purer white, the lip somewhat larger and more deeply lobed, the colour richer, and the white margin broader.

PHAIOCALANTHE IRRORATA, var. ROSEA.

Gard. Chron. 1895, vol. xvii. p. 337 (Report of R.H.S. Orchid Committee); Orchid Review, 1895, vol. iii. p. 128; Jour. of Hort. 1895, April 18th, p. 331, fig. 56.

Raised by Seden from Phaius grandifolius and Calanthe vestita gigantea: a rose-coloured form of the type.

PHAIOCALANTHE SEDENIANA, *Rolfe.*

Syn. *Phaius* × *Sedenianus*, Rchb.

Rolfe in Gard. Chron. 1888, vol. iii. p. 136; Veitchs' Man. Orch. Pl. vi. p. 17; Rchb. f. in Gard. Chron. 1887, vol. i. p. 174; Jour. of Hort. 1894, Dec. 6th, pp. 512, 513, fig. 80.

A beautiful plant with flowers of a shade of pale primrose-yellow, raised by Seden by crossing the large-flowered Phaius grandifolius with the beautiful Calanthe × Veitchii, the latter being the pollen parent and itself a hybrid: it is one of the best bigeneric hybrids.

PHAIOCALANTHE SEDENIANA, var. ALBIFLORA.

Gard. Chron. 1896, vol. xix. p. 88 (Report of R.H.S. Orchid Committee); Jour. of Hort. 1896, Jan. 30th, p. 99, fig. 15.

A white form of the preceding, raised at Langley.

PHAIUS × AMABILIS.

J. O'Brien in Gard. Chron. 1893, vol. xiii. p. 226, fig. 32, p. 229; Orchid Review, 1893, vol. i. p. 87.

A hybrid, raised by crossing Phaius grandifolius with the pollen of P. tuberculosus, the flowers with white sepals and petals tinged with rose on the face; lip claret-coloured, with darker purplish lines on a whitish ground on the base inside, the lower portion of the labellum yellow on the outside throughout.

PHAIUS × MACULATO-GRANDIFOLIUS.

Gard. Chron. 1891, vol. x. p. 591 (Report of R.H.S. Orchid Committee).

Raised from the two species expressed by the name, and exhibited for the first time in November 1891.

PHALÆNOPSIS × ARIADNE.

Orchid Review, 1896, vol. iv. p. 147.

Raised from Phalænopsis Aphrodite and P. Stuartiana, the latter the pollen parent. The sepals and petals white, of usual shape, the lip nearly intermediate; the side lobes less oblique than in P. Aphrodite, but the markings almost as in that species; the front lobe has the basal half densely spotted with purple, the remainder being white.

PHALÆNOPSIS × ARTEMIS.

Gard. Chron. 1894, vol. xvi. p. 49 (Report of R.H.S. Orchid Committee); Orchid Review, 1893, vol. i. p. 356.

From Phalænopsis amabilis and P. rosea, with a considerable resemblance to P. × intermedia.

PHALÆNOPSIS × CASSANDRA.

Gard. Chron. 1898, vol. xxiii. p. 106 (Report of R.H.S. Orchid Committee).

Raised by Seden from Phalænopsis rosea and P. Stuartiana.

PHALÆNOPSIS × F. L. AMES.

Rolfe in Gard. Chron. 1888, vol. iii. p. 201, fig.; Veitchs' Man. Orch. Pl. pt. vii. p. 48, fig.; Orchid Review, 1894, vol. ii. p. 106.

A hybrid from Phalænopsis × intermedia and P. amabilis, the former the pollen parent, itself a cross between P. Aphrodite and P. rosea.

The varietal name was given in honour of the late Hon. F. L. Ames of North Easton, Massachusetts, U.S.A.

PHALÆNOPSIS × HARRIETTÆ, *Rolfe.*

Rolfe in Gard. Chron. 1887, vol. ii. p. 8, with fig.; The Garden, 1890, vol. xxxviii. p. 158, t. 766; Veitchs' Man. Orch. Pl. pt. vii. p. 49, fig.; Jour. of Hort. 1901, vol. xlii. pp. 236, 237, with fig.

A beautiful and interesting hybrid from Phalænopsis amabilis with P. violacea. The only result from this cross was purchased by the late Hon. Erasmus Corning, of Albany, U.S.A., and named in compliment to his daughter, Miss Harriett Corning.

PHALÆNOPSIS × HEBE.

Gard. Chron. 1897, vol. xxi. p. 115 (Report of R.H.S. Orchid Committee); Orchid Review, 1897, vol. v. p. 79.

Raised from Phalænopsis rosea and P. Sanderiana, with some resemblance to P. × intermedia, the only difference in the parentage the replacement of P. Sanderiana by the closely allied P. Aphrodite. The flowers of P. × Hebe are blush-white, slightly veined and suffused with rose; the lip is bright rose, mottled on the side lobes, with the usual crest and very short tendrils.

PHALÆNOPSIS × HERMIONE.

Gard. Chron. 1899, vol. xxv. p. 174 (Report of R.H.S. Orchid Committee).

Raised by Seden from Phalænopsis Stuartiana and P. Lueddemanniana.

PHALÆNOPSIS × INTERMEDIA, var. VESTA.

Orchid Review, 1893, vol. i. p. 52; Gard. Chron. 1894, vol. xii. p. 343 (Report of R.H.S. Orchid Committee).

The hitherto supposed natural hybrid Phalænopsis × intermedia was proved to be such by Seden, who raised a plant identical with the type from P. Aphrodite and P. rosea. Subsequently fertilizing P. rosea leucaspis with P. Aphrodite, a second was obtained with smaller flowers and shorter broader segments, of a rose-purple colour, distinguished by the varietal name *Vesta.*

PHALÆNOPSIS × JOHN SEDEN, *Rolfe.*

Rolfe in Gard. Chron. 1888, vol. iii. p. 331, fig.; Veitchs' Man. Orch. Pl. pt. vii. p. 50, with fig. reproduced; Gard. Chron. 1898, vol. xxiii. p. 171, fig. 68.

Raised by Seden, and named at our request in his honour by Mr. Rolfe of the Herbarium, Kew.

The parents were Phalænopsis amabilis and P. Lueddemannia.

PHALÆNOPSIS × LEDA, *Rolfe.*

Rolfe in Gard. Chron. 1888, vol. iii. p. 457.

This plant, a stray seedling, was detected growing in a greenhouse where no other Phalænopsis seed had been sown, and nothing certain is known of its parentage; it is probably the result of a cross between Phalænopsis amabilis and P. Stuartiana.

PHALÆNOPSIS × LUEDDE-VIOLACEA.

Gard. Chron. 1895, vol. xviii. p. 102 (Report of R.H.S. Orchid Committee); *id.* 1898, vol. xxiv. p. 43, fig. 11; Orchid Review, 1895, vol. iii. p. 259; Jour. of Hort. 1895, July 25th, pp. 77, 87, fig. 11.

Raised and reared by Seden from the seed produced by crossing the two species indicated in the compound name.

The peculiar bar-like markings of amethyst possessed by Phalænopsis Lueddemanniana are in the hybrid transformed into distinct spots.

PHALÆNOPSIS × MRS. JAMES H. VEITCH.

Gard. Chron. 1899, vol. xxv. p. 114, fig. 43; Jour. of Hort. 1901, vol. xlii. pp. 226, 227, fig.

Raised by Seden from Phalænopsis Lueddemanniana and P. Stuartiana, and remarkable among Phalænopsis in shape and colouring. The sepals and petals are greenish-yellow, with brownish-crimson dots; the front lobe of the lip is white, and a yellow tinge is noticeable in the throat.

PHALÆNOPSIS × ROTHSCHILDIANA, *Rchb. f.*

Rchb. f. in Gard. Chron. 1887, vol. i. p. 606; Veitchs' Man. Orch. Pl. pt. vii. p. 51, with fig.

The second hybrid Phalænopsis ever raised, and dedicated to the Right Hon. Lord Rothschild, of Tring Park. The parents were Phalænopsis Schilleriana and P. amabilis.

PHALÆNOPSIS ×"JOHN SEDEN"
THE DELL, EGHAM

PRINCIPAL ORCHID HYBRIDS

PHALÆNOPSIS × STUARTIANO-MANNI.

Gard. Chron. 1898, vol. xxiii. p. 238 (Report of R.H.S. Orchid Committee).

Raised by Seden from the two species expressed by the name.

SOBRALIA × ROSEO-MACRANTHA.

Gard. Chron. 1897, vol. xxii. p. 277 (Report of R.H.S. Orchid Committee).

A hybrid Sobralia, one of the best the genus has produced, which resembles Sobralia macrantha in foliage and habit, and the flowers soft light rose-purple with cream-white margins, are fully 6 in. in diameter.

SOBRALIA × VEITCHII.

Gard. Chron. 1894, vol. xvi. p. 103 (Report of R.H.S. Orchid Committee); Orchid Review, 1894, vol. ii. p. 239; Jour of Hort. 1894, Aug. 2nd, pp. 98, 99, fig. 5; Dict. Ic. des Orchidées, Sobralia, hybr. pl. 1.

A hybrid from Sobralia macrantha and S. xantholeuca, and one of the most distinct of the few that have been raised between species of this genus. The flowers soft rosy blush, with a rich rose-lilac lip have a conspicuous orange-yellow throat, in size and shape approximating closely to the two parents.

SOPHROCATTLEYA BATEMANIANA.

Syns. *Lælia* × *Batemaniana*, Rchb.

Veitchs' Man Orch. Pl. pt. ii. p. 92, figs.; Rolfe in Jour. Linn. Soc. vol. xxiv. p. 156; Rchb. in Gard. Chron. 1886, vol. xxvi. p. 263.

The first hybrid to be raised between the genera Cattleya and Sophronitis, to the ordinary observer so widely separated, led orchid experts in early years to doubt the correctness of the assigned parentage. Raised from Sophronitis grandiflora and Cattleya intermedia, the seed was sown in June 1881, and the first flower opened in August 1886, and caused very great interest.

This remarkable plant is dedicated to Mr. James Bateman, author of Orchidaceæ of Mexico and Guatemala.

SOPHROCATTLEYA CALYPSO.

Gard. Chron. 1892, vol. xii. p. 744 (Report of R.H.S. Orchid Committee); *id.* 1895, vol. xx. p. 695, fig. 122; Orchid Review, 1903, vol. i. p. 291.

A hybrid raised by Seden from Sophronitis grandiflora and Cattleya Loddigesii Harrisonæ.

The flower of Sophrocattleya Calypso, in form and size, partakes much of Cattleya Loddigesii Harrisonæ, and is similar to that species in the firm substance of the sepals and petals, rose-purple in colour, with a number of very dark purple lines.

SOPHROCATTLEYA EXIMIA.

Gard. Chron. 1894, vol. xvi. p. 378 (Report of R.H.S. Orchid Committee); Orchid Review, 1894, vol. ii. p. 333; Jour. of Hort. 1894, Oct. 4th, p. 321, fig. 48; *id.* 1900, vol. xli. p. 373, fig. 103.

A bigeneric hybrid raised by Seden from Cattleya Bowringiana and Sophronitis grandiflora, resembling S. grandiflora in habit, but that the pseudo-bulbs are ovoid in shape and rather stout. The flowers are larger than those of Sophronitis, sepals and petals bright purplish-rose, and the lip darker, with a light yellow throat.

SOPHROCATTLEYA QUEEN EMPRESS.

Gard. Chron. 1899, vol. xxvi. p. 96 (Report of R.H.S. Orchid Committee); *id.* p. 112, fig. 43; Orchid Review, 1898, vol. vi. p. 258.

Raised by Seden from Cattleya Mossiæ and Sophronitis grandiflora, the latter being the male parent. The flowers are as Cattleya Mossiæ in form, but smaller, and in colour rosy crimson.

SOPHROCATTLEYA SAXA.

Orchid Review, 1903, vol. xi. p. 267.

Raised from Sophronitis grandiflora and Cattleya Trianæ. The flower is of a rose shade, with a crimson blotch on the lip.

SOPHROLÆLIA LÆTA.

Syns. *Sophrocattleya læta.*

Chapman in Watson's Orchids, ed. 2, p. 468, fig.; Gard. Chron. 1894, vol. xvi. p. 447 (Report of R.H.S. Orchid Committee); *id.* p. 476, fig. 63; Orchid Review, 1894, vol. ii. p. 333.

Raised by Seden from Sophronitis grandiflora and Lælia pumila Dayana, with flowers resembling more closely those of Sophronitis than those of Lælia, and similar in appearance to the Sophrocattleya Batemanniana.

In the Gardeners' Chronicle above quoted the figure is erroneously named Sophrocattleya Batemanniana and the matter applies to Sophrolælia læta.

SOPHROLÆLIA VALDA.

O'Brien in Gard. Chron. 1900, vol. xxix. p. 53; Orchid Review, 1901, vol. ix. p. 37.

Raised by Seden from Sophronitis grandiflora and Lælia harpophylla. The flower resembles the Lælia, but is larger and of a light orange-yellow colour, without the characteristic markings of the Sophronitis parent on the lip.

PRINCIPAL ORCHID HYBRIDS

SOPHROLÆLIOCATTLEYA VEITCHII, *Chapman.*

Syns. *Sophrocattleya Veitchii*, Veitch.

Chapman in Watson's Orchids, ed. 2, 1903, p. 469; Gard. Chron. 1892, vol. xii. p. 312 (Report of R.H.S. Orchid Committee); Orchid Review, 1893, vol. i. p. 356.

A hybrid obtained by Seden from the bright scarlet-flowered Sophronitis grandiflora and the bigeneric hybrid Læliocattleya elegans, involving three distinct genera in its parentage, and flowered for the first time in the year 1892.

SPATHOGLOTTIS × AUREO-VEILLARDII.

Gard. Chron. 1897, vol. xxii. p. 10 (Report of R.H.S. Orchid Committee); *id.* 1898, vol. xxiii. p. 309, fig. 115; Dict. Ic. des Orchidées, Spathoglottis, hybr. pl. 1.

The first hybrid raised between two species of this genus, a product of a cross effected by Seden, between Spathoglottis aurea and S. Veillardii.

The flowers were produced for the first time in May 1897, on which occasion the plant was exhibited at the Royal Horticultural Society's Show held in the Temple Gardens.

THUNIA × VEITCHIANA, *Rchb. f.*

Rchb. in Gard. Chron, 1885, vol. xxiii. p. 818; Veitchs' Man. Orch. Pl. pt. vi. p. 20.

First raised by Mr. Toll of Manchester, and shortly afterwards by Seden: a great advance on the type species.

Plants of both progenies were exhibited in flower at one of the Royal Botanic Society's Shows in 1885—Mr. Toll's under the name of Wrigleyana and our own as Veitchiana; as the materials for Professor Reichenbach's description were supplied by us, this name has priority of publication.

The parents are Thunia Marshalliæ and T. Bensoniæ.

ZYGOCOLAX LEOPARDINUS.

Syns. *Zygopetalum × leopardinus*, Rchb. f.

Veitchs' Man. Orch. Pl. pt. ix. p. 66; Rchb. in Gard. Chron. 1886, vol. xxvi. p. 66; Orchid Review, 1900, vol. viii. p. 12.

A hybrid raised by crossing Zygopetalum maxillare with the pollen of Colax jugosus, in which the influence of the latter seems to have been quite subordinate.

The plant flowered for the first time in 1886, when Professor Reichenbach described it, but at that time the parentage was not correctly known and could only be conjectured. More conclusive evidence has since been obtained, and the names of the two species now given as parents are undoubtedly correct.

ZYGOCOLAX VEITCHII, *Rolfe.*

Rolfe in Gard. Chron. 1887, vol. i. p. 765; Jour. Linn. Soc. vol. xxiv. p. 170; Veitchs' Man. Orch. Pl. pt. ix. p. 66 figs.; Jour. of Hort. 1893, Feb. 2nd, pp. 87, 94, fig. 13; Bot. Mag. t. 7980; Dict. Ic. des Orchidées, Zygocolax, hybr. pl. 1.

One of the most interesting hybrids ever raised, either from the botanical or horticultural aspect, and obtained by crossing two species of two different genera, Zygopetalum crinitum and Colax jugosus.

The seed was sown in September 1882, and the first flowers produced in March 1887.

This plant is interesting as having been the principal subject of Mr. R. A. Rolfe's articles on a uniform plan of naming bigeneric hybrid orchids. He followed Dr. M. T. Masters, who combined the elements of the two generic names of the plants concerned when he named the first artificial bigeneric hybrid on record, Philageria Veitchii, a cross between Lapageria rosea and Philesia buxifolia.

Zygocolax Veitchii is also found as a natural hybrid, and specimens introduced from Brazil and flowered in the gardens of Sir Frederick Wigan, Bart., closely resembled those of artificial origin.

ZYGOPETALUM × LEUCOCHILUM.

Gard. Chron. 1892, vol. xi. p. 214 (Report of R.H.S. Orchid Committee); Orchid Review, 1896, vol. iv. p. 62.

Raised by Seden from Zygopetalum Burkei and Z. Mackayii, the latter the seed-bearer. The flowers are over 2¼ in. in diameter, the sepals and petals light-green lined along the centre and spotted near the margin with dark brown. The lip is white with a few violet striations.

ZYGOPETALUM × SEDENII, *Rchb. f.*

Rchb. f. in Gard. Chron. 1874, p. 290; Veitchs' Man. Orch. Pl. pt. ix. p. 66, fig.; Fl. Mag. n.s. t. 417; Jour. of Hort. 1893, May 11th, p. 377, fig. 69.

Raised from Zygopetalum maxillare, crossed with Z. Mackayi, the first hybrid Zygopetalum to flower in this country.

STOVE AND GREENHOUSE PLANTS

STOVE AND GREENHOUSE PLANTS

ABUTILON PÆONIÆFLORA, *Hook.*

Syns. *Sida pæoniæflora*, Hook.

Bot. Mag. t. 4170.

Sent by William Lobb, from the Organ Mountains of Brazil, and first flowered at Exeter in January 1845.

The petals are of a deep red-rose colour which admirably contrasts with the tuft of bright yellow anthers occupying the centre of the flower and give the appearance of a miniature single Pæony.

ACALYPHA WILKESIANA, *Muell. Arg.*

Syns. *A. tricolor*, Seem.

Veitchs' Catlg. of Pl. 1868, p. 26.

A handsome stove plant with leaves mottled and blotched with bright red and crimson, introduced from New Caledonia through the late John Gould Veitch, and now rarely seen.

ADELASTER (ERANTHEMUM?) ALBIVENIS, *Lindl.*

Lindl. in Proc. R.H.S. 1861, vol. i. p. 568; Gard. Chron. 1861, pp. 387, 499 (advts.); l'Illus. Hort. 1862, t. 320.

A charming stove plant from Peru, with dark green leaves, the veins of which are marked with pure white; the undersurface is bright purple.

ADHATODA CYDONIÆFOLIA, *Nees.*

Bot. Mag. t. 4962.

A handsome Acanthaceous plant from Brazil, first flowered in the autumn of 1855. The flowers are two-lipped, large and showy, striking from the contrast presented by the dark purple lower lip and the pure white upper one.

ÆCHMEA VEITCHII, *Baker.*

Syns. *Chevalliera Veitchii*, Morren.

Bot. Mag. t. 6329; The Garden, 1881, vol. xix. p. 654, pl. cclxl. (*sic*); La Belg. Hort. vol. xxviii. (1878), p. 177.

A very fine Bromeliad discovered by Gustave Wallis in New Grenada in 1874, and introduced the same year to cultivation.

The spiny margined, gracefully curved leaves are pale green in colour and the cone-like inflorescence is brilliant scarlet.

ÆSCHYNANTHUS CORDIFOLIUS, *Hook.*

Fl. des Serres, 1861, tom. iv. p. 101; Bot. Mag. t. 5131.

A handsome species introduced from Borneo through Thomas Lobb. The leaves are dark bluish-green, rather fleshy, and the hairy flowers deep red in colour with a black line in the throat of the tube.

ÆSCHYNANTHUS FULGENS, *Wall.*

Bot. Mag. t. 4891.

A species with large, thick, entire leaves, and terminal umbels of bright crimson tubular flowers with a yellow throat.

Found by Thomas Lobb in Moulmein, it flowered for the first time in October 1855.

ÆSCHYNANTHUS LOBBIANA, *Hook.*

Bot. Mag. t. 4260.

One of the many species of the epiphytic genus Æschynanthus introduced through Thomas Lobb. This species is striking from the strong contrast offered by the purplish-black calyx and the brilliant scarlet hue of the corolla.

It was introduced from Java and first flowered in 1846.

ÆSCHYNANTHUS LONGIFLORUS, *Blume.*

Paxt. Mag. Bot. vol. xv. p. 25; Bot. Mag. t. 4328.

A species with erect, long, tubular flowers of a uniform puce colour, with exserted stamens.

A native of Java, it was detected by Thomas Lobb growing in woods in the provinces of Buitenzorg and Bantam, for the most part epiphytic on old or decaying trees.

It flowered for the first time in this country in August 1847.

ÆSCHYNANTHUS MINIATA, *Lindl.*

Gard. Chron. 1846, p. 599 (note on exhibit of New Plants); Lindl. Bot. Reg. 1846, vol. xix. t. 61; Paxt. Mag. Bot. vol. xvi. p. 67.

Introduced from Java through Thomas Lobb in 1845, and first flowered at Exeter in November of the following year.

ÆSCHYNANTHUS OBCONICA, *C. B. Clarke.*

Bot. Mag. t. 7336.

Imported from the Malayan Peninsula and flowered for the first time in July 1893: a quaintly neat plant.

It is closely allied to Æschynanthus tricolor, but the corolla-tube is shorter and the calyx broadly campanulate. The flowers are scarlet in colour with yellow stripes.

ÆSCHYNANTHUS PULCHRA, *G. Don.*

Syns. *Æschynanthus pulcher*, Hook.

Gard. Chron. 1846, p. 407 (Report of Exhibit of New Plants); Bot. Mag. t. 4264; La Belg. Hort. 1892, p. 13, col. pl.

A beautiful species with large rich scarlet flowers introduced from Java through Thomas Lobb. First exhibited in bloom at the Great Exhibition held by the Horticultural Society at Chiswick on June 13th 1846, it was awarded the Silver Knightian Medal as a plant of very exceptional merit.

ÆSCHYNANTHUS PURPURESCENS, *Hasck.*

Bot. Mag. t. 4236.

A native of the mountainous regions of Java, whence it was introduced through Thomas Lobb.

It is quite distinct but not as striking as many of the other species of Æschynanthus in cultivation. The flowers are small, yellow with a dotted throat; the calyx has long sepals, margined and tipped with purple.

ÆSCHYNANTHUS SPECIOSA, *Hook.*

Bot. Mag. t. 4320; Paxt. Mag. Bot. 1847, vol. xiv. p. 199; The Garden, 1897, vol. li. p. 188, pl. 1109.

Introduced through Thomas Lobb, who detected it on Mount Asapan, near Bantam, in Java, on the trunks of forest trees.

It was first exhibited in the Botanic Gardens, Regent's Park, in May 1847, and was voted "the most charming of the plants then exhibited." In the description in the Botanical Magazine, Sir William Hooker writes, "It is unquestionably the most beautiful species known to us of a genus eminent for the rich colour of its blossoms."

AGALMYLA STAMINEA, *Blume.*

Bot. Mag. t. 5747; Paxt. Mag. Bot. vol. xv. p.73; Fl. des Serres, 1848, vol. iv. t. 358.

A brilliantly-coloured stove epiphyte, native of Java, where it was discovered by Thomas Lobb growing in humid parts of mountain woods. In December 1847 it was exhibited in flower for the first time.

The flowers are produced in axillary fascicles of eight to fourteen together, are brilliant scarlet in colour, with long exserted stamens, purple at the tips, bright yellow towards the base.

HORTUS VEITCHII

AGAPETES MACRANTHA, *Hook. f.*

Syns. *Thibaudia macrantha*, Hook.

Gard. Chron. 1894, vol. xv. p. 501, fig.; *id.* vol. xxix. p. 47, fig.; Bot. Mag. t. 4566; Gard. Chron. 1863, p. 695 (advt.); Fl. des Serres, vol. vi. p. 345.

A handsome hard-wooded greenhouse plant with drooping inflated tubular flowers, of a rosy white colour, marked with wavy V-shaped lines of red. Raised from seed collected on Kola mountain, Moulmein, by Thomas Lobb, it flowered for the first time in this country in December 1830, and is still largely grown.

AGLAONEMA COSTATUM, *N. E. Brown.*

Gard. Chron. 1892, vol. xi. p. 426; *id.* vol. xiii. p. 86.

A pretty little stove aroid with ovate green leaves spotted and veined with white, introduced from the region of Perak.

AGLAONEMA MARANTÆFOLIUM, *Blume*, var. FOLIIS MACULATIS.

Bot. Mag. t. 5500.

The whole-coloured leaved type is a very old garden plant; the form with spotted leaves was introduced in 1864 from Manila.

AGLAONEMA OBLONGIFOLIUM, *Schott.*, var. CURTISII.

Gard. Chron. 1897, vol. xxi. p. 70; Veitchs' Catlg. of Pl. 1897, p. 2.

A variegated form of the type, sent to England by Charles Curtis from Malaysia.

The leaves, 15 to 18 in. long, are effectively marked with oblique silvery lines symmetrically arranged on both sides of the midrib, the ground colour bright grass-green.

AGLAONEMA PUMILUM, *Hook.*

Veitchs' Catlg. of Pl. 1894, p. 5, fig. p. 3.

A charming little stove aroid from the Malayan region, resembling in habit a miniature Dieffenbachia.

The leaves are from 4 to 6 in. long, beautifully marbled and veined with white on a deep sea-green ground.

ALLOPLECTUS PELTATUS, *Hook. f.*

Bot. Mag. t. 6333.

Introduced from Costa Rica through Endres, an addition to the family of Gesneriads remarkable in having one leaf of each pair permanently rudimentary, and further, the fully developed leaf distinctly peltate, a curious feature.

STOVE AND GREENHOUSE PLANTS

ALOCASIA × INTERMEDIA.

Veitchs' Catlg. of Pl. 1868, fig. p. 3; Gard. Chron. 1867, p. 660 (note on exhibit).

A hybrid raised by Seden at Chelsea from Alocasia longifolia fertilized with the pollen of A. Veitchii.

It received the Silver Medal of the Royal Horticultural Society, at the Exhibition held in June 1867, as the best garden hybrid then exhibited.

ALOCASIA LOWII, var. PICTA, *Hook.*

Syns. *Caladium Veitchii*, Lindl.

Bot. Mag. t. 5497; Gard. Chron. 1859, p. 740; Veitchs' Catlg. of Pl. 1862, p. 6.

A well-known stove plant remarkable for its massive leaves, which are of great substance, rich greyish-bronzy-green on the upper surface, deep purple beneath, with a beautiful metallic lustre.

It was sent from Borneo by Thomas Lobb.

ALOCASIA SCABRIUSCULA, *N. E. Brown.*

Gard. Chron. 1879, vol. xii. p. 296.

A stove aroid belonging to the large-leaved section of the genus of which Alocasia Lowii and A. Thibautiana are better known and more showy species.

ALOCASIA × SEDENII.

Veitchs' Catlg. of Pl. 1870, fig. p. 2.

A rather fine hybrid raised by Seden at Chelsea from Alocasia metallica and A. Lowii, which retains in its handsome leaves the metallic hue of one parent and the dark green and white veins of the other; a striking combination; the plant is now rarely seen.

ALOCASIA THIBAUTIANA, *Mast.*

Masters in Gard. Chron. 1878, vol. ix. p. 527; *id.* 1895, vol. xvii. p. 485, with fig.; Veitchs' Catlg. of Pl. 1880, p. 5, fig.

A noble stove aroid with leaves 2 to 2½ ft. long and from 15 to 20 in. broad, deep olive-greyish-green in colour, traversed by numerous grey veinlets, and of a lustrous metallic hue.

It is a native of Borneo, named in compliment to M. L. Thibaut, in his day one of the first continental nurserymen.

ALOCASIA ZEBRINA, *C. Koch & Veitch.*

Gard. Chron. 1862, p. 399 (advt.).

Introduced to cultivation through the late John Gould Veitch from the Philippine Islands.

The specific name is derived from the peculiar appearance of the leaf-stalks pale yellow in colour heavily marbled and banded with a shade of the darkest green.

AMASONIA PUNICEA, *Hort.*

Syns. *A. calycina*, Hook.

Bot. Mag. t. 6915; Veitchs' Catlg. of Pl. 1889, p. 27; The Garden, 1885, vol. xxvii. p. 130, pl. 479.

An interesting and showy stove plant from British Guiana through David Burke, known to science for nearly a century prior to its introduction, having been described by a German botanist, Martin Vahl, as long since as 1796.

As a horticultural plant, Amasonia punicea is of great value from the brilliantly coloured bracts which subtend the flowers, and which remain on the plant for nearly three months after these have fallen.

ANTHURIUM BROWNII, *Mast.*

Masters in Gard. Chron. 1876, vol. vi. p. 744, figs. 139 and 140.

Introduced from New Grenada, where it was discovered by Gustav Wallis, and named by Dr. Masters in honour of Mr. N. E. Brown of the Herbarium, Kew, who has done so much to elucidate the difficult order to which this genus belongs.

ANTHURIUM CUSPIDATUM, *Mast.*

Gard. Chron. 1875, vol. iii. p. 428, fig. 85.

Discovered in Columbia by Gustav Wallis, and sent to this country in 1874.

It is remarkable in having the petioles of its leaves entirely cylindrical and not at all sulcated as in the majority of species. The leaves are bold and handsome, some 1 ft. 7 in. in length and 1 ft. in breadth.

ANTHURIUM KALBREYERI, *Hort.*

Gard. Chron. 1881, vol. xvi. p. 117, with fig.

A handsome ornamental climbing aroid, introduced from New Grenada through Kalbreyer.

The leaves are palmately divided into nine oblong sinuate leaflets of various sizes, glabrous and rich deep green in colour.

ANTHURIUM VEITCHII, *Mast.*

Gard. Chron. 1876, vol. vi. p. 772, fig. 143; l'Illus. Hort. vol. xxviii. t. 406; Bot. Mag. t. 6968; Veitchs' Catlg. of Pl. 1878, fig. p. 5.

A magnificent aroid, found in Columbia and sent to England by Gustav Wallis, and probably the noblest inhabitant of European stoves.

The leaves are of extraordinary appearance, often attaining a length of

AMASONIA PUNICEA

from 4 to 5 ft., with a breadth of not less than one-third of these dimensions. The principal veins are sunk, and the waved appearance thus caused is further enhanced by a deep glossy green colour and a most brilliant metallic lustre.

ANTHURIUM WALLISII, *Mast.*

Gard. Chron. 1875, vol. iii. p. 429, fig. 86.

Discovered by Gustav Wallis in New Grenada, and through him introduced to this country.

ANTHURIUM WAROCQUEANUM, *Moore.*

Veitchs' Catlg. of Pl. 1878, p. 20, fig. p. 6; Fl. and Pom. 1878, p. 101, fig.

Introduced from Columbia, and dedicated to the late M. Warocqué, formerly an eminent horticulturist in Belgium.

It is a noble species with leaves 2 to 2½ ft. in length, rich deep green in colour, and prominent paler coloured veins.

APHELANDRA ACUTIFOLIA, *Nees.*

Bot. Mag. t. 5789.

A common plant in South America, collected in Mexico, Peru, New Grenada, and Surinam.

The figure in the Botanical Magazine was prepared from an imported plant flowered for the first time in October 1868.

APHELANDRA NITENS, *Hook.*

Bot. Mag. t. 5741; Veitchs' Catlg. of Pl. 1863, p. 2, fig.

Introduced from Guayaquil, New Grenada, through Richard Pearce, and flowered in May 1868.

It is a handsome and still favourite stove plant, with shining dark green leaves, and orange-scarlet flowers borne in erect spikes at the ends of the branches.

APHELANDRA VARIEGATA, *Moore.*

Bot. Mag. t. 4899.

An extremely handsome plant from Brazil, the foliage bold and striking; the large imbricated bracts forming a spike, resembling a fir cone, are of a rich orange-yellow colour, from between which the bright yellow flowers protrude.

ARALIA ELEGANTISSIMA, *Hort. Veitch.*

Gard. Chron. 1873, p. 782; Veitchs' Catlg. of Pl. 1876, p. 5, fig.

A charming stove species similar in habit to the beautiful Aralia Veitchii, but with larger and more deeply serrated leaves.

It was introduced from the South Sea Islands through the late John Gould Veitch.

ARALIA KERCHOVEANA, *Hort. Veitch.*

Gard. Chron. 1878, vol. ix. p. 430; Veitchs' Catlg. of Pl. 1883.

A graceful object for the stoves from the South Sea Islands, first distributed in 1883. The leaves are digitate, composed of 9 to 11 spreading leaflets, making almost a circular outline. The plant is dedicated to Count Oswald de Kerchove of Ghent, one of the most distinguished patrons of Belgian horticulture.

ARALIA OSYANA, *Hort. Veitch.*

Veitchs' Catlg. of Pl. 1870, p. 17, fig. p. 3, also col. pl.

A graceful free-growing stove plant, introduced from the South Sea Islands through the late John Gould Veitch.

The leaves are digitately compound, with 6 to 8 strap-shaped leaflets divided at the apex, bright green in colour with chocolate-coloured veins and tips.

ARALIA VEITCHII, *Hort.*

Veitchs' Catlg. of Pl. 1873, fig. p. 3; Fl. and Pom. 1874, p. 5; Nich. Dict. Gard. vol. i. fig. 143.

A very elegant, universally-cultivated, slender-growing stove plant from New Caledonia, unsurpassed, as a pot plant with ornamental foliage for house decoration and for the exhibition tables.

ARDISIA MAMILLATA, *Hance.*

Veitchs' Catlg. of Pl. 1888, p. 9.

A stove plant from Southern China found by Mr. Charles Ford, late Superintendent of the Botanic Gardens at Hong Kong, who also sent plants to the Royal Gardens, Kew, about the same time.

The specific name is derived from the teat-like processes or mamillæ, each surmounted by a bristle-like hair, thickly studded over the whole of the upper surface of the leaf.

The chief ornaments, however, are the rich coral-red berries which follow the flowers, and remain in perfection through the winter months.

ARDISIA OLIVERI, *Mast.*

Masters in Gard. Chron. 1877, vol. viii. p. 680, with fig.; Bot. Mag. t. 6357; Veitchs' Catlg. of Pl. 1878, fig. p. 7.

This stove flowering shrub of great beauty, somewhat resembling an

Ixora, and of considerable botanical interest, was introduced through Endres from Costa Rica.

Named by Dr. Masters in honour of Professor Oliver, of the Kew Herbarium, "as a trifling, but very sincere acknowledgment of the very valuable services he has, in a manner as thorough as it has been unobtrusive, rendered to horticultural botany for many years past." The plant was eventually distributed in 1878.

ARISTOLOCHIA PROMISSA, *Mast.*

Masters in Gard. Chron. 1879, vol. xi. p. 494.

Described by Dr. Masters from specimens collected in West Tropical Africa by Kalbreyer, who sent seed to Chelsea from which plants were raised.

Dr. Masters says, "It is one of the most extraordinary members of an extraordinary genus." The flowers extend into three tails, which sometimes reach a length of 2 ft.

ARISTOLOCHIA RINGENS, *Vahl.*

The Garden, 1879, vol. xvi. p. 335, fig.; Veitchs' Catlg. of Pl. 1880, p. 19.

A member of the remarkable genus Aristolochia with peculiar gaping flowers, from New Grenada, sent by Gustav Wallis, in 1877. It had, previous to the present introduction, been in cultivation in the Royal Gardens, Kew, and probably also in other botanical collections, but was still a very rare climber.

The flowers, usually some 6 in. in length, vary in size according to the strength of the plant, the ground colour is pale, netted with a venation of dark purple.

ARTHROPODIUM NEO-CALEDONICUM, *Baker.*

Baker in Jour. Linn. Soc. vol. xv. p. 352; Bot. Mag. t. 6326.

Introduced from New Caledonia, first flowered in May 1877, and interesting geographically as extending to New Caledonia, the range of other characteristic Australian and New Zealand genera.

J. G. Baker writes of this plant in the Botanical Magazine:—"I described it from a single dried specimen gathered on Mount Kanala in New Caledonia by M. Deplanche, and it is No. 1695 of the Vieillard collection distributed by the late M. Lenormand."

ASCLEPIAS VESTITA, *Hook.*

Bot. Mag. t. 4106.

Raised from seeds received from North America, and flowered for the first time in October 1843. The flowers, purple in the bud, are greenish-white when expanded, in dense heads in the axils of the uppermost leaves.

ASPARAGUS PLUMOSUS, *Baker.*

Baker in Jour. Linn. Soc. vol. xiv. 1875, p. 613; Masters in Gard. Chron. 1878, vol. ix. p. 527.

This popular plant, introduced from South Africa by Christopher Mudd, is commonly known as the Asparagus Fern, and is in great request for all floral decoration.

ASPARAGUS RACEMOSUS, *Willd.*

Veitchs' Catlg. of Pl. 1880, p. 19.

A climbing species of elegant habit, introduced from Mauritius through Charles Curtis. An admirable subject for covering pillars or trellises in the conservatory or warm greenhouse; the slender branchlets and sprays of glossy green are ever effective.

BARBACENIA SQUAMATA, *Paxt.*

Paxt. Mag. Bot. vol. x. p. 75; Bot. Mag. t. 4136; Fl. des Serres, 1847, p. 266; Gard. Chron. 1890, vol. viii. p. 409, with fig.

A pretty species of the monocotyledonous genus Barbacenia, with fine orange-red flowers produced singly on slender scapes, introduced from the Organ Mountains of Brazil through William Lobb in 1841.

BARLERIA PRIONITIS, *Lindl.*

N. E. Brown in Gard. Chron. 1883, vol. xix. p. 339.

A pretty acanthaceous soft-wooded plant, re-introduced to Chelsea from Sumatra through Curtis, with opposite lanceolate acuminate leaves with axillary spines, and terminal spikes of yellow flower: though previously in cultivation, it had been lost.

BEFARIA ÆSTUANS, *Linn.*

Syns. *Bejaria æstuans*, Mutes.

Gard. Chron. 1848, p. 119, with fig.; Bot. Mag. t. 4818.

Detected by William Lobb in Peru, in the province of Chochapoyas, at an elevation of 8,000 ft.

It was called æstuans, as the flowers glow like fire, and somewhat resemble a Rhododendron; of a beautiful deep rose-colour, they are borne in corymbs terminating the branches.

BEFARIA CINNAMOMEA, *Lindl.*

Gard. Chron. 1848, p. 175.

An ericaceous greenhouse shrub, with purple flowers, introduced through William Lobb in 1847, from the Andes of Peru, with the leaves remarkable in that they are covered on the lower side with a light-brown wool; it is named the Cinnamon Befaria.

STOVE AND GREENHOUSE PLANTS

BEFARIA COARCTATA, *Humb. & Bon.*

Syns. *Bejaria coarctata.*

Gard. Chron. 1848, p. 175, with fig.; Bot. Mag. t. 4433.

Raised from seeds sent home from Peru by William Lobb in 1847. The figure in the Botanical Magazine is from a plant which flowered in a cool greenhouse in the nursery of Messrs. Lucombe, Pince & Co., of the ancient city of Exeter.

BEFARIA MATHEWSII, *Fielding.*

Bot. Mag. t. 4981.

From seed sent from the mountains of Peru by Thomas Lobb, the cream-coloured flowers were seen for the first time in this country in March 1857.

BEGONIA × ACERIFOLIA.

Veitchs' Catlg. of Pl. 1899, p. 53.

A hybrid raised at Chelsea from the remarkable Burmese species Begonia Burkei crossed with the Malaysian B. decora.

The leaves resemble those of an Acer in outline, with silvery blotches on a dark bronzy-green ground.

There is a species B. acerifolia, in comparatively common cultivation, with which this hybrid must not be confused.

BEGONIA BOLIVIENSIS, *A. DC.*

Gard. Chron. 1867, p. 544, fig.; Bot. Mag. t. 5657; Fl. Mag. 1867, t. 354; Veitchs' Catlg. of Pl. 1868, fig. p. 6.

A very beautiful plant with drooping scarlet flowers, from Bolivia, sent by Richard Pearce, and of great interest as one of the original species from which the numerous garden varieties, so popular at the present day, have been derived.

BEGONIA BURKEI, *Hort.*

Veitchs' Catlg. of Pl. 1894, p. 39.

A singular, very distinct species, introduced from Upper Burmah through David Burke.

The leaves, large and peltate, are on foot-stalks over 1 ft. long; from their axils the peduncles bearing much-branched cymes of delicate pink flowers rise, and continue in great beauty through the late autumn and early winter months.

BEGONIA × CARMINATA.

Veitchs' Catlg. of Pl. 1896, p. 2, fig.

A handsome Begonia, the result of a cross between the Brazilian species Begonia coccinea and the South African B. Dregii. The foliage is neat and attractive, the bright carmine-pink flowers, in pendulous cymes composed of from twenty to thirty each, are massive and effective.

BEGONIA COCCINEA, *Hook.*

Bot. Mag. t. 3990; Paxt. Mag. Bot. 1843, vol. x. p. 73.

A very beautiful now well-known species, from the Organ Mountains of Brazil, sent by William Lobb in 1841.

The plant flowered for the first time in April 1842, soon after it was received, and was subsequently exhibited at the rooms of the Horticultural Society in Regent Street.

BEGONIA CRINITA, *Oliver.*

Bot. Mag. t. 5897.

A very elegant plant introduced through Pearce from South America, with a tendency in the branches to develop hairs on that surface only which faces the petiole of the leaf below.

BEGONIA DAVISII, *Hort. Veitch.*

Bot. Mag. t. 6252; The Garden, 1877, vol. xi. p. 70, with fig.; Fl. Mag. n.s. pl. 231; Veitchs' Catlg. of Pl. 1879, p. 22; Fl. and Pom. 1877, p. 85, col. pl.

Introduced from the Andes of Peru through Walter Davis, after whom it is named.

The plant of dwarf tufted habit has elegant bluish-green foliage, purple on the under surface, and handsome dazzling scarlet flowers.

This species was effectively used as a parent by Seden in obtaining a dwarf race of hybrids suitable for summer-bedding.

BEGONIA DECORA, *Stapf.*

Veitchs' Catlg. of Pl. 1893, p. 9, fig. p. 4.

A species with ornamental foliage from Penang.

The plant is of a dwarf compact habit with a creeping rootstock, from which are produced the handsome leaves, often 3 to 4 in. long, rich bronzy red-brown with yellow-green nerves, covered with thick-set papillæ terminating in short hairs.

This species has been crossed with varieties of Begonia Rex by various continental growers, and the offspring are amongst the finest ornamental-leaved foliage plants our stoves possess.

BEGONIA × EUDOXA.

Veitchs' Catlg. of Pl. 1899, p. 53, fig.

A hybrid Begonia raised at Chelsea from Begonia Burkei fertilized with the pollen of B. decora.

The oblique leaves 6 to 9 in. long, of a bronzy-green ground colour, covered with small white spots tinted with rose, are most effective; the under surface is rich carmine.

BEGONIA FALCIFOLIA, *Hook.*

Bot. Mag. t. 5707.

This lovely plant, introduced from Peru through Richard Pearce, by whom it was discovered, is a stove species with falcate-lanceolate leaves, of a deep red-purple beneath, deep green-bronze on the upper surface, and with numerous rose-pink flowers, in axillary panicles on the ends of the branches.

BEGONIA GOGOENSIS, *N. E. Brown.*

N. E. Brown in Gard. Chron. 1882, vol. xviii. p. 71; Fl. and Pom. 1882, p. 121.

A native of Gogoe in Sumatra, discovered by Curtis, through whom it was introduced.

It is a very handsome ornamental-foliaged species with peltate leaves, of a bronzy-metallic hue when young, changing to deep velvety-green when mature and intersected by a paler midrib and delicate veins; the under surface deep red.

BEGONIA × HERACLEICOTYLE.

Veitchs' Catlg. of Pl. 1895, p. 48.

A hybrid raised at Chelsea from Begonia heracleifolia and B. hydrocotylifolia, with bold attractive leaves 6 in. across, and large pyramidal panicles of pale pink flowers, which open in the early spring.

BEGONIA LINEATA, *N. E. Brown.*

N. E. Brown in Gard. Chron. 1882, vol. xvii. p. 199.

A pretty species, which dies down annually, with a tuberous rootstock, and blackish-green leaves more or less covered with silvery spots, sent home from Java by Curtis.

BEGONIA × MARGARITACEA.

Veitchs' Catlg. of Pl. 1895, p. 2, fig. p. 4.

A hybrid foliage Begonia raised at Chelsea, with leaves of a deep bronzy-green ground colour covered with irregular silver-rose spots and blotches, and numerous short crimson hairs. The leaves, of the usual oblique form, are toothed along the margin; the under surface is of a glossy vinous-red.

BEGONIA PEARCEI, *Hook.*

Bot. Mag. t. 5545; Veitchs' Catlg. of Pl. 1866, fig. 5, col. pl.

This species was introduced from La Paz through Richard Pearce.

The leaves are very beautiful, of a dark velvet-green above, dull red traversed with pale green nerves beneath, an agreeable contrast to the bright yellow flowers.

The source of all the yellow-flowered forms, it entered largely into the production of the summer-flowering tuberous varieties.

BEGONIA ROSÆFLORA, *Hook. f.*

Bot. Mag. t. 5680; Veitchs' Catlg. of Pl. 1869, p. 1, fig.; La Belg. Hort. 1868, p. 153.

This beautiful species, a native of the Bolivian Andes, closely resembles Begonia Veitchii.

It differs in the stouter red petioles and scapes, the broader, rounder leaves, and more numerous flowers, pale red in colour as those of a Briar Rose, and it flowered for the first time in July 1867.

From light-flowered forms of this species the first white tuberous Begonias were obtained, and these have steadily improved in quality for many years.

BEGONIA VEITCHII, *Hook. f.*

Hook. f. in Gard. Chron. 1867, p. 734, with woodcut; Bot. Mag. t. 5663; Veitchs' Catlg. of Pl. 1867, p. 11, fig.; Fl. des Serres, 1877, p. 119.

A superb species, due to Richard Pearce, who discovered it near Cuzco in Peru, at an elevation of from 10,000-12,000 ft.

At the time of its introduction, this, the finest species then known, was described in the Botanical Magazine. "Of all the species of Begonia known, this is, I think, the finest. With the habit of Saxifraga ciliata, immense flowers of a vivid vermilion cinnabar-red, that no colorist can reproduce, it adds the novel feature of being hardy in certain parts of England at any rate, if not in all."

It flowered with us for the first time in the open air in 1866.

BERTOLONIA PUBESCENS, *Hort. Veitch.*

Gard. Chron. 1865, p. 485 (notice of exhibit); Veitchs' Catlg. of Pl. 1886, p. 9.

This dwarf stove plant, the light green leaves with a very broad dark chocolate band down the centre, was introduced from Ecuador through Richard Pearce.

BORONIA ELATIOR, *Bartl.*

Bot. Mag. t. 6285; The Garden, 1876, pl. xxxix.; Fl. and Pom. 1877, p. 145, col. pl.

A charming hard-wooded greenhouse plant related to the older and better-known Boronia megastigma, but differs in having numerous pretty, small bell-shaped flowers of a bright rose-pink colour.

BOUCHEA PSEUDOGERVAÔ, *Cham.*

Bot. Mag. t. 6221.

Raised from seed imported from Brazil and first flowered in 1874. The terminal spikes of pale rosy-purple flowers are not unlike the common Verbena of South America, to which, moreover, the plant is closely allied.

BURBIDGEA NITIDA, *Hook. f.*

Bot. Mag. t. 6403; Gard. Chron. 1879, vol. xii. p. 398, fig. 63, p. 401; Veitchs' Catlg. of Pl. 1880, p. 8, with fig.

This very beautiful stove plant, the type of an entirely new genus, was discovered by Burbidge when in Messrs. Veitchs' employ in Borneo.

An interesting account of the fortunate discovery is given in the Gardeners' Chronicle quoted above.

The plant inhabits the shady forests of the Marut district in North-West Borneo, between the Lawas and Trusan rivers, at an altitude of 1,000-1,500 ft.

The generic name was given by Sir Joseph Hooker "in recognition of Mr. Burbidge's eminent services to horticulture, whether as a collector in Borneo, or as author of 'Cultivated Plants, their Propagation and Improvement.'"

CALATHEA LEUCOSTACHYS, *Hook. f.*

Bot. Mag. t. 6205.

Sent from Costa Rica by Endres and first flowered in October 1874. The specific name is in allusion to the pure white tips of the yellow bracts with which the flowers are subtended.

CALATHEA ORNATA, *Koern.*

Syns. *Maranta ornata*, Moore.

Gard. Chron. 1861, p. 499 (advt.).

An ornamental-leaved stove plant of some beauty from Borneo, sent by Thomas Lobb.

The leaves, elegantly marked, have the appearance of a dark green fern frond laid upon a pale greyish-green surface.

CALATHEA TUBISPATHA, *Hook.*

Syns. *Maranta tubispatha.*

Bot. Mag. t. 5542; Veitchs' Catlg. of Pl. 1869, p. 8, fig.

Introduced through Richard Pearce from Western Tropical South America.

Handsome leaves from 8 to 12 in. in length, of a pale green colour, with a row on either side of the midrib of oblong deep brown blotches, are the distinguishing characteristics.

CALATHEA VEITCHIANA, *Hook.*

Syns. *Maranta Veitchiana*, Van Houtte.

Bot. Mag. t. 5535; Gard. Chron. 1864, p. 414; *id.* 1870, p. 924, fig. 170; Veitchs' Catlg. of Pl. 1866, p. 3, fig. 1 on col. pl.; Fl. des Serres, xvi. t. 1656; The Garden, 1872, vol. ii. p. 544, fig.

A very beautiful ornamental-leaved stove plant from Western Tropical America, found by Richard Pearce.

The leaves attain a height of 2 ft. when mature with a blade 14 in. long by 9 in. broad. The colouring is very fine, the under surface purplish, the upper deep shining green blotched with conspicuous patches on either side of a yellowish-green midrib.

CALCEOLARIA ALBA, *Ruiz & Pav.*

Bot. Mag. t. 4157; Fl. des Serres, 1850, p. 319; Gard. Chron. 1897, vol. xxii. p. 140, fig. 40; Hemsley in The Garden, 1879, vol. xv. p. 259.

Introduced through William Lobb, when collecting in Chili, and seed sent from which plants were raised and flowered for the first time in September 1844.

The flower is snow-white, globular in appearance, resembling the ripe fruit of the snowberry (Symphoricarpus racemosus).

CALCEOLARIA AMPLEXICAULIS, *H. B. K.*

Bot. Mag. t. 4300; Hemsley in The Garden, 1879, vol. xv. p. 259.

Raised from seed collected near Muña, in Peru, by William Lobb, about the year 1849.

Easily recognized by soft dark green leaves, which clasp the stem, and a profusion of pale yellow flowers, formerly so familiar an object in gardens, it is now rarely found, though having as a summer-bedding plant undeniable merits.

CALCEOLARIA CRENATA, *Lam.*

Syns. *C. floribunda*, Hook.

Hook. in Bot. Mag. t. 4154; Hemsley in The Garden, 1879, vol. xv. p. 258.

Discovered in the environs of Quito by William Lobb, sent home in 1843, and probably now lost to cultivation. In this country the first flowers opened in September 1844.

CALCEOLARIA DEFLEXA, *Ruiz & Pav.*

Syns. *C. fuchsiæfolia*, Hemsl.; *C. grandis*, Hort.

Bot. Mag. t. 6431, as C. deflexa; The Garden, 1879, vol. xv. p. 258, col. pl. as C. fuchsiæfolia; Gard. Chron. 1849 (notice of exhibit as C. grandis); *id.* 1881, vol. i. p. 269.

A Peruvian species introduced through William Lobb, flowered for the first time in 1849, and exhibited as Calceolaria grandis.

Figured in The Garden as a new species in 1879, Mr. Hemsley applied the name C. fuchsiæfolia, from the resemblance of the leaves to those of species of Fuchsia. Later in the same year Sir Joseph Hooker figured it in the Botanical Magazine under the name of C. deflexa, remarking at the same time, "There are specimens in the Herbaria which were cultivated many years ago, from Messrs. Veitchs' garden (then probably at Exeter)."

CALCEOLARIA FLEXUOSA, *Ruiz & Pav.*

Bot. Mag. t. 5154; Hemsley in The Garden, 1879, vol. xv. p. 259; Fl. des Serres, 1877, p. 137.

Raised from seed sent by William Lobb from Peru.

A fine species of this extensive genus with dense massive panicles and large yellow flowers, well adapted for bedding-out during the summer months.

CALCEOLARIA LOBATA, *Cav.*

Bot. Mag. t. 6330; Hemsley in The Garden, 1879, vol. xv. p. 260.

Introduced from Peru, and figured in the Botanical Magazine.

The leaves are lobed, roundish cordate, the lip of the flower remarkably long, folded back upon itself about the middle, in colour a pale, clear, yellow, purple-red spots inside.

CALCEOLARIA PISACOMENSIS, *Meyen.*

Bot. Mag. t. 5677; The Garden, 1879, vol. xv. p. 260.

A species originally discovered by the distinguished traveller Meyen, near Arequipa, Peru, and subsequently by our collector, Richard Pearce, through whom it was introduced. A sub-shrubby perennial, with flowers of a rich orange-red to bright red, in cymes on the upper part of the stem; first flowered in August 1868.

CALCEOLARIA PUNCTATA, *Vahl.*

Bot. Mag. t. 5392; The Garden, 1879, vol. xv. p. 259.

This species belongs to the shrubby Calceolarias, the same section as C. violacea, in which the lips of the corolla are nearly equal, but not saccate. The flowers are lilac-coloured, with a bright yellow blotch on the lower lip, and the plant, a native of the southern provinces of Chili, was introduced through Pearce in 1862.

CALCEOLARIA TENELLA, *Pœp. & Endl.*

Bot. Mag. t. 6231; Hemsley in The Garden, 1879, vol. xv. p. 261.

A little elegant plant, with bright, glossy green leaves, and pale golden flowers, the corolla spotted with red within.

Seeds were sent from Chili by the collector, Downton, and plants raised in 1873. It had been, however, discovered by the German traveller Pœppig in 1823, and later by the English collector Bridges, both of whom failed to introduce the plant to cultivation.

CAMPSIDIUM CHILENSE, *Reiss & Seem.*

Syns. *Tecoma Guarume*, Hook.; *T. mirabilis*, Hort.

Gard. Chron. 1870, p. 1182, with fig.; Bot. Mag. t. 6111, and sub tab. 4896.

This beautiful greenhouse climber, a native of Chiloe and Chili, known to science from specimens collected by many travellers prior to its introduction through William Lobb, has pinnate dark green leaves, tubular scarlet flowers, and flowered for the first time in April 1874; it is now rarely met with.

CANAVALIA ENSIFORMIS, *DC.*

Bot. Mag. t. 4027.

This stove-climber, with handsome purple pea-shaped flowers, was raised from seed received with other plants from Ashantee, in which country it is known as the "Over-look." It is planted by the natives along the margin of their provision grounds, in the belief that it fulfils the part of a watchman, and, from some supposed dreaded power, protects property from plunder.

CANTUA BICOLOR, *Lindl. & Paxt.*

Syns. *Periphragmos uniflorus?* Ruiz & Pav.

Paxt. Mag. Bot. 1849, vol. xv. p. 220; Bot. Mag. t. 4729.

The credit of this introduction is probably due to Mr. Low, but seeds were sent to us by William Lobb about the same time, from which plants were raised and flowered in April 1853.

A greenhouse shrubby plant, with handsome drooping flowers, with a yellow tube and a rich scarlet limb.

CANTUA DEPENDENS, *Pers.*

Syns. *C. buxifolia*, Lam.; *Periphragmos dependens*, Ruiz & Pav.

Gard. Chron. 1848, p. 303 (notice of exhibit of New Plants); Bot. Mag. t. 4582; Fl. des Serres, vol. vii. p. 11; The Garden, 1885, vol. xxviii. p. 271, pl. 509.

A beautiful greenhouse climbing shrub, with long drooping orange-coloured flowers of great beauty from Peru sent by William Lobb, and flowered for the first time at Exeter in May 1848, on which occasion it was exhibited before the Horticultural Society, and awarded the Society's large Silver Medal.

It is the "Magic Tree" of the Peruvian Indians.

CANTUA DEPENDENS

COLESBORNE, GLOS.

STOVE AND GREENHOUSE PLANTS

CANTUA PYRIFOLIA, *Juss.*

Syns. *Periphragmos flexuosa*, Ruiz & Pav.

Bot. Mag. t. 4386; Fl. des Serres, 1848, p. 385.

This, not by any means the most showy of the species of Cantua, was raised from seed collected by William Lobb in Peru, and first flowered in March 1848.

The flowers in a dense terminal corymb, individually funnel-shaped, cream-white in colour, have a limb composed of five pure white segments.

CARAGUATA ANGUSTIFOLIA, *J. G. Baker.*

J. G. Baker in Gard. Chron. 1884, vol. xxii. p. 616; Bot. Mag. t. 7137.

A distinct species discovered by Kalbreyer when collecting in New Grenada, with green leaves, red-brown veins, and flowers bright yellow, subtended by scarlet bracts tipped with green.

First flowered in the houses at Kew in 1884.

CARAGUATA ZAHNII, *Hook. f.*

Syns. *Tillandsia Zahnii*, Hort. Veitch.

Bot. Mag. t. 6059; Veitchs' Catlg. of Pl. 1874, p. 14, fig.

Introduced through Zahn, who discovered it on his great journey to Costa Rica in 1870, in Chiriqui, Central America, shortly before he perished by drowning.

A handsome stove plant, with leaves 1 ft. long, the base yellow with crimson stripes, the middle portion bright crimson, passing into green at the tips.

The flowers, in dense panicles, on the end of erect scapes, are pale yellow in colour, subtended by bright scarlet bracts.

Flowers opened for the first time at Chelsea in May 1873.

CAVENDISHIA ACUMINATA, *Benth.*

Syns. *Thibaudia acuminata*, Hook.

Bot. Mag. t. 5752; Veitchs' Catlg. of Pl. 1869, p. 10, fig.

This showy, free-flowering greenhouse shrub, a native of the Andes of Columbia, sent by Richard Pearce, produced its brilliant scarlet tubular flowers for the first time in November 1868.

CELMISIA MUNROI, *Hook. f.*

Bot. Mag. t. 7496.

One of the handsomest of the New Zealand Daisy bushes, the Asters of other parts of the world. The flowers resemble a large Marguerite Daisy, the ray florets white and the disk yellow; the leaves strap-shaped, dark green, with an electric coloured pellicle and a silvery woolly under surface.

It is hardy in the warmer localities of the British Isles.

CELMISIA SPECTABILIS, *Hook. f.*

Bot. Mag. t. 6653.

This, the first species of a handsome genus to be cultivated in this country, is a native of New Zealand, where the family represents the Asters and Erigerons of the Old and New Worlds. With this sole exception they are absent from the Archipelago.

Celmisia spectabilis flowered for the first time in May 1882.

CELOSIA HUTTONI, *Mast.*

Gard. Chron. 1872, p. 215, with figs.; Veitchs' Catlg. of Garden and Fl. Seeds, 1873, p. 30, figs.; Nich. Dict. Gard. vol. i. p. 395.

An ornamental-leaved plant introduced in 1870 from Java, through Henry Hutton, after whom it is named.

It has deep claret-coloured leaves and scarlet flowers, and, although a stove perennial, succeeds as a half-hardy annual, and as such is a most useful subject for summer-bedding.

CENTROPOGON COCCINEUS, *Regel.*

Syns. *Siphocampylus coccineus*, Hook.

Bot. Mag. t. 4178; Paxt. Mag. Bot. vol. xii. p. 173; Fl. des Serres, 1846, pl. ix.

Sent by William Lobb from the Organ mountains of Brazil to Exeter, and first flowered in June 1845, and awarded a silver gilt medal by the Horticultural Society at Chiswick and by the Royal Botanic Society in Regent's Park the following month.

It is a stove species with tubular red flowers, unhappily seldom seen in cultivation.

CEPHÆLIS TOMENTOSA, *Willd.*

Bot. Mag. t. 6696.

A very singular plant related to the species that yields the valuable medicinal Ipecacuanha, but of different appearance. A native of tropical America, introduced from British Guiana.

CERATOSTEMA LONGIFLORUM, *Lindl.*

Lindl. in Gard. Chron. 1848, p. 87, fig.; Bot. Mag. t. 4779; Fl. des Serres, tom. ix. p. 197.

A splendid ericaceous shrub with tubular flowers nearly 2 in. long, of a rich orange-scarlet colour, introduced from the Andes of Peru through William Lobb, and first exhibited at the Chiswick Horticultural Summer Exhibition of 1853.

It is a very pretty evergreen greenhouse shrub, loving a *mélange* of peat and sand.

CEROPEGIA CUMINGIANA, *Dcne.*

Bot. Mag. t. 4349.

Sent from Java by Thomas Lobb, and first flowered in August 1847: dried specimens had previously been obtained from Manila by Cuming, and named in his honour.

A stove climber with numerous flowers of peculiar shape, coloured with transverse bands of white, red, chocolate, and yellow.

CEROPEGIA GARDNERI, *Thwaites.*

Bot. Mag. t. 5306.

A native of Ceylon, first detected by Mr. Gardner, whose name it bears, on Rambaddo, at an elevation of 4,000-5,000 ft. In this country a greenhouse climber, it has the peculiar shaped flowers typical of the genus, and ovate leaves purple on the under surface.

CHIRITA EBURNEA, *Hance.*

Journal of Botany, 1883, vol. xxi. p. 168.

A gesneraceous greenhouse plant with radical lanceolate-ovate leaves and tubular pinkish flowers on a white background, in clusters on erect stems, from the Province of Hupeh, Central China; flowered at Coombe Wood in 1903.

CHIRITA HORSFIELDII, *R. Br.*

Syns. *Liebigia speciosa*, DC.; *Tromsdorffia speciosa*, Blume.

Bot. Mag. t. 4315; Gard. Chron. 1846, p. 599 (Report of Exhibit of New Plants).

A beautiful gesneraceous plant, with flowers in corymbs in the axils of the uppermost leaves on stems 1½ to 2 ft. high resembling those of a Gloxinia.

The somewhat large coarse foliage hides many of the flowers and detracts from any horticultural merit.

Introduced from Java through Thomas Lobb, first exhibited in bloom on September 1st 1846 before the Horticultural Society, and now probably lost to cultivation.

CLEMATIS SMILACIFOLIA, *Wall.*

Syns. *C. glandulosa*, Blume.

Bot. Mag. t. 4259; Gard Chron. 1856, p. 338 (advt.).

A stove species of "Traveller's Joy," with leaves resembling some large-leaved species of Smilax, and dark purple, almost black, floral segments; the central mass of pure white anthers a striking contrast.

This climber was introduced to cultivation through Thomas Lobb, many years since, from Mount Salak in Java.

CLERODENDRON CRUENTUM, *Lindl.*

Gard. Chron. 1860, p. 339 (Report of Show); *id.* p. 456, Lindl.

Imported through Thomas Lobb, this stove species, with rich red flowers, was exhibited before the Royal Horticultural Society in April 1860, and created much interest.

CLERODENDRON ILLUSTRE, *N. E. Brown.*

N. E. Brown in Gard. Chron. 1884, vol. xxii. p. 424.

Introduced by Curtis, who discovered it in Celebes, the species is valuable for the bright vermilion scarlet flowers on red branches in dense terminal panicles.

CLIANTHUS DAMPIERI, *All. Cunn.*

Bot. Mag. t. 5051; Gard. Chron. 1858, p. 476; Fl. des Serres, 1850, tom. vi. p. 121; The Garden, 1890, vol. xxxvii. p. 299, fig.

This remarkable plant, commonly known as "Sturt's Pea," was first met with in the dry sandy islands of Dampier's Archipelago, North-West Australia, by that renowned navigator and buccaneer, whose name it commemorates. In 1817 Allan Cunningham found it in New South Wales, and later on Captain Sturt met with it in the same region, growing on sterile bleak open flats skirting Prince Regent's Lake.

It first flowered in this country at Exeter, was exhibited before the Horticultural Society of London at St. James's Hall on April 21st 1858, and awarded a Silver Medal.

COFFEA BENGHALENSIS, *Roxb.*

Bot. Mag. t. 4917.

A native of the mountains of the north-eastern frontier of India, chiefly about Silhet, and brought many years ago to Calcutta, where it was for some time much cultivated under the idea that it was the real Coffee of Arabia. The plant from which the figure in the Botanical Magazine was taken was sent to this country by Thomas Lobb.

COLEUS GIBSONII, *Verlot.*

Gard. Chron. 1866, p. 432 (advt.); Fl. Mag. 1867, t. 338; Veitchs' Catlg. of Pl. 1866, fig. 6 on col. pl.

Coleus Gibsonii was found by the late John Gould Veitch in New

Caledonia, in vast quantities, the highly coloured foliage a striking feature. It was used as a parent in the production of the numerous hybrids now in cultivation, and proved most unusually prolific.

COLEUS VEITCHII, *Hort. Veitch.*

Fl. Mag. 1867, t. 345; Veitchs' Catlg. of Pl. 1867, with fig.

Introduced from New Caledonia through the late John Gould Veitch.

The leaves, of heart-shaped outline, are of a deep chocolate colour with a lively green edge. This species, with Coleus Gibsonii, was much used by the hybridist, and some very beautiful-leaved varieties resulted, more especially those which originated in the Gardens of the Royal Horticultural Society at Chiswick.

COLOCASIA AFFINIS, *Schott.*

Syns. *Alocasia Jenningsii*, T. Moore.

l'Illus. Hort. 1869, t. 585; Fl. des Serres, t. 1818; Fl. and Pom. 1868, p. 276, with fig.; Veitchs' Catlg. of Pl. 1868, p. 4, with fig.

Introduced from India. A fine foliaged stove plant with glaucous green leaves blotched with deep blackish-green, and a prominent venation very effective when well grown.

The plant was honoured with a Silver Medal when shown before the Royal Horticultural Society in May 1867.

COLUMNEA KALBREYERI, *Hook. f.*

Bot. Mag. t. 6633; Masters in Gard. Chron. 1882, vol. xvii. pp. 44, 216, with fig.; Fl. and Pom. 1882, p. 26.

A superb discovery by Kalbreyer, whose name it bears, when travelling in the province of Antioquia, growing on trees in the forests of Ciñegetas. The striking contrast afforded by the pale-green, yellow-mottled upper surface, and the blood-red under surface of the leaves, with the golden yellow of the flowers, forms one of the most distinct of the many colour contrasts to be found in the vegetable kingdom.

CORDIA DECANDRA, *Hook. & Arn.*

Gard. Chron. 1849, p. 564; Bot. Mag. t. 6279.

A beautiful greenhouse shrub with large white flowers, sent from Chile in 1849 through William Lobb.

The excessively hard wood is much used for charcoal, and it is from this the local name "Carbon" is derived. It is also used for smelting copper.

The specimen figured in the Botanical Magazine first flowered at Chelsea in May 1875.

CORREA CARDINALIS, *Muell.*

Bot. Mag. t. 4912; The Florist, 1856, pl. 116.

This handsome hard-wooded greenhouse plant, with scarlet tubular flowers tipped with green, was originally discovered by Dr. Ferdinand (afterwards Baron von) Müller in sandy places in the sterile plain of Port Albert, Victoria, South Australia.

Raised from seed from the same locality, it flowered for the first time in England in May 1856.

CRAWFURDIA FASCICULATA, *Wall.*

Bot. Mag. t. 4838.

One of the climbing Gentians, a plant with beautiful blue flowers, raised from seed sent by Thomas Lobb from Khasia.

Plants flowered under glass for the first time in January 1855.

CROSSANDRA GUINEENSIS, *Nees.*

Bot. Mag. t. 6346.

A charming plant with dark-green leaves, golden reticulation, and spikes of rose-purple flowers, long known to science prior to introduction, but only flowered for the first time in October 1877.

CROTON ANEITUMENSIS, *Hort.*

Veitchs' Catlg. of Pl. 1881, p. 31.

The midribs, margins, and principal veins of the leaves of this variety are coloured gamboge-yellow on a bright green ground.

CROTON APPENDICULATUS, *Hort.*

Veitchs' Catlg. of Pl. 1876, p. 20, fig. p. 9; Fl. and Pom. 1879, p. 67, fig.

A peculiar variety in which the blade of the leaf is separated by a considerable interval occupied by the midrib only.

CROTON AUCUBÆFOLIUM, *Hort.*

Veitchs' Catlg. of Pl. 1879, p. 2, fig.

Introduced through the late John Gould Veitch from the South Sea Islands, the foliage bears a strong resemblance to the well-known vulgar Aucuba japonica of gardens.

CROTON AUREO-MACULATUS, *Hort.*

Veitchs' Catlg. of Pl. 1878, p. 26; Gard. Chron. 1878, vol. ix. p. 430.

A variety with neat and small foliage, bright green spotted with yellow.

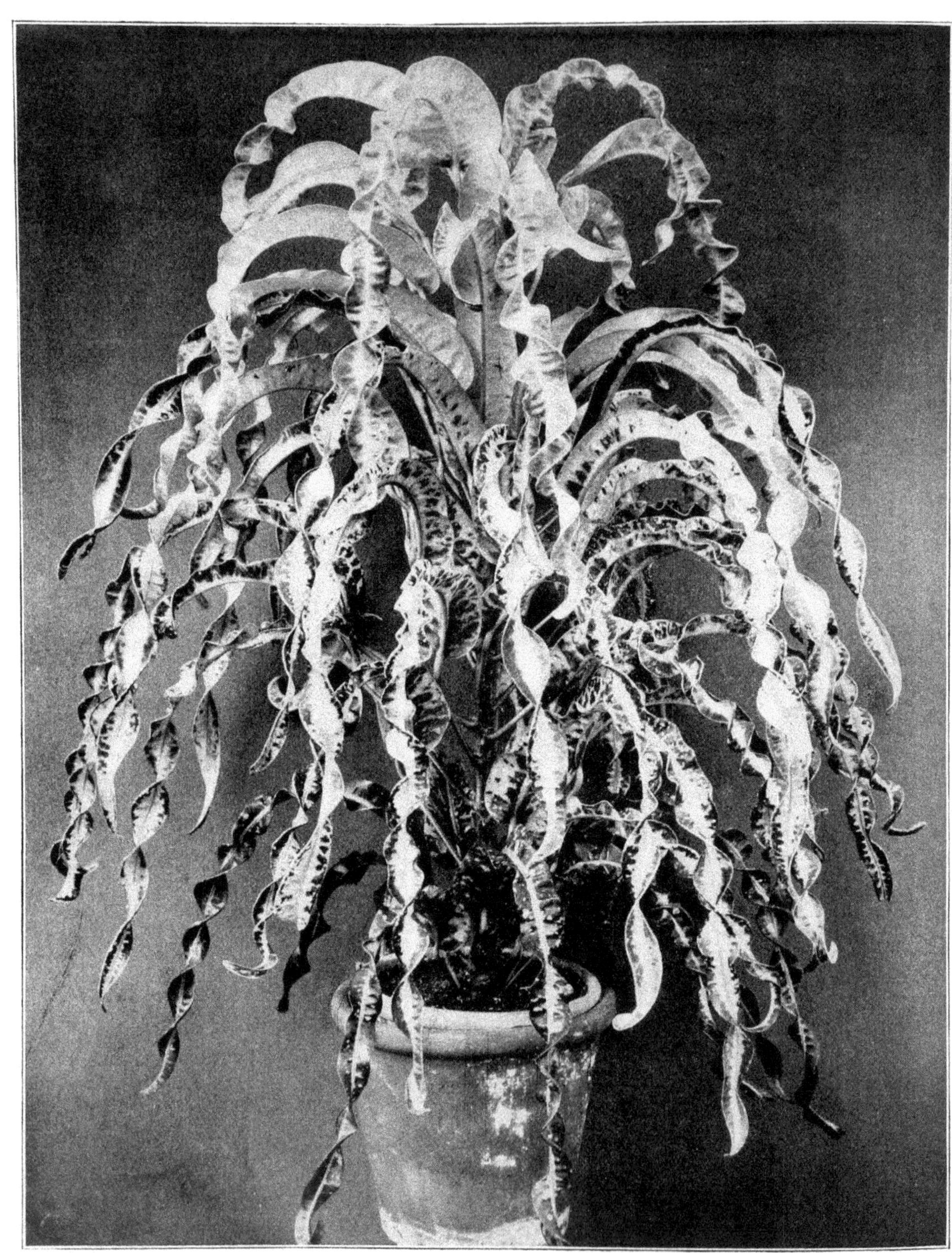

CROTON CAUDATUS TORTILIS

STOVE AND GREENHOUSE PLANTS

CROTON AUREO-MARMORATUS, *Hort.*

Fl. and Pom. 1882, p. 122; Veitchs' Catlg. of Pl. 1884, p. 13.

Introduced through Charles Moore, Esq.

The leaves, fully 1 ft. long and 3 in. broad, beautifully variegated with bright golden-yellow on a deep olive-green ground.

CROTON BISMARCK.

Veitchs' Catlg. of Pl. 1876, p. 51.

A form intermediate in the shape of the leaves between the trilobed and broadly lance-shaped varieties, spotted and blotched with yellow on a green ground.

CROTON BRAGÆANUS, *Hort.*

Veitchs' Catlg. of Pl. 1882, p. 17; Fl. and Pom. 1882, p. 122.

An elegant variety with pendulous lance-shaped leaves 18 to 21 in. long, deep olive-green, speckled and spotted in a quaint way with bright yellow of various shades.

It was dedicated to Senhor José Terceiro Da Silva Braga, formerly well known in Portugal as a distinguished and enlightened patron of Horticulture.

CROTON CAUDATUS TORTILIS, *Hort.*

Veitchs' Catlg. of Pl. 1883, p. 14, fig. p. 5.

A twisted-leaved variety, one of the best in cultivation, introduced through Charles Moore, Esq., at one time Superintendent of the Botanic Gardens, Sydney, and the last Government servant to be appointed to such a post by the home authorities.

CROTON CHALLENGER.

Veitchs' Catlg. of Pl. 1878, p. 21, fig. p. 10.

A long-leaved variety from the South Sea Islands, found by Peter C. M. Veitch, and also by Sir William MacArthur, of Camden Park, Sydney, N.S.W.

The ground colour of the leaf-blade is light green, blotched and streaked with yellow, assuming a rosy tinge with age.

CROTON CHRYSOPŒCILUS, *Hort.*

Veitchs' Catlg. of Pl. 1885, p. 34.

The foliage of this variety offers a striking instance of the peculiar coloration to which the foliage of Crotons is subject. The midrib, foot-stalks, and sometimes two-thirds of the length of the entire leaf, is coloured pale canary-yellow, the remainder a deep olive-green.

CROTON COMTE DE GERMINY.

Veitchs' Catlg. of Pl. 1880, p. 20.

A handsome variety with broad leaves, coloured crimson along the midrib and blotched with light golden-yellow on the blade, dedicated to the Comte de Germiny in recognition of that noble's continuous patronage of Horticulture.

CROTON COOPERI, *Hort.*

Veitchs' Catlg. of Pl. 1875, p. 11, fig. p. 6.

A variety with large oblong slightly wavy leaves, conspicuously veined and blotched with yellow, the markings, as the foliage ages, developing shades of red.

CROTON CORNUTUM, *Hort.*

Gard. Chron. 1868, p. 884; Veitchs' Catlg. of Pl. 1870, p. 18, fig. p. 6.

An introduction through the late John Gould Veitch from one of the South Sea Islands.

A compact growing variety with curious horn-like processes at the apex of the leaves caused by an extension of the midrib.

CROTON CRONSTADTII, *Hort.*

Veitchs' Catlg. of Pl. 1882, p. 17, fig. p. 8; Fl. and Pom. 1882, p. 122.

Leaves twisted and crisped, glossy green margined and variegated with yellow—lost to cultivation.

CROTON DAYSPRING.

Veitchs' Catlg. of Pl. 1881, p. 32.

A variety introduced through Charles Moore, Esq., late of the Botanic Gardens, Sydney, N.S.W.

CROTON DISRAELI.

Gard. Chron. 1875, vol. iv. p. 420, figs.; Veitchs' Catlg. of Pl. 1876, p. 20, fig. p. 10.

A variety with trilobed leaves marked and blotched with yellow on a green ground, sent to us by A. H. C. Macafee, Esq., of Sydney, N.S.W.

CROTON EARL OF DERBY.

Veitchs' Catlg. of Pl. 1878, p. 27.

A trilobed-leaved form of the Disraeli type, with nearly the whole central portion of the leaf-blade yellow, the margins and tips of the lobes a grassy green.

CROTON EVANSIANUS, *Hort.*

Veitchs' Catlg. of Pl. 1879, p. 23, fig. p. 7.

A very handsome form remarkable for the deep colouring of its trilobed leaves, and still to-day a garden favourite.

When first formed these are light olive-green with midribs and veins of golden-yellow; when mature the ground colour deepens to bright bronzy crimson and the yellow to an orange-scarlet.

It was introduced from the South Sea Islands through Peter C. M. Veitch.

CROTON FORDII, *Hort.*

Veitchs' Catlg. of Pl. 1880, p. 20.

A dwarf-growing form with richly coloured trilobed leaves.

CROTON HANBURYANUS, *Hort.*

Veitchs' Catlg. of Pl. 1879, p. 33, fig.

A variety with leaves of varied coloration, introduced through Charles Moore, Esq., of Sydney, N.S.W.

CROTON HARWOODIANUS, *Hort.*

Veitchs' Catlg. of Pl. 1878, p. 26.

A form with dimorphous foliage of bright and varied colour, the larger leaves are 10 in. in length and the smaller little more than 6 in.

CROTON HAWKERI, *Hort.*

Veitchs' Catlg. of Pl. 1879, p. 23, fig. p. 8.

For this variety we are indebted both to Lady Robinson, of Government House, Sydney, N.S.W., and to Charles Moore, Esq.

The leaves have the middle and lower portion together with the footstalks, coloured bright yellow, the margins and tips tending to a bright green.

CROTON HILLIANUM, *Hort.*

Veitchs' Catlg. of Pl. 1869, p. 12, fig. p. 2.

Introduced through the late John Gould Veitch from the South Sea Islands, and remarkable for the reddish-yellow effect of the foliage.

CROTON HOOKERI, *Hort.*

Veitchs' Catlg. of Pl. 1871, p. 15, fig. 5; Gard. Chron. 1868, p. 943; Fl. and Pom. 1871, p. 199, fig.

A beautiful form with leaves marked in the same manner as the Milkmaid Holly, introduced from the South Sea Islands through the late John Gould Veitch.

CROTON INTERRUPTUM, *Hort.*

Veitchs' Catlg. of Pl. 1868, p. 16, fig. p. 8; Gard. Chron. 1868, p. 844; *id.* 1870, p. 137, fig.

Introduced from the South Sea Islands through the late John Gould Veitch.

The leaves, prettily variegated with red, exhibit the peculiar phænomenon of being in separate portions, connected only by an unusually powerful midrib.

CROTON IRREGULARE, *Hort.*

Veitchs' Catlg. of Pl. 1868, p. 17, fig. p. 9.

The leaves of this variety are of variable size and shape, dark green, more or less spotted with yellow.

It was introduced through the late John Gould Veitch from the South Sea Islands.

CROTON JOHANNIS, *Hort.*

Syns. *C. angustissimum.*

Gard. Chron. 1868, p. 844; *id.* 1871, p. 612, fig. 123; Veitchs' Catlg. of Pl. 1871, p. 14, fig. p. 3.

A variety from the South Sea Islands through the late John Gould Veitch, after whom it was named, with long narrow leaves of a glossy green colour, the centre and margins flaked with bright yellow.

CROTON LACTEUM, *Hort.*

Veitchs' Catlg. of Pl. 1872, p. 12, fig. p. 4.

A distinct variety introduced through the late John Gould Veitch from the South Sea Islands.

The leaves, of dark shining green, have broad lines of milky or yellowish-white colour.

CROTON MACARTHURI, *Hort.*

Veitchs' Catlg. of Pl. 1877, p. 21, fig. p. 7.

Sent to us by Sir William MacArthur of Camden Park, Sydney, N.S.W., a well-known Australian amateur, in whose honour it was named.

The leaves are effectively variegated with large blotches and flakes of yellow on a bright-green ground.

CROTON MACULATUS KATONII, *Hort.*

Gard. Chron. 1878, vol. ix. p. 430; Veitchs' Catlg. of Pl. 1878, p. 21, fig. p. 11.

Also due to Sir William MacArthur of Camden Park, Sydney, N.S.W.

It is a trilobed form of the Disraeli type, with numerous bright yellow spots scattered over the deep but bright green leaf-blade.

CROTON MAXIMUM, *Hort.*

Veitchs' Catlg. of Pl. 1871, p. 36; Gard. Chron. 1870, p. 1668, fig.; l'Illus. Hort. p. 534.

Introduced through the late John Gould Veitch from the South Sea Islands; a species with large leaves of oblong form, bright golden-yellow in colour, blotched on either side of the centre with olive-green bands.

CROTON MOOREANUS, *Hort.*

Veitchs' Catlg. of Pl. 1876, p. 21, fig. p. 11.

Sent to us by Charles Moore, Esq.; the midrib and margin of the leaf bright orange-yellow on an olive-green ground.

CROTON MULTICOLOR, *Hort.*

Veitchs' Catlg. of Pl. 1871, p. 15, fig. p. 4; Fl. and Pom. 1872, p. 89, fig.

Introduced from the South Sea Islands through the late John Gould Veitch.

The leaves, of irregular shape, are light green in colour blotched with yellow, reddish-yellow, and red.

CROTON NEVILLIÆ, *Hort.*

Veitchs' Catlg. of Pl. 1880, p. 21, fig. p. 9.

A native of the South Sea Islands, named in compliment to Lady Dorothy Nevill, as a tribute to this great lady's interest in gardening.

The leaves, variegated green, yellow, and crimson, are suffused with a metallic hue peculiar to this plant.

CROTON NOBILIS, *Hort.*

Veitchs' Catlg. of Pl. 1877, p. 22, fig. 8; Fl. and Pom. 1878, p. 133, fig.

A beautiful variety with long, lanceolate, weeping leaves, found in the South Sea Islands.

CROTON OVALIFOLIUM, *Hort.*

Veitchs' Catlg. of Pl. 1874, p. 16, fig. p. 6.

A very distinct form with large oval leaves, from the South Sea Islands through the late John Gould Veitch.

CROTON PRINCESS OF WALES.

Veitchs' Catlg. of Pl. 1880, p. 26, fig.

A very graceful variety with pendant foliage. The leaves are some 24 in. in length, coloured light yellow in the centre, margined with light olive-green profusely spotted with yellow.

CROTON RECURVIFOLIUS, *Hort.*

Veitchs' Catlg. of Pl. 1881, p. 37, fig. p. 34.

A very fine broad-leaved variety, the foliage dense and gracefully recurved at the tips, the yellow-crimson and deep olive-green variegation unusually brilliant.

CROTON REGINÆ, *Hort.*

Veitchs' Catlg. of Pl. 1878, p. 22, fig. p. 12; Fl. and Pom. 1879, p. 59, fig.

Sent to us by J. R. Young, Esq., of Sydney, N.S.W.; this handsome form has crimson midribs and veins, and a leaf-blade deep olive-green sparingly spotted with yellow.

CROTON SINITZINIANUS, *Hort.*

Veitchs' Catlg. of Pl. 1881, p. 37, fig. p. 35.

This variety, for which we are indebted to the kindness of Sir William Macarthur, Camden Park, Sydney, N.S.W., has narrow lance-shaped leaves of a deep olive-green streaked with straw.

It is named in compliment to M. Peter Sinitzin, an accomplished Russian amateur.

CROTON TORTILIS, *Hort.*

Veitchs' Catlg. of Pl. 1877, p. 22, fig. p. 9.

A peculiar and remarkable form with the leaf-blade twisted in a spiral manner around the midrib.

CROTON UNDULATUM, *Hort.*

Veitchs' Catlg. of Pl. 1870, p. 19, fig. p. 7.

Introduced from the South Sea Islands through the late John Gould Veitch.

Of unusually free growth, the margins of the leaves are beautifully undulated and wavy.

CROTON VARIABILIS, *Hort.*

Veitchs' Catlg. of Pl. 1877, p. 22.

Sent by A. H. C. Macafee, Esq., of Sydney, N.S.W.

It has long leaves, marbled, blotched and flaked with various shades of orange, bronze-yellow, and crimson.

CROTON VEITCHIANUM, *Hort.*

Veitchs' Catlg. of Pl. 1870, p. 19, fig. p. 7.

This introduction from the South Sea Islands, through the late John Gould Veitch, has leaves which attain a large size effectively variegated with rose, carmine-purple, and creamy yellow.

CROTON WEISMANNI, *Hort.*

Veitchs' Catlg. of Pl. 1872, p. 12, fig. p. 3; Fl. and Pom. 1873, p. 55, fig.

A very handsome form introduced from the South Sea Islands, with long and narrow leaves not unlike a Dracæna, splashed with bright golden-yellow on a dark shining green ground.

STOVE AND GREENHOUSE PLANTS

CROTON YOUNGII, *Hort.*

Veitchs' Catlg. of Pl. 1873, p. 12, fig. p. 4.

This variety, sent through J. R. Young, Esq., of Sydney, N.S.W., has leaves coloured with creamy yellow and bright rosy red on a dark green ground.

CUPHEA CORDATA, *Ruiz & Pav.*

Bot. Mag. t. 4208; Gard. Chron. 1846, p. 477 (Notice of Exhibition); Fl. des Serres, 1846, pl. vii.

A native of the hills and woods of Peru, about Huassahuassi, Chacalla, Acomayo, and Huanuco. From the last-named locality seed was sent to Exeter by William Lobb in 1842; plants raised and flowered for the first time in 1845.

This beautiful species has remarkable flowers, a scarlet tubular calyx and two large petals held erect as banners, and is much valued by the Peruvians, who credit it with various medicinal properties.

CURCUMA AUSTRALASICA, *Hook.*

Bot. Mag. t. 5620.

Introduced from Cape York, North Australia, through the late John Gould Veitch, and first flowered at Chelsea in August 1866.

Prior to this discovery no species belonging to the extensive genus Curcuma had been known to inhabit the Southern hemisphere.

CURCUMA SUMATRANA, *Miq.*

N. E. Brown in Gard. Chron. 1882, vol. xviii. p. 393.

This plant, introduced from Sumatra through Curtis, is a stove species with dense spikes of yellow flowers and large deep orange-red bracts.

DARWINIA FIMBRIATA, *Benth.*

Syns. *Genethyllis fimbriata*, Kipp.

Bot. Mag. t. 5468; Veitchs' Catlg. of Pl. 1869, p. 7, fig.

A beautiful inhabitant of the greenhouse from South-West Australia, first flowered in June 1864.

The flowers are small and insignificant, but the bright rose-coloured fringed scales, resembling a large drooping bell-shaped flower, are very attractive, and for this reason alone it is cultivated.

DENDROSEROS MACROPHYLLA, *Don.*

Bot. Mag. t. 6353.

A handsome shrubby greenhouse plant of the peculiar group of tree Composites, now lost to cultivation.

This genus confined to the Juan Fernandez group of Islands, whence it was first imported through Downton, flowered in August 1877.

DIANTHERA CILIATA, *Ruiz & Pav.*

Syns. *Belperone ciliata*, Hook.; *Jacobinia ciliata*, Nees.

Bot. Mag. t. 5888; Gard. Chron. 1870, p. 1567.

A pretty winter-flowering stove plant with violet-purple flowers, raised from seed received from Venezuela, and flowered for the first time in November 1870; it had also been collected in Panama and Peru.

DIDYMOCARPUS CRINITA, *Jack.*

Bot. Mag. t. 4554; Fl. des Serres, 1850, p. 303.

An interesting stove plant, its beauty dependent on the rich velvety appearance of the leaves, purple on the lower surface; the flowers are white, sparsely produced.

Sent from Singapore by Thomas Lobb, it flowered for the first time in June of 1846.

DIDYMOCARPUS LACUNOSA, *Hook.*

Bot. Mag. t. 7236; Gard. Chron. 1893, vol. xiv. p. 120, fig. 38, p. 211.

A lovely little Gesneraceous plant with flowers similar to those of a Streptocarpus, discovered by Charles Curtis in the Island of Langkawi, on the west side of the Malayan Peninsula, first flowered at Chelsea in July 1891, and it was from this material the plate in the Botanical Magazine was prepared.

DIDYMOCARPUS MALAYANUS, *Hook.*

Gard. Chron. 1896, vol. xx. p. 123, fig. 24; Bot. Mag. t. 7536; Veitchs' Catlg. of Pl. 1897, p. 6, fig.

A charming little plant due to Charles Curtis, when in charge of the Botanic Gardens, Penang, through whom it was introduced.

The flowers, similar in appearance to those of Streptocarpus Rexii, are of a pure primrose-yellow with a darker yellow blotch on the lowermost segment.

DIEFFENBACHIA BOWMANNI, *Hort. Veitch.*

Veitchs' Catlg. of Pl. 1871, p. 6, fig.; Rev. Hort. 1872, p. 198; The Garden, 1874, vol. v. p. 416, fig.

A fine species discovered by David Bowman during his trip to South Brazil.

It is distinguished by very large leaves, of a pleasing light-green colour spotted with dark and nearly black green blotches.

DIEFFENBACHIA JENMANI, *Veitch.*

Veitchs' Catlg. of Pl. 1884, p. 8, fig.; Fl. and Pom. 1884, p. 58.

A species from British Guiana sent to us by the discoverer, Mr. G. S. Jenman, Superintendent of the Botanic Gardens, Georgetown, in whose honour it is named.

A handsome stove plant with bold foliage banded and spotted with cream-white on a bright green ground.

DIEFFENBACHIA PEARCEI, *Hort. Veitch.*

Veitchs' Catlg. of Pl. 1867, fig.

A stove plant with ornamental foliage from Ecuador found by and named after Richard Pearce.

The large leaves have a broad creamy-white midrib, and irregular blotches of the same colour scattered over the surface.

DIEFFENBACHIA PICTA, *Schott.*

Syns. *D. braziliensis*, Hort.

Veitchs' Catlg. of Pl. 1875, p. 7, fig.

A handsome species from Brazil, with large leaves effectively spotted with white or yellowish-white on a pale green base.

DIPLADENIA ACUMINATA, *Hook.*

Syns. *D. magnifica*, Hort.

Bot. Mag. t. 4828; Gard. Chron. 1854, p. 455; The Florist, 1854, col. pl.

A native of Brazil, a beautiful stove climber with large deep rose-coloured flowers, flowered for the first time in this country at Chelsea in July 1854, and still popular.

DIPLADENIA ATROPURPUREA, *A. DC.*

Syns. *Echites atropurpurea*, Lindl.

Lindl. Bot. Reg. 1843, t. 27; Paxt. Mag. Bot. 1842, vol. ix. p. 199; Fl. des Serres, tom. i. p. 70; The Garden, 1893, vol. xliv. p. 488, col. pl. 937.

A handsome stove climber from Brazil, first flowered in 1842, when it was exhibited before the Horticultural Society of London, and awarded a Banksian Medal as a new plant of exceptional merit.

Soon lost to cultivation, it was not re-introduced until it appeared as a seedling on a clump of Cattleya imported in 1889 by Mr. Russell Clarke of Croydon.

DIPLADENIA BOLIVIENSIS, *Hook. f.*

Bot. Mag. t. 5783; Veitchs' Catlg. of Pl. 1869, p. 6, fig.

Introduced from Bolivia through Richard Pearce, and flowered at Chelsea for the first time in June 1868.

The pure white flowers have a rich orange-yellow throat, and are usually less than 1 to $1\frac{1}{2}$ in. across.

DIPLADENIA SPLENDENS, *A. DC.*

Syns. *Echites splendens*, Hook.

Bot. Mag. t. 3976; Paxt. Mag. Bot. vol. x. p. 25; Fl. des Serres, tom. i. p. 74.

A beautiful stove climber described in Curtis's Botanical Magazine in the following terms:—

"This is unquestionably the most beautiful of the many handsome species of this genus, and may vie with the choicest productions of Flora which have been of late years introduced to our gardens. It was sent from the Organ Mountains of Brazil by William Lobb in 1841. Only three plants were met with, and these all reached Exeter in a living state."

DIPLADENIA UROPHYLLA, *Hook. f.*

Bot. Mag. t. 4414; Paxt. Mag. Bot. vol. xvi. p. 67.

A stove species with salmon-coloured flowers inclined to purple, obtained from seed sent from the Organ Mountains of Brazil by Thomas Lobb.

DIPTERACANTHUS SPECTABILIS, *Hook.*

Bot. Mag. t. 4494; Fl. des Serres, 1850, tom. vi. p. 49.

A stove plant with charming blue flowers, raised from seed sent from the Andes of Peru by William Lobb, and flowered for the first time in August 1849.

DRACÆNA ALBICANS, *Hort.*

Veitchs' Catlg. of Pl. 1870, p. 10, fig.

A distinct form of Dracæna terminalis, with large, somewhat undulated leaves, of a pleasing green colour, variegated with white when mature, and occasionally developing some others in which all colouring matter is entirely absent.

DRACÆNA ALBO-VIRENS, *Hort.*

Veitchs' Catlg. of Pl. 1876, p. 21, fig. p. 12.

A graceful form with bright green leaves, margined creamy-white and light crimson: white sometimes predominates, others have light crimson or pink most in evidence.

DRACÆNA AMABILIS, *Hort.*

Veitchs' Catlg. of Pl. 1873, p. 14, fig. p. 7; Fl. and Pom. 1874, p. 75, fig.

The leaves of this beautiful variety are from 24 to 30 in. in length and 4 to 5 in. in breadth, bright glossy green in colour, marked and suffused with pink and creamy white, when mature distinctly rose-coloured.

STOVE AND GREENHOUSE PLANTS

DRACÆNA BALMOREANA, *Hort.*

Veitchs' Catlg. of Pl. 1875, p. 8, fig. p. 12.

An interesting variety introduced from the islands in the South Pacific Ocean through the late John Gould Veitch.

The foliage, as it ages, has a metallic lustre, and is conspicuously marked with rich rose-coloured stripes of variable length and width.

The venation is very distinct and regular.

DRACÆNA BAPTISTI, *Hort.*

Fl. and Pom. 1875, p. 53, fig.; Veitchs' Catlg. of Pl. 1874, p. 50, fig. p. 49.

A distinct and highly ornamental variety received by us from our correspondents, Messrs. John Baptist & Sons of Sydney, N.S.W., in compliment to whom it is named.

Bold foliage, leaves measuring from 2 to 3 ft., margined and striped with yellow and pink, and the additional charm of a stem similarly variegated; a very valuable stove plant.

DRACÆNA CHELSONI, *Hort.*

Veitchs' Catlg. of Pl. 1870, p. 21, fig. p. 11.

Introduced through the late John Gould Veitch from the South Sea Islands.

The ground colour of the leaves deep glossy green, almost black, which, as the plant attains age, become mottled and suffused with a deep crimson, a broad line of the same colour bordering the leaves on both edges.

DRACÆNA ELEGANTISSIMA, *Hort.*

Veitchs' Catlg. of Pl. 1876, p. 21, fig. p. 13.

A narrow-leaved form of a deep bronze hue with a metallic lustre, very distinctly margined with bright crimson; in the young leaves crimson of a lighter shade largely predominates.

DRACÆNA GUILFOYLEI, *Hort.*

Veitchs' Catlg. of Pl. 1870, p. 21, fig. p. 12.

An introduction from the South Sea Islands.

The plant is of rather small growth, with leaves 15 to 18 in. long, striped their entire length with creamy white, in age a rosy tint, always deeper at the margin.

DRACÆNA HENDERSONI, *Hort.*

Veitchs' Catlg. of Pl. 1874, p. 17, fig. p. 10; Fl. and Pom. 1875, p. 53, fig.

A variety of graceful and elegant form, with leaves 1½ to 2 ft. in length, light green in colour, marbled with white and rosy pink lines.

DRACÆNA IMPERIALIS, *Hort.*

Veitchs' Catlg. of Pl. 1873, p. 12, fig. p. 8.

The leaves of this robust variety are $1\frac{1}{2}$ to 2 ft. in length, of a dark green ground colour, suffused with bright red over the whole of the older leaves, the younger of a lighter shade. The plant has a peculiar bronzy metallic lustre, the leathery texture of the leaves rendering it valuable for decorative purposes.

DRACÆNA × INTERMEDIA, *Hort.*

Syns. *D. hybrida*, Hort.

Fl. and Pom. 1876, p. 17, fig.; Veitchs' Catlg. of Pl. 1875, p. 12, fig. p. 8.

A fine form obtained at Chelsea by crossing the two varieties Dracæna magnifica and D. albicans, the foliage clearly indicating the parentage.

DRACÆNA JAMESII, *Hort.*

Veitchs' Catlg. of Pl. 1894, p. 6.

A distinct and beautiful variety sent to us by Charles Moore, Esq., late of the Sydney Botanic Gardens, N.S.W.

In growth of dwarf habit, the leaves rarely exceeding 1 ft. in length and 1 in. in width.

The colour is a rich maroon carmine-crimson.

DRACÆNA LEVANGERI, *Hort.*

Veitchs' Catlg. of Pl. 1875, p. 12.

An introduction from the South Pacific Ocean, through the late John Gould Veitch. The leaves, 12 to 14 in. long, are gracefully curved and distinctly marked with deep rose, almost crimson, on an emerald-green ground.

DRACÆNA MACARTHURI, *Hort.*

Veitchs' Catlg. of Pl. 1877, p. 23, fig. p. 11.

Sent to us by Sir William Macarthur, after whom it is named.

The leaves, scarcely 1 ft. in length, are elegant in form and colour, especially brilliant and attractive.

DRACÆNA MACLEAYI, *Hort.*

Veitchs' Catlg. of Pl. 1869, p. 13, fig. p. 5.

A very beautiful variety brought home by the late John Gould Veitch from the South Sea Islands.

A dwarf, robust plant, with leaves 15 to 18 in. long, of a dark, bronzy brown tint, and a decided gloss over the upper surface.

DRACÆNA MAGNIFICA, *Hort.*

Fl. and Pom. 1871, p. 273, fig.; Veitchs' Catlg. of Pl. 1871, p. 16, fig. p. 8.

A magnificent variety introduced from the South Sea Islands through the late John Gould Veitch.

The leaves attain a length of from 1½ to 2 ft., are of a bronzy pink colour, changing, when mature, to a deeper shade.

DRACÆNA MOOREANA, *Hort.*

Fl. and Pòm. 1872, p. 233; Veitchs' Catlg. of Pl. 1870.

Discovered by the late John Gould Veitch in the South Sea Islands, this magnificent variety has foliage 2 to 3 ft. in length, beautifully undulated. The colouring of the leaf is rich, glossy bronze, with a bright reddish-crimson midrib and leaf-stalk; it was dedicated to Charles Moore, Esq.

DRACÆNA NIGRO-RUBRA, *Hort.*

Veitchs' Catlg. of Pl. 1869, p. 13, fig. p. 5.

A narrow-leaved form from the South Sea Islands, with leaves 16 to 20 in. in length, dark blackish-brown in colour, with a bright rosy crimson centre, the latter colour entirely predominating in the young foliage.

DRACÆNA PRINCESS MARGARET.

Veitchs' Catlg. of Pl. 1879, p. 24, fig. p. 11.

A very fine variety of bold habit and distinct markings, introduced from the South Sea Islands by Peter C. M. Veitch.

When the leaves first unfold they are of a creamy white, slightly shaded with green, with oblique stripes of a deeper shade, and with a pale crimson artery.

The light-coloured portions of the midrib are suffused with delicate rosy pink.

DRACÆNA PORPHYROPHYLLA, *Hort.*

Veitchs' Catlg. of Pl. 1871, p. 12, fig. p. 9.

This charming variety, from the South Sea Islands, where it was found by the late John Gould Veitch, attains really noble proportions. The bold leaves, of a deep bronzy tint on the upper surface, effectively contrast with the glaucous hue of the under.

DRACÆNA REGINA, *Hort.*

Veitchs' Catlg. of Pl. 1869, p. 21, fig. p. 18; Fl. and Pom. 1872, p. 63, fig.

A magnificent Dracæna discovered by the late John Gould Veitch in the South Sea Islands; the leaves, large and broad, are exquisitely marked with creamy white on a green ground.

DRACÆNA ROSEO-PICTA, *Hort.*

Fl. and Pom. 1878, p. 29, fig.; Veitchs' Catlg. of Pl. 1877, p. 23, fig. p. 12.

A sub-erect leaved variety of robust habit and bold aspect, raised at Chelsea. The leaves, from 18 to 20 in. in length, are beautifully coloured with a delicate rose tint, deepening with age to a bright carmine.

DRACÆNA SPECIOSA, *Hort.*

Veitchs' Catlg. of Pl. 1877, p. 23, fig. p. 13.

The foliage of this tricoloured variety is broadly striped or margined with yellowish-white, stained edged with light rosy pink; the ground colour is *eau-de-nile.*

DRACÆNA × TAYLORI, *Hort.*

Veitchs' Catlg. of Pl. 1876.

Raised at Chelsea in early days from the two varieties Dracæna magnifica and D. Mooreana, a handsome hybrid, with deeply coloured foliage, and a very decided metallic lustre tinged with crimson.

The petioles of the leaves are of a light crimson hue.

DRYANDRA CALOPHYLLA, *Br.*

Bot. Mag. t. 7642.

A shrubby plant from the Antipodes, of the Order Proteaceæ, first flowered in this country in the Royal Gardens, Kew, in 1898, but raised by us from seed received in 1891.

ECHEVERIA × GLAUCO-METALLICA.

Veitchs' Catlg. of Pl. 1870, p. 22, fig. p. 14.

A hybrid raised by Seden at Chelsea from Echeveria metallica and E. secunda glauca. It was in its early days much used for summer-bedding on account of a dwarf character and large leaves, approaching in size those of E. metallica; of a bronzy glaucous blue-green hue.

EPISCIA ERYTHROPUS, *Hook. f.*

Bot. Mag. t. 6219.

Introduced from New Grenada, and flowered for the first time in March 1874, this stove herbaceous plant has handsome oblanceolate leaves, bright green above, pale and suffused with red beneath. The flowers, white with a yellow throat, are clustered at the base and more or less hidden.

STOVE AND GREENHOUSE PLANTS

ERANTHEMUM ASPERSUM, *Hook. f.*

Bot. Mag. t. 5711; Veitchs' Catlg. of Pl. 1869, p. 6, fig.

Introduced by the late John Gould Veitch from the Solomon Islands, where he discovered it in 1866.

The flowers, white speckled with purple, one lobe almost wholly purple, give the corolla limb the appearance of a member of the orchidaceous family.

ERANTHEMUM BORNEENSE, *Hook. f.*

Bot. Mag. t. 6701.

Discovered in North-West Borneo by Curtis, and flowered in England for the first time in May 1882.

The flowers, collected in a dense terminal cone-like inflorescence, are pure white in colour with a faint lemon tinge on the lower segment of the corolla.

ERANTHEMUM COOPERI, *Hook.*

Bot. Mag. t. 5467; Fl. Mag. 1864, t. 182; Fl. des Serres, 1880, p. 293.

A handsome species of Eranthemum raised from seed sent by Sir Daniel Cooper, Bart., from New Caledonia.

The flowers, pure white with purple spots on the lower lip opened for the first time in this country in September 1863.

ERANTHEMUM LAXIFLORUM, *A. Gray.*

Bot. Mag. t. 6336; Veitchs' Catlg. of Pl. 1878, p. 29.

A handsome stove flowering plant introduced from the New Hebrides through Peter C. M. Veitch. The flowers, of a rich purple colour, are borne in many-flowered cymes in the axils of the uppermost leaves, and produced continuously during the autumn months.

ERANTHEMUM SANGUINOLENTUM, *Hort. Veitch.*

Fl. des Serres, 1862-1865, tom. xiv. p. 157.

An Acanthaceous plant from Madagascar, with opposite leaves beautifully marked along the midrib and veins with bright carmine on an emerald-green ground.

ERANTHEMUM TUBERCULATUM, *Hook.*

Bot. Mag. t. 5405.

This very floriferous species, raised from seed sent by Sir Daniel Cooper, flowered for the first time in June 1863.

ERYTHRINA INDICA, *Lam.*, var. MARMORATA.

Veitchs' Catlg. of Pl. 1879, p. 12, with fig.

A beautiful stove shrub with variegated leaves, found in several of the South Sea Islands.

The leaves, broader than the variety commonly cultivated, are effectively variegated with white and blotched with orange-yellow.

EUPHORPIA PULCHERRIMA, *Willd.*, var. PLENISSIMA.

Syns. *Poinsettia pulcherrima*, var. *plenissima*, Hort.

Gard. Chron. 1876, vol. v. p. 16, with fig.; The Garden, 1876, vol. ix. p. 288, pl. xiii.; Fl. Mag. 1876, pl. 200; Veitchs' Catlg. of Pl. 1876, p. 1, with figs.

The discovery of this remarkable plant is due to Herr Benedict Roezl, who met with it in an Indian village in Mexico.

It differs from the ordinary Poinsettia in having branched inflorescences, so called "double," and as a consequence produces two or three times as many brilliantly coloured bracts as does the type.

The plant, distributed for the first time in 1876, proved of difficult culture, and is now lost to cultivation.

EURYGANIA OVATA, *Hook. f.*

Bot. Mag. t. 6393.

A very beautiful evergreen greenhouse shrub, allied to Thibaudia, with brilliant glossy green leaves relieved by bright red petioles.

A native of the Andes of Peru, it was introduced to England through William Lobb.

FICUS PARCELLI, *Hort. Veitch.*

Veitchs' Catlg. of Pl. 1874, p. 17, figs. pp. 8 and 9.

A handsome stove plant received through Messrs. Baptist & Sons, of Sydney, whose collector Mr. Parcell discovered it in the South Sea Islands.

The leaves are handsome, as large as those of Ficus elastica, but thinner in texture, and profusely blotched with irregular patches of cream-white on a green ground; it is still in use as a stove decorative plant.

FUCHSIA DEPENDENS, *Hook.*

Gard. Chron. 1847, p. 734 (Notice of Exhibit of New Plants); Hooker's Ic. Pl. t. 65.

A beautiful species from Quito, in which locality it had previously been met with by Dr. Jamieson, from whose dried specimens the figure in Hooker's Icones was prepared. This Fuchsia first flowered and was exhibited at Exeter in 1847.

The drooping tubular flowers, 2 to 3 in. long, are soft rosy scarlet, the petals of a deeper shade. It is now seldom met with outside Botanical Gardens.

FUCHSIA × DOMINIANA.

The Florist, 1855, pl. 96; Fl. des Serres, 1854-1855, tom. x. pl. 1004.

Raised by John Dominy in 1852 from seed obtained by crossing the two fine species Fuchsia spectabilis and F. serratifolia multiflora.

The flower tube is some 2 in. in length, coloured bright scarlet; the spreading calyx lobes of a pinkish colour on the inner surface.

FUCHSIA MACRANTHA, *Hook.*

Bot. Mag. t. 4233; Gard. Chron. 1846, p. 238 (Report of Exhibition); Paxt. Mag. Bot. vol. xiii. p. 97; Fl. des Serres, 1846, pl. xix.

This handsome species was first discovered by Mr. Mathews climbing on trees in lofty mountains at Andinamarca, Peru, and he sent home Herbarium specimens, and later detected by William Lobb, in woods near Chasula, Columbia, at an altitude of 5,000 ft.

Plants, flowered at Exeter for the first time in 1846, were exhibited at the Horticultural Society's rooms in April of that year.

The flowers are without petals, but the long tubular calyx is brightly coloured rose-red, changing to cream-white at the apices of the four-cleft limb.

FUCHSIA SERRATIFOLIA, *Ruiz & Pav.*

Bot. Mag. t. 4174; Lindl. Bot. Reg. 1845, t. 41; Fl. des Serres, 1849, p. 447; The Garden, 1877, vol. xi. p. 70.

This species, one of the handsomest for greenhouse culture, raised from seed collected in moist shady situations at Muña in Peru by William Lobb, flowered for the first time in April 1845.

The young deep red shoots and large flowers, 2 to 3 in. long, single on long stalks from the axils of the leaves, are most striking, and when first exhibited before the Horticultural Society was awarded a Silver-gilt Medal as a plant of exceptional merit.

FUCHSIA SIMPLICICAULIS, *Ruiz & Pav.*

Bot. Mag. t. 5096; The Garden, 1877, vol. xi. p. 70; Fl. des Serres, 1858, tom. iii. p. 179.

Introduced from Peru through William Lobb, and flowered for the first time at Chelsea in 1858.

The calyx tube, sepals, and petals, are brilliant red in colour, the showy blooms in dense drooping clusters at the termination of the branches.

FUCHSIA SPECTABILIS, *Hook.*

Bot. Mag. t. 4375; Gard. Chron. 1848, pp. 238, 270, and 319, with fig.; Paxt. Mag. Bot. vol. xvi. p. 225; Fl. des Serres, 1848, pp. 359-360.

A truly beautiful species introduced from the Andes of Cuenca, Peru, through William Lobb, who described it in his letters as the "loveliest of the lovely, found in shady woods and growing from 2 to 4 ft. high."

It flowered for the first time at Exeter, was exhibited before the Horticultural Society in their rooms in Regent Street on April 18th, 1848, and awarded the large Silver Medal of the Society.

The plant forms a shrub of moderate size, with young wood of a blood-red colour, glabrous and shining; the large rich velvety green leaves are rich purple on the under surface, and the bright red flowers, 4 in. in length, have petals of a deeper tint. Dr. Lindley describes it as a magnificent thing:—the "Queen of Fuchsias."

GARDENIA FLORIDA, *L.*, var. RADICANS FOLIIS VARIEGATA.

Syns. *G. radicans*, Thunb., *foliis variegata.*

Gard. Chron. 1861, p. 499 (advt.).

A beautiful form of this well-known plant with leaves effectively variegated; from Japan through the late John Gould Veitch.

GESNERA DONKLARII, *Hort.*

Bot. Mag. t. 5070.

Figured in the Botanical Magazine from a specimen which flowered in 1858, and a native of Columbia, a region rich in species belonging to this genus. The leaves, large and cordate, rich green above and velvety, have a purple under surface; the tubular, drooping flowers dull red in colour, a yellow bronze throat.

GESNERA POLYANTHA, *DC.*

Bot. Mag. t. 3995.

Plants raised from seed sent from the Organ Mountains of Brazil by William Lobb, produced their rich and copiously flowered panicles for the first time in August 1842: now lost to cultivation.

GLOBBA ALBO-SANGUINEA, *N. E. Brown.*

Syns. *G. atrosanguinea*, Teijsm. & Binn.; *G. coccinea*, Hort.

Bot. Mag. t. 6626; Gard. Chron. 1881, vol. xv. (Report of R.H.S. Floral Committee); N. E. Brown in Gard. Chron. 1882, vol. xviii. p. 71; Veitchs' Catlg. of Pl. 1882, p. 10, fig.; The Garden, 1882, vol. xxi. p. 361, fig.; Fl. and Pom. 1882, p. 89.

A stove plant of the Ginger family (Zingerberaceæ), found by Curtis in Borneo.

The plant has slender erect or gracefully arched stems with deep glossy green leaves terminated by a dense inflorescence. The flowers, of peculiar structure, are pale yellow in colour with bright scarlet bracts.

GLOXINERA BRILLIANT.

Gard. Chron. 1895, vol. xvii. p. 145, fig. 22.

A bigeneric hybrid raised at Chelsea by the foreman West, by crossing a florist's variety of Gloxinia, Radiance with the pollen of Gesnera pyramidalis.

The brilliant scarlet flowers, tinged with magenta in the shadows, are of intermediate character, and the foliage follows more the Gloxinia than the Gesnera type.

GLOXINIA SPECIOSA, *Lodd.*, var. MACROPHYLLA VARIEGATA.

Bot. Mag. t. 3934.

This, the finest variety of Gloxinia known at the time of its introduction, has rich purple drooping flowers and large green leaves variegated along the veins with greenish-white.

Raised from seed sent by William Lobb from the Organ Mountains of Brazil, the first of the Veitchian introductions to be figured in 1842 in Curtis's Botanical Magazine.

GRAVESIA GUTTATA, *Triana.*

Syns. *Bertolonia guttata*, Hook. f.

Bot. Mag. t. 5524; Veitchs' Catlg. of Pl. 1866, p. 2, fig. 7 on col. pl.; Fl. Mag. 1867, t. 347.

A variegated-leaved stove plant received from Madagascar, though the habitat is probably Brazil.

The dark olive-green leaves, splashed at regular intervals with bright rose spots, offer a charming contrast.

HEBECLADUS BIFLORUS, *Miers.*

Syns. *Atropa biflora*, Ruiz & Pav.

Bot. Mag. t. 4192; Fl. des Serres, 1846, pl. iv.

A native of the Andes of Peru, about Tarma, and Canta Cullnay, where it was collected by Ruiz and Pavon, and by Mathews, but it was not introduced till William Lobb sent specimens in a living state to Exeter, where it flowered in 1845.

HEDYSCEPE CANTERBURYANA, *W. & D.*

Syns. *Veitchia canterburyana*, Wend.; *Kentia canterburyana*, Sec.

Gard. Chron. 1872, p. 327, fig. 116; Veitchs' Catlg. of Pl. 1872, p. 10, fig.

A useful Palm for decorative purposes, of robust constitution and dwarf habit, introduced by the Veitchian firm from Lord Howe's Island.

HELICONIA AUREO-STRIATA, *Hort.*

Veitchs' Catlg. of Pl. 1881, p. 37.

A handsome foliage plant, with Canna-like leaves, and midrib and veins of a light golden-yellow, from the South Sea Islands through Charles Moore, Esq., late of the Botanic Gardens, Sydney, N.S.W.

HEMICHÆNA FRUTICOSA, *Benth.*

Bot. Mag. t. 6164.

A rock plant with handsome yellow Mimulus-like flowers, sent from Costa Rica by Endres, and first flowered in 1873; not hardy in the British Isles.

HETEROTRICHUM MACRODON, *Planch.*

Bot. Mag. t. 4421.

A remarkable and very handsome Melastromaceous plant, with beautiful velvety green leaves and pure white flowers in terminal corymbs, raised from seed sent by William Lobb from New Grenada.

HIBISCUS ROSA-SINENSIS, *L.*, var. COOPERI.

Syns. *H. Cooperi*, Hort.

Fl. des Serres, 1862-1865, t. xiv. p. 109.

Sent by Sir Daniel Cooper from Australia. The leaves, beautifully variegated, have irregular blotches of white, rose-carmine, and greenish-white, on a dark green ground.

HIBISCUS SCHIZOPETALUS, *Hook. f.*

Syns. *H. rosa-sinensis*, var. *schizopetalus*, Mast.

Masters in Gard. Chron. 1879, vol. xii. p. 272, fig. 45; Boulger in Gard. Chron. 1879, vol. xii. p. 372; Bot. Mag. t. 6524; Veitchs' Catlg. of Pl. 1880, p. 14, with fig.

This singular and beautiful plant discovered by Dr. Kirk, H.B.M. Consul at Zanzibar, who found it at Mombasa and in various other localities in East Tropical Africa, and sent seeds to Kew.

It grows both in dry rocky slopes, damp mountain glades, and in dense shade, with Bignonias, Balsams, and Ferns.

STOVE AND GREENHOUSE PLANTS

Plants brought to us from Eastern Africa by the Rev. J. A. Lamb, of the Church Missionary Society, in 1878, attracted much attention from the beauty of the flowers and a peculiar drooping elongated structure and curiously laciniated petals.

HINDSIA LONGIFLORA, *Benth.*

Syns. *Rondeletia longiflora*, Cham.

Paxt. Mag. Bot. vol. ix. p. 217; Bot. Mag. t. 3977; Lindl. Bot. Reg. 1843, pl. 42; *id.* 1844, t. 40.

A beautiful stove plant, with long salver-shaped bright blue flowers in dense terminal corymbs, from the Organ Mountains of Brazil, flowered for the first time at the Mount Radford Nursery, near Exeter, in August 1842.

HINDSIA VIOLACEA, *Benth.*

Bot. Mag. t. 4135; Lindl. Bot. Reg. 1844, t. 40; Paxt. Mag. Bot. vol. xi. p. 198; Fl. des Serres, tom. i. p. 19.

A species somewhat resembling Hindsia longiflora, with large flowers of a violet-blue colour.

Imported from the Organ Mountains of Brazil, through William Lobb, it was exhibited in flower for the first time in May 1843, on which occasion the Horticultural Society awarded a large Silver Medal.

HOYA BELLA, *Hook.*

Bot. Mag. t. 4402; Paxt. Mag. Bot. 1848, vol. xv. p. 243; Fl. des Serres, 1848, p. 399.

This plant, a native of the Talung Kola Mountain, Moulmein, found by Thomas Lobb, is described in the Botanical Magazine as "The most lovely of all the Hoyas, resembling an amethyst set in silver." It was exhibited for the first time in June 1848.

HOYA CINNAMOMIFOLIA, *Hook.*

Bot. Mag. t. 4347; Fl. des Serres, 1848, p. 310.

A handsome stove climber, from Java through Thomas Lobb, flowered for the first time in July 1847.

The flowers in a globular head are effective, a strong contrast to the deep purple blood-colour of the staminal crown and the pale yellow-green of the corolla.

HOYA CORIACEA, *Blume.*

Bot. Mag. t. 4518; Fl. des-Serres, 1850, tom. vi. p. 143.

Discovered by Dr. Blume in mountain woods on the western side of Java, and detected in the same island by Thomas Lobb on Mount Salak, who transmitted living plants to Exeter: first bloomed in 1849.

This twining stove plant has thick, almost fleshy, leaves, and dense umbels of honey-yellow flowers.

HOYA CORONARIA, *Blume.*

Bot. Mag. t. 4969.

Blume was apparently the first to discover this Hoya in the moist woods and shady banks of Western Java, though it was sent to this country by Thomas Lobb, and produced pale sulphur-yellow flowers for the first time in November 1856.

HOYA FRATERNA, *Blume.*

Bot. Mag. t. 4684; Fl. des Serres, tom. viii. p. 179.

A fine stove climber, discovered by Blume in Java, and later in the same locality by Thomas Lobb, through whom it was introduced.

The honey-yellow or buff-coloured flowers opened for the first time at Exeter during the summer of 1851.

HOYA LINEARIS, *Wall.*

Bot. Mag. t. 6682.

A curious species, with hairy linear leaves and corymbs of creamy white flowers, first flowered at Chelsea in April 1883.

HOYA PURPUREO-FUSCA, *Hook.*

Bot. Mag. t. 4520; Fl. des Serres, 1850, p. 143.

Introduced through Lobb, who describes it as a handsome climber common to the woods at Panarang, this denizen of the stoves, with ovate leaves and umbels of peculiar brownish-purple blossoms, flowered in September 1849.

HYPOCYRTA PULCHRA, *N. E. Brown.*

N. E. Brown in Gard Chron. 1894, vol. xvi. p. 244; Bot. Mag. t. 7468.

A pretty little stove Gesneraceous plant, detected in New Grenada by David Burke, through whom the introduction was made. The flowers, singly in the axils of the uppermost leaves, are about 1 in. in length, hairy, of a bright orange colour.

HYPOCYRTA STRIGILLOSA, *Mart.*

Bot. Mag. t. 4047.

A Gesneraceous plant with bright scarlet and yellow flowers in the axils of the leaves; the remarkable tubular corolla with an inflation on the underside resembling the breast of a "pouter" pigeon.

Grown from seed sent from the Organ Mountains of Brazil by William Lobb, and first flowered in May 1843.

HYPOESTES ARISTATA, *Soland.*

Bot. Mag. t. 6224.

A native of Extra-tropical South Africa, detected by Forbes when travelling for the Horticultural Society in 1822, though flowered with us for the first time in February 1874. The flowers, freely produced, have a rose-purple outside, and the inside striped with white—they are showy.

HYPOESTES SANGUINOLENTA, *Hort. ex Veitch.*

Syns. *Eranthemum sanguinolentum*, Van Houtte.

Bot. Mag. t. 5511.

A pretty little Acanthaceous plant, conspicuous for the broad pale-purple bands that mark each vein: a native of Madagascar.

IMPATIENS JERDONIÆ, *Wight.*

The Florist, 1854, n.s. vol. iv. pl. 82; Bot. Mag. t. 4739.

Sent from the Neilgherry Hills, British India, by Mr. McIvor in 1852, and also about the same time to the Royal Gardens, Kew.

This Balsam has flowers of singular shape and strikingly-contrasted colours—brilliant red, bright yellow, and green.

A flowering plant, exhibited for the first time at the Horticultural Society's rooms in Regent Street on October 18th 1853, was honoured with a Knightian medal "in testimony of its singular beauty and usefulness."

IMPATIENS MIRABILIS, *Hook.*

Bot. Mag. t. 7195.

Sent to us by Curtis, by whom it was discovered in Langkawi Island, off the east coast of Sumatra. In Curtis's Botanical Magazine Sir Joseph Hooker writes as follows:—

"It would be difficult to conceive a wider departure from the habit of its genus than this remarkable plant presents. It is an undoubted species of Impatiens, but, whereas the other species of that large genus are weak succulent annuals or branched perennials, Impatiens Mirabilis possesses an erect naked trunk that attains a height of 4 ft. in its native country and the thickness of a man's leg crowned with a tuft of many leaves, from the axils of which spring erect racemes of golden flowers, larger by far than in most of the genus known to me, but slightly uncouth in form."

ISOLOMA HYPOCYRTIFLORUM, *Benth. & Hook. f.*

Syns. *Gloxinia hypocyrtiflorum*, Hook.; *Hypocyrta brevicalyx*, Hort. Veitch.

Bot. Mag. t. 5655; Veitchs' Catlg. of Pl. 1867, with fig.

An interesting Gesneraceous plant, introduced through Richard Pearce from Ecuador, with ornamental foliage and attractive flowers.

In the Botanical Magazine above quoted, Dr. Hooker writes of it as follows:—"In its habit, fibrous roots, and the presence of propagula it is a Gloxinia; in the corolla a Hypocyrta; in the glands a Gesnera; whilst in the small calyx it differs from all these genera."

The leaves are of a pleasing green with silvery white veins; the flowers scarlet with a yellow gibbous portion.

IXORA ACUMINATA, *Roxb.*

Gard. Chron. 1857, p. 378 (advt.).

A handsome species with large pure white trusses, introduced from India through Thomas Lobb.

IXORA BURBIDGEI, *Hort. Veitch.*

Gard. Chron. 1881, vol. xvi. p. 153 (Report of R.H.S. Floral Committee); Veitchs' Catlg. of Pl. 1883, p. 14.

Sent from Borneo by F. W. Burbidge, and like Ixora salicifolia in habit and foliage. It differs from that species in having inflorescenses axillary in the uppermost leaves, as well as terminal, and the flowers bright orange-scarlet in dense trusses.

IXORA FLORIBUNDA, *Griseb.*

Gard. Chron. 1855, p. 315 (advt.).

A very neat and compact-growing species with bright reddish-scarlet flowers, abundantly produced in large trusses, from Java through Thomas Lobb.

IXORA FULGENS, *Roxb.*

Syns. *I. salicifolia*, DC.

Bot. Mag. t. 4523.

A scarlet-flowered species with narrow leaves, a native of Java, in which island it was first found by Blume, and later introduced to cultivation through Thomas Lobb from Mount Seribu: exhibited for the first time in July 1850.

IXORA LOBBII, *Loudon.*

Gard. Chron. 1853, p. 497 (advt.); Loudon's Ency. Pl. supp. ii. 1853.

A stove species with bright orange-scarlet flowers discovered in the

Seribu Mountains, Java, by Thomas Lobb, through whom it was introduced.

Apparently but a short time in cultivation, it is now not to be found in any garden collection.

IXORA MACROTHYRSA, *Teijsm. & Binn.*

Syns. *I. Duffii*, T. Moore.

Bot. Mag. t. 6853; Fl. and Pom. 1878, p. 76, with fig.

Discovered by Mr. Duff of the Sydney Botanic Gardens, in Ualan or Strong Island, one of the Caroline group in the Pacific; the plant produces a large truss of pure bright scarlet flowers, and is one of the most imposing of all cultivated species of this genus.

Specimens from the Sydney Botanic Gardens first flowered in England in 1878.

IXORA SALICIFOLIA, *Blume;* var. VARIEGATA, *N. E. Brown.*

N. E. Brown in Gard. Chron. 1882, vol. xviii. p. 71.

This variety, with a feathered silvery grey band down the centre of the leaf, was found on the island of Sumatra by Curtis, through whom it was introduced.

The type is widely spread among the islands of the Malay Archipelago, but the variety appears to be restricted to the one on which Curtis made the fortunate discovery.

IXORA × WESTII.

Gard. Chron. 1881, vol. xvi. p. 281 (Report of International Hort. Show at Manchester); The Garden, 1892, vol. xlii. p. 496, pl. 886; Veitchs' Catlg. of Pl. 1882, p. 18.

A hybrid raised at Chelsea by West from Ixora odorata, a white flowered species from Madagascar, and I. amboinensis, one with rich vermilion-scarlet blooms.

The combination of white and scarlet has resulted in flowers of a rich rose-pink, a character peculiar to I. Westii; the noble size of the flower trusses and their distinct and novel appearance make this hybrid an important gain.

JASMINUM GRACILLIMUM, *Hook. f.*

Gard. Chron. 1881, vol. xv. p. 9, fig.; Bot. Mag. t. 6559; The Garden, 1881, vol. xix. p. 628, pl. 279; Veitchs' Catlg. of Pl. 1881, p. 8, fig.; Fl. and Pom. 1881, p. 23, fig.

A charming white flowered stove species introduced from Borneo through Burbidge, one of the numerous species found in Eastern Asia and its many islands, the type of which, Jasminum pubescens, is a native of India and China. J. gracillimum is the most distinct in its graceful habit and in the abundance of large pure white sweet-scented blooms, more copiously produced than by any other cultivated species.

A favourite with the natives of Borneo, it is used as a perfume for the hair. Plants cultivated in gardens or in open grassy plots near houses are browsed by goats during the dry season and denuded of their leaves and young branches. On the return of the wet season the plants break into leaf as if by magic, and become fountains of pure white deliciously fragrant flower.

KÆMPFERIA VITTATA, *N. E. Brown.*

N. E. Brown in Gard. Chron. 1882, vol. xviii. p. 264.

A distinct species discovered at Loboe, in Sumatra, by Curtis, through whom it was introduced.

The flowers insignificant, but foliage attractive, much resembling that of some species of Calathea.

LAPAGERIA ROSEA, *Ruiz & Pav.*

Bot. Mag. t. 4447; Fl. des Serres, 1849, p. 491; The Garden, 1878, vol. xiv. p. 376, pl. cli.; *id.* 1896, vol. xlix. p. 174, pl. 1056; Flora and Sylva, vol. ii. p. 221.

This beautiful greenhouse climber, commonly known as the Chilian Bell-flower, from the bell-like appearance of the pendant blooms, was first introduced by Mr. Richard Wheelwright, who sent a plant to the Royal Gardens, Kew, in 1847.

In the following year plants were received at Exeter from William Lobb, then collecting in Chili, and a coloured drawing of the flower, from a plant in its native habitat. From this drawing the coloured plate in the Botanical Magazine was prepared, and when the plants flowered later, the colouring was found to be unusually faithful.

LAPAGERIA ROSEA, *Ruiz & Pav.*, var. ALBIFLORA, *Hook.*

Syns. *L. alba*, Decn.

Bot. Mag. t. 4892; l'Illus. Hort. 1864, t. 406; Fl. Mag. t. 199; The Garden, 1878, vol. xiv. p. 376, pl. cli.; *id.* 1896, vol. xlix. p. 174, pl. 1056; Flora and Sylva, vol. ii. p. 221, fig.

This charming companion to the rose-coloured type flowered for the first time in Europe in the Jardin des Plantes, Paris, in 1855, a living plant having been sent from Chili by M. Abadi.

In 1860 Richard Pearce sent home seed and living specimens from Chili, where the plant is rare, and from this source a stock raised was afterwards distributed.

It flowered with Messrs. Veitch for the first time in 1862, and was exhibited with Lilium auratum, at that time rare, before the Royal Horticultural Society on July 2nd of that year.

LAPAGERIA ROSEA

STOVE AND GREENHOUSE PLANTS

LATUA VENENOSA, *Philippi.*

Syns. *Lycioplesium pubiflorum*, Griseb.

Bot. Mag. t. 5373; Gard. Chron. 1863, p. 388, fig.

An extremely handsome half-hardy Solanaceous shrub sent from Valdivia, South Chili, by Richard Pearce; William Lobb had procured specimens in the island of Chiloe in 1848, but failed to introduce to cultivation.

The habit of the shrub is like that of Cestrum (Habrothamnus), the shape of the flowers resembles that of Cestrum fasciculatus, though they are larger.

According to Dr. Philippi the inhabitants of Chili and Chiloe regard this plant with superstition: it is to them the Latué, Palo-mato, and Palo de los bruyos, or the tree of the magician.

LEEA AMABILIS, *Mast.*

Veitchs' Catlg. of Pl. 1882, p. 11, with fig.; Jour. of Hort. April 6th, 1882; Gard. Chron. 1882, vol. xvii. p. 493, with fig.; The Garden, 1882, vol. xxi. p. 352, with fig.; Fl. and Pom. 1882, p. 75.

A handsome stove foliage plant, collected in Borneo by Curtis.

When the leaflets first expand they are a bright crimson tinged with a rich shade of brown, a central midrib of pale rose; when mature they assume a dark bronzy-green, and a broad silvery white band develops on either side of the central midrib, from which short branches are given off at each nerve; the under surface claret-red.

LESCHENAULTIA BILOBA, *Lindl.*

Gard. Chron. 1842, p. 95; Lindl. Bot. Reg. 1842, p. 2; Paxt. Mag. Bot. vol. viii. p. 151; The Garden, 1884, vol. xxvi. pl. 460.

This beautiful blue-flowered Australian plant of difficult culture, now rarely met with, enjoyed a high degree of popularity at the time "hard-wooded" plants were more generally cultivated: it first flowered at Exeter, and when exhibited before the Royal Horticultural Society obtained a large Silver Medal.

LHOTSKYA ERICOIDES, *Schauer.*

Bot. Mag. t. 7753.

A small-growing, hard-wooded greenhouse shrub with white flowers, of the Myrtle family, having the general appearance of an Erica, and now quite lost to cultivation.

Raised from seed collected in Western Australia by James H. Veitch, it first flowered at Kew in June 1900.

LICUALA VEITCHII, *Watson.*

Syns. *Pritchardia grandis*, Veitch.

Veitchs' Catlg. of Pl. 1885, p. 54; Gard. Chron. 1886, vol. xxv. p. 139; Bot. Mag. t. 7053.

A beautiful Palm with a short stem and spreading, rounded, much-plaited bright green leaves with short petioles, forming a compact crown.

The species, unfortunately distributed as Pritchardia grandis, under the erroneous impression that it was that much-sought-for Palm, was on flowering found to be of the genus Licuala. Difficult to cultivate, few, if any, now exist in collections.

Curtis discovered Licuala Veitchii in Sarawak, North Borneo, and made a successful introduction.

LOMATIA FERRUGINEA, *R. Br.*

Gard. Chron. 1854, p. 355 (advt.).

A fine evergreen shrub with cut foliage covered by rusty tomentum, usually cultivated as a cool greenhouse or conservatory plant, though in favoured localities in this country plants have attained considerable dimensions in the open air.

It was introduced from Chiloe and Patagonia through William Lobb in 1851, and is still grown.

MACLEANIA PUNCTATA, *Hook.*

Bot. Mag. t. 4426.

Sent from the Andes of El Ecuador by William Lobb, and first flowered at Exeter in November 1848.

A greenhouse shrub of the Vaccinium family, with neat evergreen leaves punctuated with dots, bearing numerous bright scarlet tubular flowers tipped with white.

MANDEVILLA HISPIDA, *Hemsl.*

Syns. *Echites hirsuta*, Ruiz & Pav.

Bot. Mag. t. 3997.

A stove climber with delicate pale yellow and rose-coloured flowers, from the Organ Mountains of Brazil through Thomas Lobb in 1842, and flowered for the first time at Exeter in September 1843.

MANETTIA BICOLOR, *Paxt.*

Syns. *M. luteo-rubra*, Benth.

Paxt. Mag. Bot. 1843, vol. x. p. 27; Fl. des Serres, vol. ii. p. 445, t. 6; The Garden, 1899, vol. lvi. p. 6, pl. 1229; Bot. Mag. t. 7776.

This charming little trailing plant, well known in our stoves, is seldom out of flower, and the brilliant vermilion-scarlet and yellow blossoms are always appreciated.

It was introduced to Exeter through William Lobb from the Organ Mountains of Brazil.

MANETTIA MICANS, *Pœpp.*

Bot. Mag. t. 5495.

Originally discovered by Pœppig in Peru, and afterwards by McLean, Mathews, and others. This plant was not introduced until Pearce met with it near Muña, at an elevation of from 3,000-4,000 ft., and sent home seed, from which plants were raised and flowered for the first time in December 1864.

MARANTA STRIATA, *Veitch.*

Gard. Chron. 1864, p. 671 (advt.); Nich. Dict. Gard. ii. p. 327.

A charming little plant from the Philippine Islands due to the late John Gould Veitch: the leaves, of a bright green ground colour, are profusely marked with streakings of pale yellow.

MARIANTHUS DRUMMONDIANUS, *Benth.*

Bot. Mag. t. 5521.

A West Australian climber, producing pretty pale blue flowers, first opened in May 1865, from which the figure in the Botanical Magazine was prepared.

MEDINILLA CUMINGII, *Naud.*

Syns. *M. speciosa,* Hook.

Gard. Chron. 1847, p. 485; Bot. Mag. t. 4321, as M. speciosa.

First discovered by Mr. Cuming in the Philippine Islands and later introduced to cultivation through Thomas Lobb, when, as the handsomest species then known, it attracted much attention at the Chiswick Horticultural Fête held in July 1847.

The plant bears delicate rose-coloured flowers in large, drooping panicles, with handsome dark green ample leaves, in opposite pairs.

MEDINILLA CURTISII, *Hook. f.*

Gard. Chron. 1883, vol. xx. p. 621, fig. 108; Bot. Mag. t. 6730; Veitchs' Catlg. of Pl. 1884, p. 9, fig.

A native of the western coast of Sumatra, discovered by Curtis when collecting in that region. Though not so striking a plant as Medinilla amabilis or M. magnifica, it is worthy of a place in any stove. The graceful habit, ivory-white flowers, purple anthers and coral-red flower-stalks are of a distinct order.

MEDINILLA MAGNIFICA, *Lindl.*

Syns. *M. bracteata*, Hort. Veitch (*non* Blume).

Paxt. Fl. Gdn. vol. i. t. 12; Bot. Mag. t. 4533; Fl. des Serres, 1850-1851, t. 572.

A magnificent species, first found in Manila by Thomas Lobb, who sent plants, first flowered in April 1850.

Numerous heads are produced in a dense drooping panicle 1½ ft. long, and their value is greatly enhanced by the addition of large delicately coloured bracts; at its best, perhaps, before the full perfection of the flowers, when the large imbricated bracts separate and allow the buds to be partially seen. As the expansion of the blossoms advances the upper bracts fall, but the lower ones remain and reflex. It has proved to be one of the most showy ornamental stove plants ever imported, and is still largely grown.

MICONIA HOOKERIANA, *Triana.*

Syns. *M. pulverulenta*, Hook.

Gard. Chron. 1863, p. 693 (advt.); Bot. Mag. t. 5411.

Introduced through Richard Pearce from Peru in 1862, this plant attains a height of from 3 to 4 ft., is furnished with elliptic rugose dark green leaves 12 to 15 in. long, marked by a broad central silvery bar. The flowers, white, are followed by bunches of red berries.

MITRARIA COCCINEA, *Cav.*

Gard. Chron. 1848, p. 350 (Report of Exhibit of New Plants); Bot. Mag. t. 4462; Paxt. Mag. Bot. vol. xv. p. 148; Fl. des Serres, 1848, p. 385; The Garden, 1883, vol. xxiv. p. 17, with fig.

An interesting greenhouse shrub, remarkable as monotypic and as confined to a group of islands off the coast of Chili, of which Chiloe is the principal; introduced through William Lobb, and shown for the first time at an exhibition held by the Horticultural Society in May 1848. The plant, with neat evergreen foliage, tubular brilliant scarlet flowers with yellow base depending from drooping tubercled peduncles, is attractive, but does not thrive in this country—the perennial mists of Chiloe are essential to its existence.

MONOPYLE RACEMOSA, *Benth.*

Bot. Mag. t. 6233.

A handsome Gesneraceous plant with terminal racemes of pure white flowers, the first species of the genus cultivated in this country; raised from seed from New Grenada, and flowered for the first time in July 1875.

STOVE AND GREENHOUSE PLANTS

MUSA BASJOO, *Sieb. & Zucc.*

Syns. *M. japonica*, Hort.

Bot. Mag. t. 7182; Gard. Chron. 1900, vol. xxviii. p. 456.

Introduced through Charles Maries from Japan, where it is cultivated for fibre, as is Musa textilis in the Philippines. This plant is hardy in favoured localities, with slight protection from mats or netting in severe winters.

MUTISIA DECURRENS, *Cav.*

Bot. Mag. t. 5273; The Garden, 1876, vol. x. p. 134; col. pl.; *id.* 1883, vol. xxiv. p. 552, col. pl.; Fl. des Serres, 1877, p. 101.

A thinly branched greenhouse climber, with narrow undivided leaves remarkable for a blade running down the stem in the form of a wing. The flower-heads, deep orange or almost vermilion in colour and from 4 to 5 in. in diameter, resemble a single dahlia in appearance.

Introduced from the Chilian Andes through Richard Pearce, and first flowered in 1861, but now rarely met with, as the necessary climatic requirements are difficult to reproduce.

MYRMECODIA BECCARII, *Hook.*

Bot. Mag. t. 6883.

Imported from the Gulf of Carpentaria in 1884, Sir Joseph Hooker states:—"This plant is one of the most singular ever imported in a living state to this country, and it belongs to a genus, or rather to one of a group of genera, of epiphytic Rubiaceæ, which have been long known from their singular habit of forming often spinous toothed tubers of great size, the interior of which is galleried by ants of various species, and of which insects these are the native homes."

Named after Dr. Beccari, the eminent botanist and traveller, and the author of a work the bulk of which is devoted to the four Rubiaceous genera, Myrmecodia, Hydnophytum, Myrmephytum, and Myrmedoma, under which their botany and the economy of their growth, and the insects they harbour, are described with a fulness and ability that are quite admirable.

OPLISMENUS BURMANNII, *Beauv.*, var. VARIEGATA.

Syns. *Panicum variegatum*, Hort.

Gard. Chron. 1867, p. 458, Veitchs' Catlg. of Pl. 1867, fig.; Fl. des Serres, t. 1715.

A prettily variegated grass, common in glass-houses, introduced from New Caledonia through the late John Gould Veitch, and exhibited for the first time in April 1867.

ORTHOSIPHON STAMINEUS, *Benth.*

Gard. Chron. 1869, p. 941 fig.; Bot. Mag. t. 5833; Fl. Mag. 1871, p. 546; Veitchs' Catlg. of Pl. 1870, p. 15; also col. pl. fig. 10.

A remarkable plant with an inflorescence more resembling a Clerodendron than that of the Labiate family, to which it belongs.

From Assam and Burma to the Philippine Islands it is widely distributed, and from the Nicobars and Siam to Java, Borneo, and Cape York in North-East Australia. The late John Gould Veitch sent it from the last-named locality to Chelsea, where it flowered for the first time in July 1869.

OSBECKIA RUBICUNDA, *Arnott.*

Masters in Gard. Chron. 1866, p. 562; Hooker's Companion to Bot. Mag. vol. ii. p. 309.

A handsome Melastomaceous undershrub, native of Ceylon, recalling in appearance a species of Pleroma, but with stamens equal. The peculiar calycine scales, with the purple flowers and yellow anthers, give a very rich appearance.

OUVIRANDRA FENESTRALIS, *Poiret.*

Bot. Mag. t. 4894; Fl. des Serres, 1856, tom. i. 2nd ser. p. 65.

This very remarkable aquatic, known as the "Lattice-leaf Plant," was discovered by Aubert du Petit-Thouars, in Madagascar, and described by him in a work on the plants of that island.

For its introduction to this country we are indebted to the Rev. William Ellis, who was living in Madagascar in 1855, and brought home plants in a satisfactory state.

Specimens were presented to the Royal Gardens, Kew, to various other botanical establishments in this and other countries, and the remainder placed in Messrs. Veitchs' hands for propagation. For a long time after its introduction Ouvirandra fenestralis was a source of great interest, and plants found their way all through the principal gardens, but good specimens are now rare; they are in excellent condition in the Chatsworth gardens.

PALICOUREA NICOTIANÆFOLIA, *Cham. & Schlecht.*

Syns. *P. discolor*, Hort.

Veitchs' Catlg. of Pl. 1867, p. 7; Bot. Mag. t. 7001.

A stove plant with variegated foliage sent from Peru by Richard Pearce.

The leaves are 5 to 9 in. long and 2 to 3 in. broad, bright green above with yellow midrib and veins; the under surface is yellow-green.

OUVIRANDRA FENESTRALIS

THE ROYAL GARDENS, KEW

STOVE AND GREENHOUSE PLANTS

PALISOTA BICOLOR, *Mast.*

Masters in Gard. Chron. 1878, vol. ix. p. 527.

This stove species, closely allied to Palisota Barteri, imported from Fernando Po, West Africa, has very ornamental foliage.

PANAX FRUTICOSUM, *Linn.*, var. LACINIATUM.

Syns. *P. laciniatum*, Hort.

Veitchs' Catlg. of Pl. 1877, p. 24, fig. p. 15.

A stove plant with a fern-like aspect, the leaves cut in fine segments of variable shape and size.

A native of the South Sea Islands, sent to us by Mr. A. H. C. Macafee, of Sydney, N.S.W.

PANDANUS BAPTISTI, *Hort. Veitch.*

L'Horticulture Belge, 1893, p. 166, fig. 35.

Introduced from the Botanic Garden, at Sydney, N.S.W.

The plant has long, gracefully disposed, linear leaves, effectively marked with white or cream colour on a green ground; an absence of spines an additional recommendation.

PANDANUS PACIFICUS, *Veitch.*

Gard. Chron. 1892, vol. xi. p. 664 (Report of Show); Veitchs' Catlg. of Pl. 1893, p. 11.

A species introduced from the Pacific Islands through Charles Moore, Esq., late of the Sydney Botanic Gardens, N.S.W., with bright glossy green leaves 15 to 20 in. in length, fringed with short spines along both margins.

PANDANUS VEITCHII, *Hort.*

Veitchs' Catlg. of Pl. 1871, p. 11, fig., also col. pl.; Regel's Gartenflora, 1872, p. 310; Fl. and Pom. 1871, p. 177, fig.

The well-known stove plant with sword-shaped leaves, 2 to 4 ft. in length, with serrated edges, pale green in the centre, margined with clearest bands of white.

Introduced from the South Sea Islands by the late John Gould Veitch, and the most popular of all stove plants in all civilized countries.

PASSIFLORA ACTINIA, *Hook.*

Bot. Mag. t. 4009; Fl. des Serres, 1846, pl. x.; Jour. R.H.S. 1872, vol. iv. p. 142; Gard. Chron. 1902, vol. xxxii. p. 15, with fig.

Seeds of this species, known as "The Sea Anemone Passion-flower," were sent from Brazil in 1841, plants raised and flowered for the first time at Exeter in 1842.

From the resemblance the flowers bear to the Sea Anemone, Sir Joseph Hooker named the Passion-flower Actinia after a genus of low marine animals.

PASSIFLORA BILOBATA, *Juss.*

Masters in Gard. Chron. 1875, vol. iv. p. 420.

Introduced from Costa Rica, and flowered for the first time in September 1875; belonging to the small section (Cieca) with no petals, it can boast but little beauty.

PASSIFLORA ORGANENSIS, *Gardn.*, var. MARMORATA.

Gard. Chron. 1869, p. 1158.

The form with spotted leaves introduced through Bowman from the Organ Mountains, Brazil.

The flowers small, greenish-white, the coronal threads violet and tipped with white; the leaves green, prettily mottled with cream-white or yellow.

PAULLINIA THALICTRIFOLIA, *Juss.*

Bot. Mag. t. 5879; Veitchs' Catlg. of Pl. 1872, p. 8, fig.; The Garden, 1873, vol. iii. p. 84, fig. p. 87.

A native of the Rio de Janeiro province of Brazil, introduced through Bowman; a pretty semi-scandent stove foliage plant with finely divided feathery leaves resembling the fronds of some species of Davallia.

The flowers, greenish and insignificant, were produced for the first time in October 1870.

PENTAPTERYGIUM RUGOSUM, *Hook.*

Syns. *Vaccinium rugosum*, Hook. and Thoms. MS.

Bot. Mag. t. 5198; Gard. Chron. 1860, p. 384; Fl. des Serres, 1874, p. 131.

A hard-wooded greenhouse plant introduced through Thomas Lobb from the Khasia Hills, where it was originally discovered by Griffiths, and later by Drs. Hooker and Thomson.

The drooping tubular flowers with transverse lines of red on a white ground, a peculiarity which gained for the plant the trivial name of "Chinese Lantern Flower," were first produced and exhibited as a species of Thibaudia, in April 1860, before the Royal Horticultural Society.

PHILAGERIA VEITCHII, *Mast.*

Masters in Gard. Chron. 1872, p. 358, with fig.

A more than remarkable bigeneric hybrid raised at Chelsea from Lapageria rosea crossed with the pollen of Philesia buxifolia, in habit intermediate between the two parents, though rather more akin to the female than to the male.

In the character of the flower-stalk, calyx, and corolla the plant more closely follows Philesia than Lapageria, but in stamens it resembles the male parent, and in colour the mother. The few plants still existing are sickly subjects—Dame Nature will have none of it!

The compound name given by Dr. Masters, derived from the two generic names of the parents, formed a precedent since followed in naming all bigeneric hybrids.

PHILODENDRON ANDREANUM, *Devans.*

Veitchs' Catlg. of Pl. 1887, p. 11, fig. p. 4.

A striking stove Aroid, first discovered by M. André of Paris, after whom it is named, and subsequently introduced direct from New Grenada, the native country.

The large leaves, which resemble those of Anthurium Veitchii, often attain a length of from 4 to 5 ft., and are deflected vertically from a stout erect foot-stalk. When young they are of a decided scarlet colour tinged with brown, when older of a bronzy red-brown before finally changing to the bright velvety-green of the mature leaf. The midrib and veins of a whitish colour through all stages of development.

PHORMIUM TENAX, *L. f.*, var. VARIEGATUM.

Veitchs' Catlg. of Pl. 1870, p. 33.

A variegated form of the New Zealand Flax sent from New Zealand by the late John Gould Veitch.

The long strap-shaped leaves of a dark green ground colour are ornamented with broad stripes of yellow of varying breadth, the whole length of the leaf.

The plant is ornamental for indoor decoration, and a striking object when planted out-of-doors during the summer months.

PHORMIUM TENAX, *L. f.*, var. VEITCHII, *Hort.*

Veitchs' Catlg. of Pl. 1866, p. 13.

A handsome form of the New Zealand Flax with shorter narrower leaves than the type, effectively variegated along the whole length with broad stripes of creamy white on a pea-green ground.

It was introduced to cultivation through the late John Gould Veitch, is a much-valued plant in the class to which it belongs, admirable for conservatories or winter-gardens, and a striking subject for sub-tropical bedding.

PHRYNIUM VARIEGATUM, *N. E. Brown.*

Syns. *Maranta arundinacea*, L., var. *variegata.*

Veitchs' Catlg. of Pl. 1887, p. 11, fig. p. 5.

An elegant stove plant allied to Maranta and Calathea with leaves on foot-stalks about 1 ft. high, most effectively variegated with cream-white on a green ground.

PHYLLANTHUS ROSEO-PICTUS, *Hort.*

Veitchs' Catlg. of Pl. 1877, p. 24, fig.; Fl. and Pom. 1878, p. 13, fig.

A stove shrubby plant of graceful habit introduced from the South Sea Islands. The small rounded alternate leaves on slender branches arch gracefully, the variegation is rich, no two blades showing the same marking; some are delicate rose colour, some pure white, others a mixture of the two on a dark olive-green ground.

PHYSOSTELMA WALLICHI, *Wight.*

Syns. *Hoya campanulata*, Blume.

Bot. Mag. t. 4545; Lindl. Bot. Reg. 1847, t. 54; Fl. des Serres, 1850, p. 283.

A very remarkable stove plant, native of the copses in the mountainous districts of Java, first detected by Blume, and later introduced to cultivation through Thomas Lobb.

It produced a somewhat curious waxy pale buff-coloured flower for the first time in this country in October 1846, on which occasion, exhibited before the Royal Horticultural Society, it was awarded a Silver Banksian Medal.

PINANGA VEITCHII, *Wendl.*

Veitchs' Catlg. of Pl. 1880, p. 16, with fig.; The Garden, 1880, vol. xvii. p. 435, fig.; Fl. des Serres, 1877, p. 93.

A remarkable distinct Palm distinguished by the unusual colour of the foliage, a native of Borneo, sent by F. W. Burbidge.

The fronds in an early stage are pale green, blotched and stained with spots of a deeper shade, the under surface dull crimson, deepening with age to a bronzy hue. Of dwarf shrubby habit, shoots from the rhizome rise freely in a young state, the chief feature as a decorative plant the foliage coloration.

PIPER BORNEENSE, *N. E. Brown.*

N. E. Brown in Gard. Chron. 1882, vol. xvii. p. 108.

A species of Pepper sent from Western Borneo through Curtis, by whom it was discovered. The leaves are of a rich dark green with silvery-grey stripes between the nerves; the flowers, as in the majority of this genus, inconspicuous.

PIPER ORNATUM, *N. E. Brown.*

Veitchs' Catlg. of Pl. 1884, p. 13; Gard. Chron. 1884, vol. xxii. p. 424.

A stove climbing plant introduced from the island of Celebes in the Malay Archipelago through Curtis.

The leaves are heart-shaped with pale rose foot-stalks and numerous rose-coloured spots on a bright glossy green blade.

PIPER PORPHYROPHYLLUM, *N. E. Brown.*

Syns. *Cissus porphyrophyllus*, Lindl.; *C. cordifolius?* Hort.

Nich. Dict. Gard. vol. iii. fig. 189; Gard. Chron. 1860, p. 482 (Report of Show); Fl. des Serres, 1861, tom. iv. p. 263.

A handsome stove climber raised from seed sent from India by Thomas Lobb as Cissus cordifolius? and first exhibited before the Royal Horticultural Society in May 1860.

The dark olive-green leaves with salmon-pink spots along the principal veins of the upper surface, purple beneath, are very handsome.

PIPTOSPATHA INSIGNIS, *N. E. Brown.*

N. E. Brown in Gard. Chron. 1879, vol. xi. p. 138, fig.; Bot. Mag. t. 6598.

An interesting little Aroid, discovered in Borneo by F. W. Burbidge while plant-collecting in that island.

On this species Mr. N. E. Brown of Kew founded the new genus Piptospatha, a small Malayan group of the tribe Philodendreæ, closely related to Schismatoglottis, from which it however differs in wanting the terminal spike of neuter florets so conspicuously terminating the spadix in that genus.

PLUMBAGO ROSEA, *Lindl.*, var. COCCINEA.

Bot. Mag. t. 5363.

A handsome variety raised from seed sent by a correspondent in the Neilgherries.

Previously cultivated at Kew, but not in commerce, the type species, a very old garden plant, was figured in Curtis's Botanical Magazine in the year 1794, t. 230.

The variety *coccinea* produces panicles of bright brick-red or scarlet flowers often more than 2 ft. in length, when well cultivated a very striking object.

PODOLASIA STIPITATA, *N. E. Brown.*

N. E. Brown in Gard. Chron. 1882, vol. xviii. p. 70; Fl. and Pom. 1882, p. 123.

A remarkable and interesting Aroid, introduced through Curtis when collecting in Borneo.

On this species Mr. N. E. Brown of Kew founded the genus, which previously had received the provisional name of Lasia, to which it is closely allied. The leaves are arrow-shaped on prickly petioles; the spathe boat-shaped, open to the base, brownish-red, 3 to 4 in. long.

POTHOS CELATOCAULIS, *N. E. Brown.*

N. E. Brown in Gard. Chron. 1880, vol. xiii. p. 200; Veitchs' Catlg. of Pl. 1880, p. 23.

Introduced from Borneo by F. W. Burbidge, this interesting climber, flat on any surface it can feel, holds its position by numerous adventitious roots; the leaves are oblique, of a dark velvety green, produced under almost all conditions in any stove.

PRIMULA OBCONICA, *Hance.*

Syns. *P. poculiformis*, Hook.

Bot. Mag. t. 6582; The Garden, 1881, vol. xix. p. 655, fig.; *id.* 1884, vol. xxvi. p. 206, pl. 456, *id.* 1897, vol. li. p. 316, pl. 1116; Gard. Chron. 1883, vol. xix. p. 121, fig. 19.

This pretty greenhouse Primula, now well known, has undergone great improvement since first introduced through Charles Maries, who found it in the gorges of the Yangtsze, in the Ichang district. Plants flowered at Chelsea for the first time in September 1880.

The colour in the virgin species is an undecided lilac, but many seminal forms show a wide range, from a pure white to a deep rosy purple.

The form of the flower has much changed, is now more circular, and, in some forms, the margins are deeply fimbriated. It has, after many unsuccessful attempts, been made to cross with another species of the genus —Primula megasæfolia; the result first shown at the Temple Show of 1905.

PROUSTIA PYRIFOLIA, *Lay.*

Bot. Mag. t. 5489; Gard. Chron. 1898, vol. xxiv. p. 142, fig. 37.

A woody greenhouse climber, a Composite, with unattractive small white flowers and holly-like foliage. As the fruit approaches maturity it is a singular object, the pappus of slender rose-purple hairs collectively forming a plumose mass of great length and breadth.

It is a native of Chili, introduced through Richard Pearce, and flowered for the first time at Chelsea in July 1864.

PSYCHOTRIA TABACIFOLIA, *Muell.*

Syns. *Palicourea nicotianæfolia*, Cham. & Schlecht; *P. discolor*, Hort.

Veitchs' Catlg. of Pl. 1866, p. 4; Gard. Chron. 1866, p. 432 (advt.); Bot. Mag. t. 7001.

An ornamental-leaved stove plant introduced from Peru through Richard Pearce in 1865, lost to cultivation.

The figure in the Botanical Magazine, above quoted, was prepared from a specimen flowered at Kew, but the details of origin have been lost.

PTYCHOSPERMA MACARTHURII, *Wendl.*

Syns. *Kentia Macarthurii*, Hort.

Veitchs' Catlg. of Pl. 1880, p. 72, with fig.

An elegant Palm with sub-erect leaves and graceful semi-pendulous leaflets introduced from the neighbourhood of the Katan River, New Guinea.

The stems of the leaves smooth and slender, the leaflets, from 4 to 8 in. in length and ½ in. in breadth, are effective.

REHMANNIA ANGULATA, *Hemsl.*

Gard. Chron. 1903, vol. xxxiii. p. 290, suppl. illus.; Flora and Sylva, 1904, vol. ii. p. 280, col. pl.; Rev. Hort. 1905, Dec. 16th, col. pl.

This perennial herbaceous plant, too tender for the open border in the winter months, but in summer suitable for German bedding, attains a height of 6 ft. A native of Central China, *très répandu*, it was first detected by Dr. Henry, and afterwards introduced to cultivation through Wilson.

The flowers in shape as a large Mimulus, with a broad spreading lip and reflexed standard; in colour a shade of deep rosy-pink with a cream-white throat spotted with purple, flowered for the first time at Coombe in May 1902.

RHODODENDRON BROOKEANUM, *Low.*

Gard. Chron. 1855, p. 404, fig.; Bot. Mag. t. 4935; Fl. des Serres, 1857, tom. ii. p. 111.

An East Indian species with golden-yellow flowers, from Sarawak, Borneo, through Thomas Lobb, named after Sir James Brooke, the distinguished Rajah of Sarawak.

The plant flowered for the first time and was exhibited in July 1855.

RHODODENDRON BROOKEANUM, *Low*, var. FLAVUM.

Veitchs' Catlg. of Pl. 1872, p. 9, fig.

A form of Rhododendron Brookeanum with clear yellow flowers and thick dark green very glossy leaves, from Borneo.

RHODODENDRON BROOKEANUM, *Low*, var. GRACILIS.

Veitchs' Catlg. of Pl. 1871, p. 12, fig.

A variety introduced from Borneo through Thomas Lobb, differing from the type in a more slender, graceful habit, and in having larger flowers of a pale yellow colour.

RHODODENDRON JASMINIFLORUM, *Hook.*

Gard. Chron. 1851, p. 183 (Report of Show); Bot. Mag. t. 4524.

A native of Mount Ophir, Malacca, at elevations of 5,000 ft., sent to Exeter by Thomas Lobb in 1848.

In the Botanical Magazine above quoted, the Editor remarks:—"At the first, and truly splendid, exhibition of flowers at the Chiswick Gardens of the present year (1850), few plants excited greater attention among the visitors most distinguished for taste and judgment, than the one here figured, from the nursery of Messrs. Veitch of Exeter. Many excelled it in splendour, but the delicacy of form and the colour of the flowers (white with a deep pink eye), and probably their resemblance to the favourite Jessamine (some compared them to the equally favourite Stephanotis), attracted general notice. So unlike indeed are they to the ordinary form of Rhododendron blossoms that the Gardeners' Chronicle, in recounting the prizes of the day, seemed to imply that it was probably no Rhododendron at all!"

This species has entered largely into the production of the race known as the javanico-jasminiflorum hybrids.

RHODODENDRON JAVANICUM, *Benn.*

Gard. Chron. 1847, p. 374 (Notice of Exhibit of New Plants); Bot. Mag. t. 4336; Paxt. Mag. Bot. vol. xv. p. 217.

A native of Java, extremely handsome in foliage and in the brilliant colouring of the flowers. Introduced through Thomas Lobb from Java, an imported plant was exhibited in flower for the first time before the Royal Horticultural Society in June 1847.

In reporting the Show, the Gardeners' Chronicle states it "promises to be a great acquisition as well on account of its own intrinsic merit as for the purposes of hybridization." That the species has justified the high opinion held when first exhibited, the race of Greenhouse Rhododendrons, known as javanico-jasminiflorum hybrids, afford to-day ample evidence. The history of these hybrids, in the production of which Rhododendron javanicum took a large share, is given in a special part of this work devoted to the genus Rhododendron.

RHODODENDRON JAVANICUM, *Benn.*, var. AURANTIACUM.

Fl. des Serres, 1850, tom. vi. p. 135.

Introduced through Thomas Lobb from Java.

The flowers of the typical species vary considerably in colour, from citron to red-orange—the colour of this variety is golden-yellow.

RHODODENDRON JAVANICUM, *Benn.*, var. FLAVUM.

Gard. Chron. 1848, p. 303 (Report of Exhibit of New Plants).

A form with flowers of a paler yellow colour than the type, introduced through Thomas Lobb from Java.

RHODODENDRON JAVANICUM, *Benn.*, var. TUBIFLORA.

Bot. Mag. t. 6850.

A variety differing from Rhododendron javanicum in having much larger flowers of a citron-yellow colour, with broad corolla-lobes and a more vigorous habit of growth.

It was sent from Sumatra by Curtis, and flowered for the first time in June 1885.

RHODODENDRON LOBBII, *Veitch.*

Veitchs' Catlg. of Pl. 1870, p. 16, fig., also col. pl.; Gard. Chron. 1871, p. 1323; Fl. Mag. 1861, pl. 10.

A very distinct species of the East Indian Rhododendrons with bright scarlet tubular flowers in trusses of from 8 to 12 blooms.

Introduced from Borneo through Thomas Lobb in 1861, it has proved useful in the production of that beautiful race of garden forms known as the javanico-jasminiflorum hybrids.

RHODODENDRON MALAYANUM, *Jack.*

Bot. Mag. t. 6045.

Introduced by Thomas Lobb in 1854 from Mount Ophir, it was originally discovered by Dr. William Jack, of the East India Company's service, on the summit of Gunong Bunko, an insulated mountain in the interior of Bencoolen, Sumatra, commonly called by Europeans the Sugar-loaf.

The flowers are small, of a beautiful cerise-crimson colour, and the habit of the plant neat and compact. It has been successfully used as a parent in the production of the beautiful greenhouse hybrid Rhododendrons, and has materially influenced the whole strain.

RHODODENDRON MOULMEINENSE, *Hook.*

Bot. Mag. t. 4904.

A species detected by Lobb at Moulmein, on the Gerai Mountains, at an elevation of 5,000 ft., and by him introduced to Chelsea, flowered for the first time in January 1856; flowers pure white with a tinge of yellow in the centre.

RHODODENDRON MULTICOLOR, *Miquel.*

Bot. Mag. t. 6769.

A small-flowered species with blooms of various colours, native of the island of Sumatra, where first discovered by Curtis, it was through him introduced.

The type species, with bright yellow flowers, bloomed for the first time in February 1884.

RHODODENDRON MULTICOLOR, *Miquel*, var. CURTISII.

Syns. *R. Curtisii*, Hort.

Fl. and Pom. 1883, p. 185; *id.* 1884, t. 615.

A variety of Rhododendron multicolor with dark red flowers, first produced in November 1883, and sent in 1880 from the island of Sumatra, where it had been detected by Curtis on mountains at an elevation of some 2,000 ft.

RHODODENDRON (AZALEA) OLDHAMII, *Maximow.*

Masters in Gard. Chron. 1882, vol. xvii. p. 524.

Introduced from the island of Formosa through Charles Maries, but previously met with by Mr. Oldham, after whom it was named by Maximowicz.

This greenhouse shrub has reddish salmon-coloured flowers, suffused rosy lilac on the upper lobe, with numerous small darker blotches.

RHODODENDRON × PRINCESS ALICE.

Gard. Chron. 1862, p. 262 (Report of Horticultural Society's Show); Fl. Mag. 1865, pl. 206.

A hybrid between the two beautiful Himalayan spècies, Rhododendron Edgeworthii and R. ciliatum.

In size the flowers of the hybrid nearly equal those of R. Edgeworthii, and possess a delicate perfume. They are slightly tinted with rose in the bud, but become pure white as they expand.

RHODODENDRON VEITCHIANUM, *Hook.*

Gard. Chron. 1857, pp. 326, 347; Bot. Mag. t. 4992; Fl. des Serres, 1861, tom. iv. p. 57; *id.* 1862, tom. iv. p. 41.

A beautiful species with large flowers of the purest white, and crinkled margins, native of Moulmein, sent by Thomas Lobb, and the finest of the genus, exhibited for the first time in flower at a meeting of the Royal Horticultural Society in London, May 6th, 1857. Succeeding admirably in this country, it produces in early spring large white flowers in great profusion. Reproduction is simple, from home-saved seed.

RHODODENDRON VEITCHIANUM

RUELLIA PEARCEI, *Veitch.*

Syns. *Stemonacanthus Pearcei*, Hook.

Bot. Mag. t. 5648.

A native of Bolivia, discovered by Pearce, whose name it bears.

An erect-growing stove shrub, with lance-shaped leaves 5 to 6 in. long, green above, brown-purple below, with a lax inflorescence of scarlet tubular flowers.

SALVIA OPPOSITIFLORA, *Ruiz & Pav.*

Paxt. Mag. Bot. xv. p. 53; Fl. des Serres, 1848, p. 345.

A half-hardy sub-shrub with scarlet hairy flowers, discovered by Thomas Lobb in exposed situations in Tarma, Peru, and introduced by him to cultivation.

SANCHEZIA LONGIFLORA, *Hook. f. ex Planch.*

Syns. *Ancylogyne longiflora*, Hook.

Bot. Mag. t. 5588; Veitchs' Catlg. of Pl. 1868, with fig.; Fl. des Serres, 1880-1883, tom. xxiii. 257, t. 2460.

Undoubtedly one of the finest tropical Acanthaceæ ever sent to this country from Guayaquil through Pearce, by whom it was discovered.

The flowers in a drooping elongated branched panicle, about 2 in. long, of a rich vinous-purple, the corolla tubular, and mouth oblique, the small segments of a lighter colour.

SANCHEZIA NOBILIS, *Hook.*

Bot. Mag. t. 5594; Veitchs' Catlg. of Pl. 1867, fig.; The Florist, 1867, p. 154.

A most beautiful and interesting Acanthaceous plant, discovered by Richard Pearce in Ecuador in 1863, and flowered for the first time in this country at Chelsea in June 1866. The flowers are bright yellow, densely borne in panicles, the bracts bright red and the branches deep purple.

A form known as *variegata* has the midrib and veins of the leaves coloured yellow, and is cultivated in quantity.

SARMIENTA REPENS, *Ruiz & Pav.*

Bot. Mag. t. 6720; Fl. Mag. 1862, t. 112; Fl. des Serres, tom. xvi. t. 1646; Gard. Chron. 1901, vol. xxix. p. 303, fig.

This plant belongs to a monotypic genus, closely allied to another also monotypic from Chili, Mitraria coccinea, has larger flowers of a somewhat similar form.

Sarmienta repens inhabits the southern provinces of the main-land from Concepcion southwards, and the island of Chiloe, which is the southern limit.

It was introduced through Richard Pearce, and thrives in cool, damp conservatories among moss, stones and stumps of plants.

SCHISMATOGLOTTIS CRISPATA, *Hook. f.*

Bot. Mag. t. 6576; Nich. Dict. Gard. vol. iii. fig. 443.

A variegated-leaved Aroid, through F. W. Burbidge from Borneo, flowered for the first time in the Chelsea stoves in January 1881.

The leaves are dark green above with a broad irregular greyish-green band on either side of the midrib.

SCHISMATOGLOTTIS LAVALLEEI, *Linden*, var. PURPUREA, *N. E. Brown.*

N. E. Brown in Gard. Chron. 1882, vol. xviii. p. 298.

A variety differing from the type in the deep vinous-purple colour of its petioles, introduced through Curtis, who discovered it in Sumatra.

SCHISMATOGLOTTIS NEOGUINEENSIS, *N. E. Brown.*

Syns. *S. variegata*, Hort.

Gard. Chron. 1862, p. 399 (advt.); Veitchs' Catlg. of Pl. 1862, p. 8.

An ornamental-leaved stove plant of the Aroid family, sent from Borneo by Thomas Lobb.

The leaves resemble those of a Maranta, have a glossy green surface along the whole length of which runs a feathery line of silver-white.

SCINDAPSUS CUSCUARIA, *Prestl.*

Syns. *Aglaonema commutatum*, Schott.

Gard. Chron. 1863, p. 695 (advt.); *id.* 1863, p. 460 (Notice of Exhibit).

A stove plant belonging to the Aroid family, introduced through the late John Gould Veitch from the interior of Luzon, Philippine Islands. The plant grows to a height of about 2½ ft. and produces large leaves of a glossy green colour, effectively ornamented with flakes of creamy white and pale green.

SCINDAPSUS PICTUS, *Hassk.*

Syns. *Pothos argyreus*, Hort.

Gard. Chron. 1859, p. 426 (Report of Show); *id.* p. 603 (advt.)

A charming little stove plant of the Aroid family, introduced to cultivation from Borneo through Thomas Lobb, with ornamental leaves rivalling in their markings the Anœctochili.

SCUTELLARIA FORMOSANA, *N. E. Brown.*

N. E. Brown in Gard. Chron. 1894, vol. xvi. p. 212; Bot. Mag. t. 7458.

A pretty purple-flowered Labiate from the Island of Formosa through Mr. Ford, late Curator of the Botanic Gardens, Hong Kong.

It succeeds well in a cool greenhouse, producing freely pale violet-blue flowers during the early summer months.

SCUTELLARIA INCARNATA, *Vent.*

Bot. Mag. t. 4268.

A pretty little greenhouse plant of the same order with sage-like leaves and bright purplish rose-coloured flowers, raised from seed gathered on the western declivities of the Andes, sent by Professor Jameson.

It flowered for the first time in July 1846.

SENECIO KÆMPFERI, *DC.*, var. ARGENTEA.

Syns. *Ligularia Kæmpferi*, Sieb. & Zucc., var. *argentea*; *Farfugium Kæmpferi*, Benth., var. *argentea*.

Gard. Chron. 1863, p. 694 (advt.).

An ornamental-leaved herbaceous plant introduced from Japan through the late John Gould Veitch. The foliage resembles that of Senecio (Farfugium) grande, but the variegation is in flakes and blotches of white.

SINNINGIA CONCINNA, *Hook. f.*

Syns. *Stenogastra concinna*, Hook.

Gard. Chron. 1861, p. 530 (Notice of Exhibit); Fl. des Serres, 1862-1865, tom. xiv. p. 65; Bot. Mag. t. 5253; The Garden, 1897, vol. lii. p. 22, pl. 1126; Gard. Chron. 1898, vol. xxiii. p. 361, fig. 137.

A curious and pretty but rather small hothouse herbaceous plant with small roundish leaves forming a tuft, from among which spring numerous peduncles bearing solitary nodding tubular blossoms.

Flowered for the first time in April 1861, it was exhibited before the Royal Horticultural Society in June of that year.

SINNINGIA SPECIOSA, *Hiern.*, var. MACROPHYLLA VARIEGATA.

Syns. *Gloxinia speciosa*, Lodd., var. *macrophylla variegata*.

Bot. Mag. t. 3934.

Raised from seed from our collector in the Organ Mountains of Brazil, flowered in September 1841, and exhibited before the Royal Horticultural Society.

As with other species from which the florists' Gloxinia has been derived, the flowers droop and are more like those of Achimenes than the Gloxinias we are accustomed to at the present day.

SONERILA BENSONI, *Hook. f.*

Bot. Mag. t. 6049.

Raised from seed sent by Col. Benson, who procured it in the Western Ghauts of Malabar.

This stove plant has shining green leaves shot with brown, purple above, rose-purple beneath, and bright rose-purple flowers with yellow stamens.

SONERILA ELEGANS, *Wight.*

Bot. Mag. t. 4978.

A pretty stove plant with beautiful leaves and delicate rose-pink blooms, introduced from the Neilgherries through Thomas Lobb, and first flowered in January 1851.

SONERILA MARGARITACEA, *Lindl.*

Gard. Chron. 1854, p. 727; Bot. Mag. t. 5104; Fl. des Serres, t. 1126; The Florist, 1855, pl. 98.

A very ornamental stove plant with small rose-pink flowers and dark green leaves regularly punctuated with silvery white spots on the upper surface marked beneath with rose-purple veins. From India through Thomas Lobb, first flowered during the summer of 1854.

SONERILA SPECIOSA, *Zenker.*

Syns. *S. orbiculata*, Lindl.

Bot. Mag. t. 5026; Lindl. in Jour. R.H.S. 1853, p. 56.

A species remarkable for the beauty of deep rose-purple flowers, introduced with Sonerila elegans from the Neilgherries through Thomas Lobb in 1856.

SONERILA STRICTA, *Hook.*

Bot. Mag. t. 4394.

The first species of this genus of ornamental-leaved plants to be cultivated in Europe. The seed received from Thomas Lobb from Java, plants raised flowered in May 1848.

STENOSPERMATION POPAYANENSE, *Schott.*

Syns. *Spathiphyllum Wallisii*, Hort.; *Stenospermation Wallisii*, Mast.

Gard. Chron. 1875, vol. iii. p. 558, with figs.

An interesting and ornamental Aroid, from Columbia through Gustav Wallis.

The stems reach a height of from 2 to 3 ft. and produce alternate petiolate leaves. The spathes on long slender peduncles which bend, are boat-shaped, ivory-white in colour, and enclose an oblong spadix, which bears the same relation to the spathe as the clapper does to a bell.

STIGMAPHYLLON HETEROPHYLLUM, *Hook.*

Bot. Mag. t. 4014.

A handsome yellow-flowered stove climber, from seed sent by

Mr. Tweedie from Buenos Ayres in 1841, first flowered in the December of 1842.

This interesting plant derives its generic name from the curious green foliaceous appendage of the stigma.

STREPTOSOLEN JAMESONI, *Miers.*

Syns. *Browallia Jamesoni*, Benth.

Gard. Chron. 1847, pp. 374, 401 (Reports of Exhibitions); *id.* 1848, p. 618 (advt.); Bot. Mag. t. 4605; Fl. des Serres, 1849, p. 436; Paxt. Mag. Bot. vol. xvi. p. 5; The Garden, 1884, vol. xxvi. p. 6, pl. 447.

A well-known greenhouse climber, native of North Peru, found by William Lobb at an elevation of 6,000 ft. in woods near Monitre, in the province of Cuença.

First exhibited in flower on June 5th, 1847, as Browallia Jamesoni, and again in 1848 under the same name.

The figure in the Botanical Magazine above quoted was prepared from material supplied by Hector Munro Esq., of Druid's Stoke, near Bristol, with whom it flowered in June 1851.

After a few years the plant apparently fell out of cultivation until 1882, when re-introduced it became more common, and is now frequently met with.

STYLIDIUM SAXIFRAGOIDES, *Lindl.*

Bot. Mag. t. 4529.

Raised from seeds from the Swan River Settlement in 1849, this interesting little tufted plant has the appearance of a mossy Saxifraga —scapes of rather large white or yellowish flowers.

TACSONIA MOLLISSIMA, *H. B. & K.*

Bot. Mag. t. 4187; Paxt. Mag. Bot. vol. xiii. p. 25; Fl. des Serres, 1846, pl. v.

A beautiful greenhouse climber with lovely blossoms of a deep rose colour, a native of the woods near Quito, whence seeds were sent to Exeter by William Lobb, and plants raised flowered for the first time in 1845.

TECOMA FULVA, *Don.*

Bot. Mag. t. 4896; Fl. des Serres, 1856, tom. i. p. 83.

A greenhouse shrub apparently identical with Tecoma Smithii, said to be of garden origin, with pinnate leaves and handsome orange-yellow tubular flowers borne in terminal racemes, reared from seed received from Peru.

THIBAUDIA ACUMINATA, *Hook.*

Bot. Mag. t. 5752; Veitchs' Catlg. of Pl. 1869, p. 15.

A distinct greenhouse shrub from the Andes of Columbia and Ecuador, introduced through Richard Pearce in 1868, but previously collected by Jameson, Hartweg, and others, though not introduced. The flowers are brilliant red in colour, the plant almost constantly in bloom.

THUNBERGIA LUTEA, *T. Anders.*

Syns. *Hexacentris lutea*, Lindl.; *H. mysorensis*, Wight, var. *lutea* of Fl. des Serres.

Anders. Jour. Linn. Soc. 1867, vol. ix. p. 448; Lindl. in Gard. Chron. 1854, p. 150; Fl. des Serres, 1854, tom. ix. p. 217.

A stove climber from India through Thomas Lobb, with yellow flowers generally resembling those of Thunbergia (Hexacentris) mysorensis, but differing in important technical details.

THUNBERGIA MYSORENSIS, *T. Anders.*

Syns. *Hexacentris mysorensis*, Wight.

Bot. Mag. t. 4786; Paxt. Fl. Gdn. vol. iii. p. 88; Gard. Chron. 1852, pp. 307, 310.

A shrubby stove climber with racemes of rich yellow flowers, raised from seed received from Mr. M'Ivor of the Botanic Gardens, Ootacamund, and exhibited for the first time in flower at an Exhibition held in the Chiswick Gardens of the Royal Horticultural Society, May 8th, 1852.

THUNBERGIA NATALENSIS, *Hook.*

Bot. Mag. t. 5082.

A native of Natal, from seed received from that colony through Mr. Cuming, and flowered for the first time in July 1858.

A greenhouse Acanthaceous plant with flimsy pale blue flowers, yellow in the centre.

TIBOUCHINA ELEGANS, *Cogn.*

Syns. *Pleroma elegans*, Gardn.

Bot. Mag. t. 4262; Paxt. Bot. Mag. vol. xv. p. 27.

First discovered by Mr. Gardner in the Organ Mountains of Brazil, and later introduced to cultivation through William Lobb from the same locality.

It produced the splendid rich blue blossoms for the first time in the summer of 1846.

TIBOUCHINA GAYANUM.

Syns. *Pleroma Gayanum*, Triana.

Bot. Mag. t. 6345.

Pleroma Gayanum, one of the least conspicuous in the genus, a native of Cuzco, in Peru, was discovered by the French botanist and traveller Claude Gay, in whose honour it is named.

It was first introduced to this country through Walter Davis, and flowered in October 1874.

TIBOUCHINA ORNATA, *Baill.*

Syns. *Pleroma strigosum*, Triana; *Chætogastra strigosa*, DC.

Paxt. Mag. Bot. vol. xv. p. 265.

A native of Guadaloupe, originally discovered growing in beds of Sphagnum on the summit of Sulphur Mountain, and introduced through Thomas Lobb.

TILLANDSIA CHRYSOSTACHYS, *E. Morren.*

Bot. Mag. t. 6906; La Belg. Hort. 1881, vol. xxxi. p. 87; Rev. Hort. 1887, p. 166.

A beautiful species with bright lemon-yellow spikes of flowers from the forests of the Peruvian Andes through Walter Davis in 1881.

The specific name is in allusion to the long narrow yellow flower-spike, the colour most pronounced in the numerous overlapping leathery bracts.

TOCOCA LATIFOLIA, *Naud.*

Syns. *Sphærogyne latifolia*, Naud.

Gard. Chron. 1862, p. 399 (advt.); Veitchs' Catlg. of Pl. 1862, p. 8.

A magnificent stove plant with large leaves of a rich velvety olive-green on the upper surface, red beneath, stems and leaf-stalks thickly covered with recurved hairs.

It received the Silver Knightian Medal when exhibited before the Royal Horticultural Society in May 1862.

TRICHANTHA MINOR, *Hook.*

Hooker's Ic. Pl. t. 666; Gard. Chron. 1864, p. 172, with fig.; Bot. Mag. t. 5428.

This plant was first described in Hooker's Icones Plantarum from specimens collected in Columbia by Thomas Lobb in 1861, but plants were not obtainable till Richard Pearce sent seed from Guayaquil, and they flowered in 1863.

A stove climber, with ovate acuminate leaves, and clustered axillary flowers, the limb yellow, and the tube striped with blackish-purple, surrounded by a red hairy calyx of many segments.

TRICUSPIDARIA DEPENDENS, *Ruiz & Pav.*

Syns. *Crinodendron Hookerianum*, Miers, Gay; *C. Patagua*, Cav.; *T. hexapetala*, Turcz.

Bot. Mag. t. 7160 as T. dependens; Gard. Chron. 1849, p. 564 (Notice of New Plants) as Crinodendron Patagua; The Garden, 1880, vol. xviii. p. 542, col. pl.; *id.* 1890, vol. xxxviii. p. 273, fig., as Crinodendron Hookerianum; Nich. Dict. Gard. vol. iv. fig. 99, as T. hexapetala.

This much-named plant is a beautiful greenhouse shrub with evergreen leaves and drooping urn-shaped flowers of a brilliant scarlet colour.

A native of Chili in the Province of Valdivia, and of the island of Chiloe, introduced through William Lobb in 1848, and successively re-introduced by Downton and Pearce.

Though rarely met with, an attractive plant, not difficult to cultivate if planted in a peaty soil in a cool greenhouse or in the open in the favoured counties of Devon and Cornwall.

TROPÆOLUM CRENATIFLORUM, *Hook.*

Bot. Mag. t. 4245; Fl. des Serres, 1846, pl. iv.; Hemsley in The Garden, 1878, vol. xiii. p. 442.

Introduced through William Lobb from Pillao and Chagula, Peru, about 1845, but of little value in comparison with other species of the genus.

TROPÆOLUM UMBELLATUM, *Hook.*

Bot. Mag. t. 4337; Hemsley in The Garden, 1878, vol. xii. p. 444.

One of the most remarkable of all the Tropæola in having umbellate and not solitary flowers, as in all hitherto known species.

First discovered by Professor Jameson of Quito, who gathered it on Pilzhum, a mountain to which it is peculiar, at an elevation of 7,000 ft., but it did not reach England till sent by William Lobb in 1847.

TROPÆOLUM VIOLÆFLORUM, *A. Dietr.*

Syns. *T. azureum*, Hook.

Bot. Mag. t. 3985; Lindl. Bot. Reg. 1842, t. 65; Paxt. Mag. Bot. vol. ix. p. 247; Fl. des Serres, 1846, pl. vii.

Genera in which predominate bright red or orange-coloured flowers, seldom blossoms of a blue colour, an exception in the genus Tropæolum. Tubers of a blue-flowered species sent by William Lobb from Brazil, on being cultivated for only a few months, produced flowers at the Mount Radford Nursery, causing great interest when exhibited at the Royal Horticultural Society's meeting of October 4th, 1842, on which occasion a Silver Medal was awarded.

Figured in the name of T. azureum in the Botanical Magazine, it has now been found identical with T. violæflorum previously collected by Mr. Miers in Chili, and also by Bridges on the mountain range Campana de Quillota at an elevation of 4,000 ft.

VACCINIUM REFLEXUM, *Hook. f.*

Bot. Mag. t. 5781.

A greenhouse sub-shrub, from Bolivia through Richard Pearce; an interesting plant of pendulous habit with reflexed glossy green leaves and bright red flowers, opened for the first time in January 1869.

STOVE AND GREENHOUSE PLANTS

VEITCHIA JOHANNIS, *Wendl.*

Gard. Chron. 1883, vol. xx. p. 205, fig. 32.

A specimen commemorating the late John Gould Veitch, who introduced the plant to European gardens from Fiji.

First discovered by Dr. Seemann, it was sent to the Sydney Botanic Gardens, but an attempt to introduce it alive to England proved a failure.

The plant is of slender, elegant growth, in a young state with some resemblance to Kentia Exorrhiza, with which it has been confused. The fruits, at first green, gradually turn to a bright orange, ultimately red at the base, the much-branched panicle is highly ornamental, and in cultivation this Palm is rare.

VERTICORDIA NITENS, *Schauer.*

Syns. *Chrysorrhoe nitens*, Lindl.

Bot. Mag. t. 5286.

An interesting plant of the Myrtle family, first made known from specimens collected in Western Australia by Captain Mangles, and described by Dr. Lindley as "the magnificent Chrysorrhoe nitens, whose yellow flowers of metallic lustre form masses of golden stars some feet in diameter." Long a desideratum, it was at last raised from seed sent by a correspondent in Australia, and flowered for the first time in August 1861.

Under cultivation it did not attain that perfection expected, and is now apparently lost to British gardens.

VITIS ENDRESII, *Hort.*

Syns. *Cissus Endresii*, Hort. Veitch.

Veitchs' Catlg. of Pl. 1876, p. 8, fig.

A stove climber collected in Costa Rica by Endres, named in compliment.

The upper surface of the leaf is of a rich velvety green, reddish prominent veins and midribs, the younger leaves and tendrils strongly tinged with a rich purple-crimson.

WORMIA BURBIDGEI, *Miquel.*

Bot. Mag. t. 6531; Veitchs' Catlg. of Pl. 1885, p. 13.

Discovered by F. W. Burbidge in Northern Borneo, related to the Hibbertias of our glasshouses, and interesting as the first species of the genus to flower in Europe. The blossoms golden-yellow in colour, are 4 to 5 in. across.

XERONEMA MOOREI, *Brongn.*

Gard. Chron. 1878, vol. x. p. 8, fig. 3; Veitchs' Catlg. of Pl. 1889, p. 12, fig. p. 8.

A singular Liliaceous plant with the habit of an Iris, re-introduced through the late John Gould Veitch from the South Sea Islands, at the time lost to cultivation.

The brilliant crimson flowers, turned to one side of the rachis, are very brightly coloured; each is about 1 in. in length, an erect tube, from the centre of which protrude the stamens fully ½ in. beyond the mouth, and which, from their crowded position, impart a striking effect to the inflorescence.

ZAMIA MONTANA, *A. Braun.*

Veitchs' Catlg. of Pl. 1876, p. 34.

Introduced from New Grenada through Wallis.

The sharp-pointed pinnæ near the extremities of erect leaf-stalks, deeply furrowed throughout, are about 1 ft. in length, very beautiful in form and colour.

ZAMIA OBLIQUA, *A. Braun.*

Dyer in Gard. Chron. 1882, vol. xvii. p. 460, fig. 72; Bot. Mag. t. 7542.

Originally introduced through Gustav Wallis, by whom it was discovered in New Grenada, this species of rather small size, but neat habit, produces bright green sharp-pointed leaflets, 6 to 8 in. in length, on the extremities of slender leaf-stalks.

ZAMIA WALLISII, *A. Braun.*

Bot. Mag. t. 7103.

One of three new species discovered by Gustav Wallis when collecting in New Grenada in 1873, and first flowered in this country in May 1889.

ZINGIBER COLORATUM, *N. E. Brown.*

N. E. Brown in Gard. Chron. 1879, vol. xii. p. 166.

This very interesting plant, of little horticultural value, was introduced from Borneo through Burbidge. The showy inflorescence at the base is to a great extent concealed.

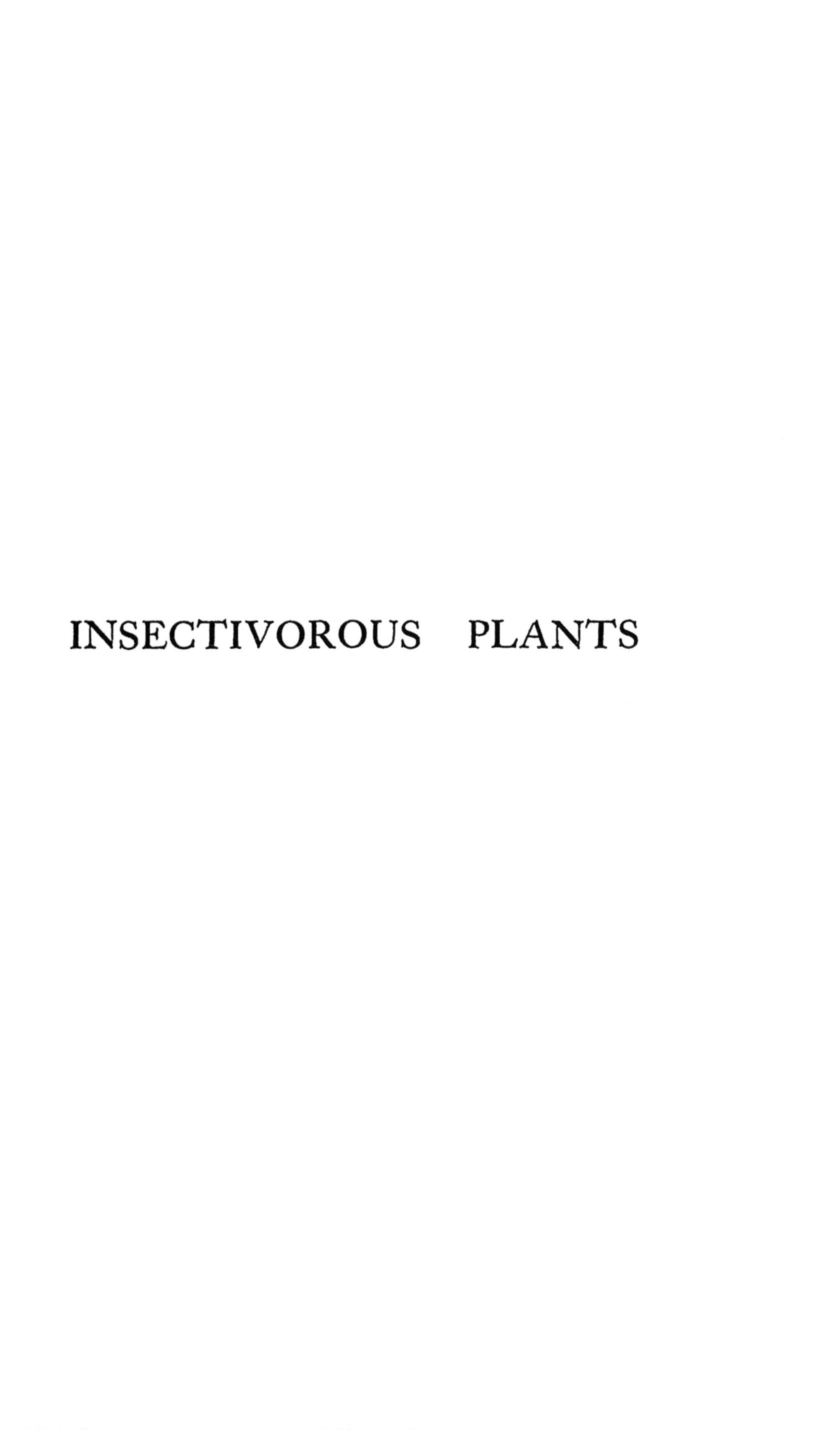

INSECTIVOROUS PLANTS

HELIAMPHORA NUTANS

CAMBRIDGE BOTANIC GARDENS

INSECTIVOROUS PLANTS

DROSERA CAPENSIS, *Linn.*

Bot. Mag. t. 6583; La Belg. Hort. 1880, p. 311, pl. xvi.

Introduced by us from the Cape in 1874, and flowered for the first time in a cool greenhouse in the Royal Gardens, Kew, in July 1881.

This interesting plant, allied to the Sundew, Drosera rotundifolia, has strap-shaped leaves from 4 to 8 in. long.

HELIAMPHORA NUTANS, *Benth.*

Bot. Mag. t. 7093; Gard. Chron. 1905, vol. xxxvii. p. 194, fig. 79.

This most remarkable plant discovered by the brothers Schomburgk, the energetic travellers, the first to visit the mountain Roraima, in British Guiana, at the base of which this pitcher plant was found.

Sir Robert Schomburgk made an excellent drawing and sent it, with dried specimens, to Mr. Bentham of Kew, who published a full account in the Trans. of the Linn. Soc. vol. xviii. p. 432, and on this species founded the new genus Heliamphora.

Re-discovered in the same locality by the collector, David Burke, in 1881, plants sent home flowered for the first time in June 1889.

NEPENTHES ALBO-MARGINATA, *Lobb.*

Gard. Chron. 1849, p. 580, with fig.; The Garden, 1880, vol. xvii. p. 542, col. pl. fig. 3; Fl. des Serres, 1877, p. 165.

Sent to Exeter by Thomas Lobb from Borneo with several other fine pitchers, this small, handsome species takes its specific name from the peculiar white band surrounding the throat of the urn, below the peristome. The base is green, the upper portion bright rosy carmine dotted with creamy white on the interior as on the lid.

NEPENTHES ANGUSTIFOLIA, *Mast.*

Masters in Gard. Chron. 1881, vol. xvi. p. 524.

A species found in Sarawak by both Curtis and Burbidge, and raised from seed sent home by the former; the pitchers small, not very ornamental.

NEPENTHES BICALCARATA, *Hook. f.*

Masters in Gard. Chron. 1880, vol. xiii. p. 200, fig.; Burbidge in Gard. Chron. 1880, vol. xiii. p. 264; The Garden, 1880, vol. xxvii. p. 542, pl. 237; *id.* 1888, vol. xxxiii. pp. 29, 78.

A remarkable Pitcher plant, first described by Sir Joseph Hooker from specimens collected in Borneo by Low and others, but not introduced to this country till Burbidge succeeded in sending home living specimens from Borneo to Chelsea.

The pitchers, very singular, and distinct from those produced by any other species in cultivation, are furnished with two sharp incurved spurs which project downwards over the mouth of the urn, and resemble the fangs of a snake with the head lifted to strike.

A possible use for these fangs is suggested by Burbidge in the Gardeners' Chronicle above quoted, where he observes that they serve to prevent the contents of the pitchers being rifled by a little creature known as the "Tamperlilie" or "Spectre Tarsier," an insectivorous quadruped in the habit of emptying the urns of their captured insects.

NEPENTHES BURKEI, *Mast.*

Masters in Gard. Chron. 1889, vol. vi. p. 492, fig. 69; Veitchs' Catlg. of Pl. 1890, p. 6, fig.

A handsome wingless species from the Philippine Islands by David Burke.

The pitchers, some 8 in. in length, are cylindrical, contracted in the middle, green in colour spotted with red; the top bordered by a deep red rim, divided at the margin into irregularly sharp-pointed lobes.

NEPENTHES BURKEI, *Mast.*, var. EXCELLENS.

Jour. R.H.S. vol. xxi. pt. ii., fig. 47; Veitchs' Catlg. of Pl. 1891, p. 66.

A variety of the variable Nepenthes Burkei, introduced from the Philippines through David Burke.

Distinguished from the type by much larger and somewhat more cylindrical pitchers more richly coloured; the sharp-pointed, lobed rim of the aperture is broader, of a rich chestnut-red, the spots on the pitchers larger and more numerous.

NEPENTHES BURKEI, *Mast.*, var. PROLIFICA.

Masters in Gard. Chron. 1890, vol. viii. p. 184.

A variety differing from the type in slender growth, narrower leaves, and smaller, less highly coloured pitchers, produced in such profusion as to suggest the name prolifica.

NEPENTHES CINCTA, *Mast.*

Masters in Gard. Chron. 1890, vol. viii. p. 240; *id.* 1884, vol. xxi. p. 576, with fig.; Veitchs' Catlg. of Pl. 1886, p. 5, with fig.

Introduced from Borneo in an importation of Nepenthes Northiana by David Burke. The seed was collected from plants of N. Northiana growing in company with N. albo-marginata, and the plants to which the name N. cincta is applied in all probability are natural hybrids between these species.

The tubular pitchers, 9 in. long from the lid to the base, are pale yellowish-green stained with crimson marked with numerous deep purple blotches. The rim oblique, deep purple, finely ribbed, has several angular lobes as in N. Northiana; beneath the rim is a pale band as in N. albo-marginata, from which the name is derived.

NEPENTHES CURTISII, *Mast.*

Masters in Gard. Chron. 1887, vol. ii. p. 681, fig. p. 689; Veitchs' Catlg. of Pl. 1888, p. 6, fig.; Bot. Mag. t. 7138.

A very interesting, attractive species from Borneo through the collector whose name it bears.

The peculiar, long, narrow pitchers are yellowish-green, thickly mottled with large blotches of purple, wider at the mouth than at the base.

The operculum is remarkable in having two horns projecting from the inner or lower surface, one towards the apex and one at the junction of lid and pitcher. The horn at the back in the majority of cases projects from the base of the lid, and is some distance beneath it.

NEPENTHES CURTISII, *Mast.*, var. SUPERBA.

Gard. Chron. 1889, vol. vi. p. 661, with fig.; Veitchs' Catlg. of Pl. 1895, p. 68.

The urns of this variety are not only much larger than those of the type, but the colour is more rich; the ground tint a brilliant sanguineous-red with longitudinal yellow-green streaks and markings.

NEPENTHES GRACILIS, *Korth*, var. MAJOR.

Veitchs' Catlg. of Pl. 1867, fig.

A variety from Borneo with pitchers of medium size, of elegant outline, green, spotted and marked with deep brown, larger than those of the type.

NEPENTHES KENNEDYANA, *F. Muell.*

Gard. Chron. 1882, vol. xvii. p. 257, with fig.; Veitchs' Catlg. of Pl. 1879, p. 53.

A native of Cape York, North Australia, sent to Messrs. Veitch through the Botanic Gardens, Sydney.

It is distinguished by handsome sub-cylindrical crimson pitchers, from 3 to 4 in. long, inflated below the middle, contracted above, narrow wings in front and a prominent rib on the sides.

NEPENTHES LÆVIS, *Lindl.*

Gard. Chron. 1848, p. 655, with fig.

A species imported from Java and Singapore, described as follows in the Gardeners' Chronicle :—

"It is readily distinguished from Nepenthes phyllamphora by its narrow, shining, leathery leaves, wholly destitute of fringed teeth, and by its smaller pitchers inflated near the base. The crests of the pitchers are sometimes fringed and sometimes naked."

NEPENTHES LANATA, *Hort.*

Gard. Chron. 1854, p. 375 (Note on Exhibit of New Plants); l'Illus. Hort. 1876, t. 261; Hooker's Monograph in the Trans. Linn. Soc.; Masters in Gard. Chron. 1882, vol. xvii. p. 178.

Introduced from Sarawak, Borneo, where it was discovered by Thomas Lobb.

There has been confusion regarding this species, which had been mistaken for Nepenthes Veitchii and *vice versâ*. By an oversight the plate in l'Illustration Horticole, Gand, quoted above, bears the name N. lanata, but the accompanying text is headed N. Veitchii, and the description applies to that species.

The pitchers are 6 in. long, greenish, not in the least ventricose, winged, and like its supporting tendril thickly covered with coarse hairs.

NEPENTHES MADAGASCARIENSIS, *Poiret.*

Masters in Gard. Chron. 1881, vol. xvi. p. 685, fig. 139.

This handsome Pitcher plant, native of the exposed swamps of Madagascar, and probably the earliest Nepenthes to be made known in gardens, was discovered by Comerson, the first European traveller in Madagascar, in 1661.

Nothing further was heard for 136 years, when Poiret, in 1797, published a description in Lamarck's Encyclopædia, and it is remarkable that this species was almost the last to be introduced, nor was it sent home till Curtis made a collecting mission in Madagascar in 1878-1879.

The pitchers, only of medium size, are unusually elegant, and richly coloured.

First exhibited on October 11th, 1881, before the Royal Horticultural Society, a First Class Certificate was awarded.

NEPENTHES NORTHIANA, *Hook. f.*

Masters in Gard. Chron. 1881, vol. xvi. p. 717, fig. 144 and suppl.

The existence of this noble species was first made known to science through Miss North, the well-known botanical artist.

Nepenthes Northiana formed the subject of a sketch by this lady, made in Sarawak, North-West Borneo, from specimens brought to her by Mr. Herbert Everett of the Borneo Company, who "traversed pathless forests amid snakes and leeches, to find and bring them down to her."

The sketch having been shown to Mr. Harry Veitch, was at once recognized as a new and desirable species, and Curtis, about to start at the time on a collecting expedition to the East, was instructed to especially search for the plant. After some difficulty he succeeded, sent seed to Chelsea in 1881, from which plants were raised.

NEPENTHES RAJAH, *Hook. f.*

Gard. Chron. 1881, vol. xvi. p. 492, fig. 91; Veitchs' Catlg. of Pl. 1883, p. 53, fig.; Bot. Mag. t. 8017.

This remarkable species first made known in 1851 by Sir Hugh Low, who discovered it on Mount Kina Balu, in Borneo, though he failed to introduce to cultivation, and it was Burbidge, collecting in 1878, who first succeeded in procuring seed from which plants were raised at Chelsea.

The huge bag-shaped pitchers are thus described by Sir J. D. Hooker:— "The broad ampullaceous pitcher is 6 in. in diameter and 12 in. long; it has two fimbriated wings in front, is covered with long rusty hairs above, and studded with glands within. The margin is scolloped into bold undulations, and the lid is sub-orbicular 10 in. long and 8 in. broad."

Unfortunately this noble species is unusually *difficile*, and now exceedingly rare.

NEPENTHES SANGUINEA, *Lindl.*

Gard. Chron. 1849, p. 580, fig.; Veitchs' Catlg. of Pl. 1873, p. 43.

A magnificent species, with pitchers of a dark crimson, blood-red colour, with a peculiar milk-white interior. Introduced to Exeter through Thomas Lobb, who met with it on Mount Ophir; the pitchers often 12 in. in length by 2½ to 3 in. in width, are of cylindrical shape with a dilatation at the middle and a broad margin round the aperture, and have been known to hold a pint of water.

NEPENTHES STENOPHYLLA, *Mast.*

Masters in Gard. Chron. 1890, vol. viii. p. 243; *id.* 1892, vol. xi. p. 401, fig.

A species from Borneo, at first thought to be but a mere form of

Nepenthes Curtisii, from which, however, it differs in several important particulars.

The pitchers, 6 to 7 in. in length by about $1\frac{3}{4}$ in. at their broadest part, are green, thickly mottled with longitudinal dark reddish-purple blotches, cylindric, narrowed in the middle, with two moderately deep laciniate wings.

NEPENTHES VEITCHII, *Hook. f.*

Hook. f. in Trans. Linn. Soc. vol. xxii. p. 421; Burbidge in Fl. Mag. t. 265; Masters in Gard. Chron. 1872, p. 541; *id.* 1881, vol. xvi. p. 780, fig. 152; The Garden, 1880, vol. xxvii. p. 542, pl. 237; W. W. in Gard. Chron. 1896, vol. xx. p. 239.

A very handsome species sent to Exeter by Thomas Lobb in 1847, from Mount Kina Balu in Sarawak, confused with one, Nepenthes villosa, probably from the fact that the pitchers and leaves are densely covered with hair: figured in the Botanical Magazine under that name.

Though destitute of the rich colouring possessed by many, the broad reflexed finely ribbed peristome, of a creamy olive or reddish colour, and the pale green hairy body of the pitcher, render it a striking object.

Never common in cultivation, it is now rare.

NEPENTHES × BALFOURIANA.

Gard. Chron. 1899, vol. xxvi. pp. 90, 91, fig. 39.

A very remarkable hybrid named in compliment to Professor Bailey Balfour of the Edinburgh Botanic Gardens, raised at Chelsea from Nepenthes × mixta and N. × Mastersiana, the descendant of four distinct species.

NEPENTHES × CHELSONI.

Veitchs' Catlg. of Pl. 1874, p. 11, fig.; The Garden, 1877, vol. xi. p. 429, fig.

A hybrid raised by Seden at Chelsea from a cross between Nepenthes × Dominii and N. Hookeri.

The pitchers, of a peculiar urn-shape, broad at the base, narrowing towards the rim, with two broad wings ciliate along the margin, are heavily blotched with crimson.

NEPENTHES × CHELSONI, var. EXCELLENS.

Veitchs' Catlg. of Pl. 1903, p. 59, with fig.

A hybrid raised at Chelsea by Tivey from the original Nepenthes × Chelsoni and N. Rafflesiana, the latter the seed parent and a very magnificent urn.

In shape resembling N. × Chelsoni, but the wings are broader, and the colouring of the profuse crimson blotches richer.

NEPENTHES × COURTII.

Veitchs' Catlg. of Pl. 1878, p. 13, fig.; Masters in Gard. Chron. 1881, vol. xvi. p. 844, fig. 160.

A hybrid from seed of an unnamed species from Borneo fertilized with pollen of Nepenthes × Dominii, raised at Chelsea by Court.

NEPENTHES × CYLINDRICA.

Gard. Chron. 1887, vol. ii. p. 521, fig.; Veitchs' Catlg. of Pl. 1888, p. 6, fig.

A distinct hybrid raised at Chelsea by George Tivey from Nepenthes hirsuta glabrescens (N. zeylanica rubra of gardens) crossed with N. Veitchii.

NEPENTHES × DICKSONIANA.

Gard. Chron. 1888, vol. iv. p. 543, fig. 541; Veitchs' Catlg. of Pl. 1889, p. 11, fig. p. 6.

The offspring of Nepenthes Rafflesiana flowering in the Botanic Garden at Edinburgh, fertilized with the pollen of N. Veitchii sent from Chelsea. The cross effected by Mr. Lindsay, the progeny was dedicated to the late Professor Dickson.

NEPENTHES × DOMINII.

Gard. Chron. 1862, p. 398.

This, the first hybrid Nepenthes ever raised by man, was obtained at Exeter by John Dominy, by crossing N. Rafflesiana with an unnamed species with green-coloured pitchers from Borneo.

NEPENTHES × F. W. MOORE.

Jour. of Hort. 1904, Nov. 10th, p. 414, fig.

A hybrid raised at Chelsea by Tivey from Nepenthes × mixta with N. × Dicksoniana, and of the same parentage as N. × Sir W. Thiselton-Dyer and × N. picturata.

The pitchers are green, more or less globular in shape, with a fine red margin to the mouth and deep fimbriated wings.

NEPENTHES × INTERMEDIA.

Gard. Chron. 1875, p. 118 (Report of Exhibit of New Plants); Veitchs' Catlg. of Pl. 1875, p. 9, with fig.; The Garden, 1877, vol. xi. p. 429, with fig.; Masters in Gard. Chron. 1882, vol. xvii. p. 179, with fig.

Raised at Chelsea by William Court from Nepenthes Rafflesiana and an unnamed Bornean species with small spotted pitchers.

NEPENTHES × MACULATA.

Syns. *N. × hybrida maculata.*

Veitchs' Catlg. of Pl. 1866, p. 4; Gard. Chron. 1866, p. 432 (advt.); Fl. Mag. 1868, pl. 409.

Raised by Dominy from a cross between Nepenthes distillatoria and an unnamed species from Borneo.

NEPENTHES × MASTERSIANA.

Masters in Gard. Chron. 1881, vol. xvi. p. 748; Veitchs' Catlg. of Pl. 1883, p. 6, with fig.; Flora and Sylva, vol. ii. p. 112.

A hybrid raised at Chelsea by Court from Nepenthes sanguinea crossed with the pollen of N. Khasiana (N. distillatoria of gardens).

NEPENTHES × MIXTA.

Gard. Chron. 1893, vol. xiii. p. 46, fig. 9.

An interesting hybrid from Nepenthes Curtisii and N. Northiana, the first named the pollen parent.

First exhibited under the name of Northisii, a compound of the names of the two parent species, but subsequently changed to mixta.

NEPENTHES × MIXTA, var. SANGUINEA.

Gard. Chron. 1894, vol. xvi. p. 318 (Report of R.H.S. Meeting).

A very dark form of the type with the ground colour suffused with deep sanguineous-red.

NEPENTHES × PICTURATA.

Gard. Mag. 1903, Oct. 10th, pp. 670, 677, with fig.; Flora and Sylva, 1904, vol. ii. p. 69, fig.

Raised at Chelsea by Tivey from Nepenthes × mixta and N. × Dicksoniana, of the same parentage as the hybrid N. × Sir William T. Thiselton-Dyer. A singularly handsome seedling.

NEPENTHES × RATCLIFFIANA.

Veitchs' Catlg. of Pl. 1880, p. 53, 1881, p. 56, fig.; Masters in Gard. Chron. 1882, vol. xvii. p. 178, fig. 28.

A hybrid dedicated to Alfred E. Ratcliff Esq., Edgbaston, Birmingham, a distinguished amateur of this interesting class of plants, and obtained at Chelsea from Nepenthes phyllamphora and N. Hookeriana.

NEPENTHES × RUBRO-MACULATA.

Veitchs' Catlg. of Pl. 1880, p. 53; Masters in Gard. Chron. 1882, vol. xvii. p. 143, fig. 24.

A hybrid raised at Chelsea from Nepenthes × hybrida maculata and a Bornean species which, at that time, had not been identified (probably N. lanata).

SARRACENIA × COURTII

INSECTIVOROUS PLANTS

NEPENTHES × RUFESCENS.

Masters in Gard. Chron. 1888, vol. iv. p. 669, with fig.; Veitchs' Catlg. of Pl. 1889, p. 49.

A hybrid obtained by Court at Chelsea from Nepenthes zeylanica rubra (N. hirsuta glabrescens) crossed with the pollen from a hybrid named Courtii.

NEPENTHES × SEDENII.

Veitchs' Catlg. of Pl. 1871, p. 10, fig.

A hybrid raised at Chelsea by Seden from a cross between an unnamed species with deep-coloured pitchers and Nepenthes distillatoria.

NEPENTHES × SIR WILLIAM T. THISELTON-DYER.

Gard. Chron. 1900, vol. xxviii. p. 137 (Report of R.H.S. Floral Committee), pp. 256, 257, fig. 76.

A very fine hybrid won by George Tivey, the result of crossing the two fine hybrids Nepenthes × mixta and N. × Dicksoniana, the descendant of four distinct species.

NEPENTHES × TIVEYI.

Gard. Chron. 1897, vol. xxii. p. 187 (Report of R.H.S. Floral Committee); Gard. Chron. 1897, vol. xxii. pp. 200, 201, figs. 59, 60.

Raised by George Tivey at Chelsea from Nepenthes Veitchii and N. Curtisii superba, the latter the seed-bearer.

NEPENTHES × WRIGLEYANA.

Veitchs' Catlg. of Pl. 1880, p. 54; Masters in Gard. Chron. 1882, vol. xvii. p. 143, fig. 23.

A hybrid named in compliment to Oswald Wrigley Esq., of Bridge Hall, Lancashire, a distinguished amateur of Pitcher plants, and raised at Chelsea from Nepenthes phyllamphora crossed with the pollen of N. Hookeriana.

SARRACENIA × CHELSONI.

Veitchs' Catlg. of Pl. 1879, p. 18, with fig.; Gard. Chron. 1880, vol. xiii. p. 722, fig. 125; *id.* 1881, vol. xv. p. 817, fig. 148a.

A beautiful hybrid raised by Court at Chelsea from Sarracenia rubra and S. purpurea, the latter the pollen parent.

The pitchers intermediate in form between those produced by the two species, are of a bright rich crimson hue.

SARRACENIA × COURTII.

Veitchs' Catlg. of Pl. 1885, p. 8, with fig.; Gard. Chron. 1881, vol. xvi. p. 381 (Report of R.H.S. Floral Committee).

A hybrid raised at Chelsea by Court from Sarracenia purpurea and S. psittacina, the last named the pollen parent.

The procumbent pitchers, as those of both the parents, in a regular radiating rosette, bright crimson-purple from the middle upwards, are veined and reticulated with a deeper tint: they change with age to a deep blood-red with blackish-purple veins.

SARRACENIA × FORMOSA.

Masters in Gard. Chron. 1881, vol. xvi. p. 41; Veitchs' Catlg. of Pl. 1883, p. 55.

A beautiful hybrid from Sarracenia psittacina and S. variolaris in which the characteristics of the two are intimately blended in the ascending pitchers, intermediate in form and length, while the beak-like lid is altogether that of S. psittacina.

All the upper portion of the pitcher has bright crimson reticulated nervation, with the characteristic spotting of S. variolaris; the basal portion pale fulvous-green.

SARRACENIA × MELANORHODA.

Masters in Gard. Chron. 1881, vol. xvi. p. 40; Veitchs' Catlg. of Pl. 1883, p. 54, with fig.

A hybrid raised by Court at Chelsea between Sarracenia purpurea and S. × Stevensii, the latter itself a hybrid by the late Mr. Stevens, one time gardener to the Duke of Sutherland, at Trentham.

The pitchers are semi-decumbent and about intermediate in position between the erect ones of S. × Stevensii and the prostrate ones of S. purpurea, funnel-shaped with a broad wing on the upper side, when mature blood-red veined with crimson. The lid erect and crisped, is beautifully veined with blackish crimson on a reddish-yellow ground.

SARRACENIA × WILLISII.

Gard. Chron. 1894, vol. xv. p. 761 (Report of R.H.S. Floral Committee).

Raised at Chelsea from Sarracenia × melanorhoda and S. × Courtii—derived from four distinct species.

The green pitchers with purple veining are some 8 in. in length.

UTRICULARIA ENDRESII, *Rchb. f.*

Rchb. in Gard. Chron. 1874, vol. ii. p. 582; Bot. Mag. t. 6656; The Garden, 1880, vol. xxviii. p. 432, col. pl.; Veitchs' Catlg. of Pl. 1879, p. 20, with fig.

A pretty epiphytal Bladder-wort, introduced in 1868 through Endres from Costa Rica.

The flowers, of a light lilac or delicate mauve colour with a creamy white lip and yellow palate, resemble those of Utricularia montana in shape.

EXOTIC FERNS

EXOTIC FERNS

ACROSTICHUM AUREUM, *L.*

Gard. Chron. 1866, p. 432 (advt.); Veitchs' Catlg. of Pl. 1866, p. 1.

A bold-habited semi-aquatic fern, introduced by the late John Gould Veitch from Cape York, Australia. The tall erect fronds of pinnate form, the pinnæ oblong, emarginate, sometimes almost cuspidate, the lower entirely barren, the upper wholly fertile; it is usually found near the sea.

ACROSTICHUM LECHLERIANUM, *Hook.*

Syns. *Polybotrya Lechleriana*, Mettenius.

Moore in Gard. Chron. 1886, vol. xxv. pp. 394, 400, 401, figs. 79, 80.

Introduced through Walter Davis from Peru, this fine stove species, of scandent habit, has finely divided fronds exceeding 2 ft. in length.

At first taken to be Cyathea microphylla, it was found, on fructification, to belong to the genus Polybotrya, now included under Acrostichum by Hooker.

ADIANTUM ÆMULUM, *Moore.*

Moore in Gard. Chron. 1877, vol. viii. p. 584, with figs.

Introduced from Brazil, this handsome species, less rigid in habit than Adiantum decorum, and less drooping than A. cuneatum, is of about equal stature, but a bluer green.

ADIANTUM ÆTHIOPICUM, *L.*, var. CHILENSE, *Kaulf.*

Syns. *A. chilense*, Kaulf.

Gard. Chron. 1862, p. 399 (advt.); Veitchs' Catlg. of Pl. 1862, p. 11.

For this Chilian Maiden-hair Fern, a very beautiful species, we are indebted to Richard Pearce, who discovered it in Chili; the smooth-looking glaucous-green broad pinnuled fronds are distinct and remarkably handsome.

ADIANTUM ÆTHIOPICUM, *L.*, var. SCABRUM, *Kze.*

Syns. *A. scabrum*, Kze.

Gard. Chron. 1862, p. 399 (advt.); Veitchs' Catlg. of Pl. 1862, p. 11.

This Silver Maiden-hair Fern, a native of Chili, sent by Richard Pearce, is of dwarf habit, with oblong fronds and black stipes; the pinnules, oblong or trapeziform, are sprinkled on both surfaces with farinose powder.

ADIANTUM COLPODES, *Moore.*

Gard. Chron. 1865, p. 530.

Raised from spores collected in Ecuador by Richard Pearce, the fronds tinted with rose-colour in the young state like those of Adiantum tinctum, somewhat resemble A. Capillus-veneris in outline, but the sori are very distinct. The margin of the pinnule has a crenated appearance with bay-like recesses as the specific name implies.

ADIANTUM CONCINNUM, *H. B. K.*, var. LATUM, *Moore.*

Veitchs' Catlg. of Pl. 1868, p. 14, fig. p. 2.

A beautiful form, differing from the type in having broader fronds, a more erect habit and stronger growth.

It was introduced from Muña, a Province of Peru, through Richard Pearce, and exhibited for the first time before the Royal Horticultural Society in May 1867.

ADIANTUM CUNEATUM, *L. & F.*, var. GRANDICEPS, *Hort.*

Gard. Chron. 1881, vol. xvi. p. 685.

A variety of garden origin, a multifid form of the well-known Adiantum cuneatum, of the same habit of growth as the type, differing only in tasselled apices.

ADIANTUM DIGITATUM, *Presl.*, var. SPECIOSUM, *Hook.*

Syns. *A. speciosum*, Hort.

Veitchs' Catlg. of Pl. 1875, p. 10, fig. p. 1; Fl. and Pom. 1875, p. 272, fig.

A variety from Peru, with the whole of the pinnules covered with dense short hairs, a woolly feeling to the touch.

The semi-scandent habit and the deciduous fronds are uncommon characters in the genus to which the Maiden-hair Ferns belong.

ADIANTUM HENSLOVIANUM, *Hook. f.*

Syns. *A. sessilifolium*, Hook.

Veitchs' Catlg. of Pl. 1874, p. 15; Moore in Fl. and Pom. 1873, p. 277.

A distinct greenhouse species of Adiantum with fronds 2½ to 3 ft. in length, of a semi-pendulous character, sent from Peru through Richard Pearce.

FERNS

ADIANTUM MACROPHYLLUM, *Sw.*, var. GLAUCUM, *T. Moore.*

Moore in Gard. Chron. 1875, vol. iii. p. 621, figs. 126 and 127.

Imported both from Peru and New Grenada.

As compared with the typical species, this form may be distinguished by a pronounced glaucous-green hue, and by the narrower pinnæ with a strong tendency to be straight-edged.

ADIANTUM MONOCHLAMYS, *Eaton.*

Veitchs' Catlg. of Pl. 1886, p. 9; Hooker's Sp. Fil. p. 125.

A dwarf species of Maiden-hair Fern from Japan, closely allied to the beautiful Himalayan Adiantum venustum, from which, however, the dwarf habit and glaucous foliage are distinct.

ADIANTUM MOOREI, *Baker.*

Syns. *A. amabile*, Moore.

Baker in Gard. Chron. 1873, p. 811; Moore in Gard. Chron. 1868, p. 1090; Fl. and Pom. 1872, p. 287, fig.

Introduced through Richard Pearce from the Andes of Peru.

Unfortunately the specific name amabile which Mr. Moore selected had already been used by Liebmann for a Mexican species, and in the 2nd edition of Synopsis Filicum, Mr. Baker changed the name to Moorei in compliment to Moore of Chelsea, who had done so much to elucidate the nomenclature of Ferns.

ADIANTUM PALMATUM, *T. Moore.*

Moore in Gard. Chron. 1877, vol. vii. p. 40, fig. 5; Veitchs' Catlg. of Pl. 1877, p. 20, fig. p. 4.

One of the most beautiful and attractive of Maiden-hair Ferns, from very high elevations on Chimborazo, South America: the fronds deciduous.

ADIANTUM PERUVIANUM, *Klotzsch.*

Gard. Chron. 1870, p. 457, fig.; Hooker and Baker, Synop. Fil. ii. 35, t. 81; Veitchs' Catlg. of Pl. 1873, p. 10, fig. p. 3.

A noble Maiden-hair first made known to science from specimens collected in Peru by Ruiz and Pavon, later in the same locality by Mathews, and introduced to cultivation through Richard Pearce.

A remarkably handsome species, the fronds spread and hang gracefully dependent as the boughs of a weeping willow.

ADIANTUM PRINCEPS, *T. Moore.*

Veitchs' Catlg. of Pl. 1876, p. 18, fig. p. 4; Moore in Gard. Chron. 1875, vol. iv. p. 197, figs. 43, 44; Fl. and Pom. 1877, p. 243, figs.

Introduced to cultivation from New Grenada through Gustave Wallis, and remarkable for the size of the individual pinnules and for the graceful pendant form of the frond.

ADIANTUM RUBELLUM, *T. Moore.*

Gard. Chron. 1868, p. 866; Veitchs' Catlg. of Pl. 1870, p. 1. fig.

Introduced from Peru through Richard Pearce. A dwarf and compact plant belonging to the Capillus-veneris section, characterized by a roseate hue with which the young fronds are more or less decorated, and which suggested the specific name.

ADIANTUM SEEMANNI, *Hook.*

Syns. *A. Zahnii*, Hort.

Veitchs' Catlg. of Pl. 1875, p. 10, fig. p. 2; Moore in Gard. Chron. 1875, p. 396; Fl. and Pom. 1876, p. 218, fig.

A noble species of Maiden-hair, first exhibited under the name of Adiantum Zahnii, but afterwards found to be identical with A. Seemanni, described in 1851 by Sir William Hooker, from dried specimens collected by Dr. Seeman at Veraguas, Central America; introduced to cultivation from the same locality through Zahn.

ADIANTUM SULPHUREUM, *Kaulfuss.*

Gard. Chron. 1862, p. 399; Veitchs' Catlg. of Pl. 1862, p. 12.

A charming greenhouse Maiden-hair Fern with the whole under surface of the frond covered with golden-coloured powder, as in several of the Gymnogrammes and other gold ferns. Introduced from Chili through Richard Pearce.

ADIANTUM TINCTUM, *T. Moore.*

Gard. Chron. 1862, p. 932.

A beautiful Maiden-hair Fern from Peru, apparently closely related to Adiantum concinnum. The fronds, however, appear to be only bipinnate, the habit of growth quite different. It is especially remarkable for the rose-red tint of the young fronds, approaching in outline the larger forms of A. Capillus-veneris, but the fructification has no resemblance.

ADIANTUM VEITCHIANUM, *T. Moore.*

Moore in Gard. Chron. 1868, p. 1090; Veitchs' Catlg. of Pl. 1871, p. 1, fig.

One of the Peruvian introductions of Richard Pearce, by whom it was found at Muña, in the Peruvian Andes.

The young fronds are very beautiful, of a fine red colour, the mature pale green.

FERNS

ADIANTUM WAGNERI, *Mett.*

Syns. *A. decorum*, T. Moore.

Gard. Chron. 1869, p. 582.

Introduced from the Andes of Peru through Richard Pearce.

In aspect this species is suggestive both of Adiantum concinnum and A. cuneatum, having much the same outline of frond as the latter, and pinnules of about the same size, but more variable in form. In habit stiffer and more erect than either of the two named, due to the stouter stipites and rachides.

ADIANTUM WEIGANDII, *Moore.*

Veitchs' Catlg. of Pl. 1885, p. 9; Moore in Gard. Chron. 1883, vol. xx. p. 748; Fl. and Pom. 1884, p. 9.

A pretty distinct fern of garden origin, obtained from Mr. Weigand of Astonia, Long Island, New York, for our houses.

ASPIDIUM ACULEATUM, *Sw.*, var. TSUS-SIMENSE, *Hook.*

Syns. *Polystichum Tsus-simense*, Schott.

Hooker's Sp. Fil. t. ccxx.; Veitchs' Catlg. of Pl. 1891, p. 60.

A useful decorative fern, from Japan, of dwarf, compact habit, the fronds rarely exceeding 1 ft. in length, spread gracefully on all sides; the pinnæ of leathery texture are of a deep lustrous green.

ASPIDIUM ARISTATUM, *Sw.*, var. VARIEGATUM, *Hort.*

Syns. *Lastrea aristata*, Moore, var. *variegata*.

Veitchs' Catlg. of Pl. 1880, p. 75.

An elegant greenhouse fern received from Japan, differing from the type in bands of yellowish-green running through the bases of the pinnules along the course of the rachis.

ASPIDIUM FALCATUM, *Sw.*, var. PENDULUM, *Hort.*

Syns. *Cyrtomium falcatum*, Pappe & Raws, var. *pendulum*.

Veitchs' Catlg. of Pl. 1892, p. 8.

A pendulous form of the type with more narrow pinnæ, but with the same glossy-green colour, well adapted for cultivation as a basket plant in the cold greenhouse.

ASPIDIUM FLEXUM, *Kze.*

Syns. *Polystichum flexum*, Remy.

Gard. Chron. 1862, p. 399 (advt.); Veitchs' Catlg. of Pl. 1862, p. 13.

A fine hardy fern, sent from Chili by Richard Pearce, with tripinnate fronds, oblong toothed segments of a thick leathery texture.

ASPIDIUM SETOSUM, *Blume.*

Syns. *Polystichum setosum*, Schott.

Veitchs' Catlg. of Pl. 1862, p. 14; Gard. Chron. 1862, p. 399 (advt.); Nich. Dict. Gard. (Century Suppl.), fig. 89.

Introduced to cultivation from Japan through the late John Gould Veitch.

A hardy species with bipinnate fronds 2 to 3 ft. or more in height; the pinnules remarkable for a fringe of setæ or stiff hairs, stand erect from the plane of the frond and have a bristly appearance.

ASPIDIUM TRIPTERON, *Kunze.*

Syns. *Polystichum tripteron*, Kunze.

Gard. Chron. 1881, vol. xv. p. 74; Veitchs' Catlg. of Pl. 1881,p. 17.

Introduced from Japan through Charles Maries, found growing in rocky places on the shady hills of that country.

It is a hardy or cool greenhouse fern of great elegance, with fronds from 12 to 18 in. in length.

ASPLENIUM BELANGERI, *Kunze.*

Syns. *A. Veitchianum*, Moore.

Lowe's British and Exotic Ferns, vol. v. pl. v.

A native of Java, and various parts of the Malay Peninsula; the fronds 1 ft. to 18 in. in length, are bipinnate, with numerous pinnæ 1 to 1½ in. long by ½ in. broad. It is an elegant species, a stove temperature essential for successful cultivation.

ASPLENIUM CETERACH, *L.*, var. AUREUM, *Link.*

Syns. *Ceterach aureum*, Link.

Veitchs' Catlg. of Pl. 1874, p. 34.

An introduction from Teneriffe, resembling the common Ceterach officinarum, but the fronds covered behind with silvery scales assume a yellowish-brown hue as they develop.

ASPLENIUM CONSIMILE, *Gay.*

Gard. Chron. 1863, p. 695 (advt.).

Introduced from Chili by Richard Pearce, this dwarf tufted species, with pinnated deep green fronds, rising from a dense crown of brownish semi-transparent scales, is remarkable for its enduring properties.

FERNS

ASPLENIUM FERULACEUM, *Moore.*

Veitchs' Catlg. of Pl. 1877, p. 44; Nich. Dict. Gard. (Century Suppl.), fig. 103.

A beautiful stove fern from New Grenada by Gustave Wallis, with fronds unusually light and pleasing in appearance, and pinnules delicate and fine in texture.

ASPLENIUM LATIFOLIUM, *Don.*

Syns. *Athyrium latifolium*, Presl.

Gard. Chron. 1860, p. 432 (advt.); *id.* p. 634; Veitchs' Catlg. of Pl. 1866, p. 1.

A distinct, hardy fern, not unlike Asplenium lanceolatum in appearance, introduced from Chili through Richard Pearce.

ASPLENIUM LONGISSIMUM, *Blume.*

Veitchs' Catlg. of Pl. 1873, p. 27.

A useful basket subject from Java, with fronds 2 to 8 ft. in length, in some respects resembling a species of Nephrolepis.

ASPLENIUM OBTUSILOBIUM, *Hook.*

Hooker's Ic. Pl. t. 1000; Gard. Chron. 1861, p. 696.

A handsome fern brought to this country by Sir Daniel Cooper, Bart., from the New Hebrides, and placed in the hands of the Veitchian people for distribution. Dark green smooth shining fronds form a thick tuft, out of which grow numerous stolons, on which at intervals young fern plants are produced.

ASPLENIUM SCANDENS, *J. Sm.*

Gard. Chron. 1887, vol. i. p. 639.

Introduced from Sumatra through Curtis. A stove species of climbing habit, with lanceolate fronds 1 ft. or more in length, narrowed gradually from the middle towards the base and apex.

ASPLENIUM VIELLARDII, *Mett.*

Syns. *A. Schizodon*, Moore.

Moore in Gard. Chron. 1871, p. 1004, with fig.; *id.* 1872, p. 1654; Veitchs' Catlg. of Pl. 1874, p. 15.

Introduced from New Caledonia through the late John Gould Veitch in 1868, and remarkable for the frond serration irregularly bi- or tri-serrate, and from 3 to 4 in. long.

BLECHNUM NITIDUM, *Presl.*, var. CONTRACTUM.

Gard. Chron. 1864, p. 670 (advt.); Hooker's Sp. Fil. iii. p. 55.

A bold habited Lomaria-like fern from the Philippine Islands, with fronds in a young state of a deep red colour.

CHEILANTHES MYSURENSIS, *Wallich.*

Hooker's Sp. Fil. ii. p. 100; Gard. Chron. 1862, p. 399 (advt.); Veitchs' Catlg. of Pl. 1862, p. 12.

An elegant, small-growing species, raised from spores collected in Nagasaki, Japan, by the late John Gould Veitch.

The fronds, from 6 in. to 1 ft. in length, are furnished with oblong pinnules; the rachis and stipes are dark purplish-brown and scaly.

CHEILANTHES UNDULATA, *Hope & C. H. Wright.*

Gard. Chron. 1903, vol. xxxiv. p. 397.

This species was first discovered in Yunnan, Central China, by Dr. Henry, and later, from the same region, Wilson sent home material from which plants were raised at the Royal Gardens, Kew.

DAVALLIA AFFINIS, *Hook.*

Syns. *Acrophorus affinis*, Moore.

Veitchs' Catlg. of Pl. 1862, p. 11; Gard. Chron. 1862, p. 399; Nich. Dict. Gard. fig. 622.

A handsome stove species sent from Borneo through Thomas Lobb.

The thick scaly rhizomes creep along the surface of the soil, and give off at intervals finely divided fronds, 2 to 3 ft. long.

DAVALLIA ALPINA, *Bl.*

Veitchs' Catlg. of Pl. 1867, fig.; Nich. Dict. Gard. fig. 623.

An interesting little stove fern of dwarf creeping habit, with neat fronds only 2 or 3 in. in length, of exceptional interest to the collector, sent from Borneo through Thomas Lobb.

DAVALLIA BULLATA, *Wall.*, var. MARIESII, *Moore.*

Veitchs' Catlg. of Pl. 1880, p. 21, fig. p. 11; Fl. and Pom. 1880, p. 151, fig.

This beautiful evergreen fern, by Charles Maries from Japan, produces flexuose rhizomes freely in all directions, suitable for basket-work, and for training in all possible shapes.

The familiar fern-balls of the London shops are made of this fern.

DAVALLIA ELEGANS, *Sw.*, var. POLYDACTYLA, *Moore.*

Moore in Gard. Chron. 1881, vol. xv. p. 562; Veitchs' Catlg. of Pl. 1882, p. 18; Fl. and Pom. 1882, p. 52.

A pretty crested form, raised at Chelsea by George Schneider from spores of Davallia elegans.

FERNS

DAVALLIA FERRUGINEA, *Desv.*

Baker in Gard. Chron. 1887, vol. i. p. 639.

Described from specimens collected in Madagascar by Curtis, through whom it was introduced to cultivation, this large sarmentose stove species has decompound fronds and sessile crowded ultimate segments.

DAVALLIA FIJIENSIS, *Hook.*, var. MAJOR, *Hort.*

Veitchs' Catlg. of Pl. 1879, p. 24.

An elegant fern, a native of the Fiji Archipelago, sent by the late Charles Moore, Esq., one time of the Botanic Gardens, Sydney, N.S.W.

A stove species with slender stipes, gracefully arching fronds, the pinnules finely cut and of a very bright green.

DAVALLIA HIRTA, *Kaulfuss*, var. CRISTATA.

Syns. *Microlepia hirta*, var. *cristata.*

Veitchs' Catlg. of Pl. 1878, p. 24, fig. p. 14.

A crested form of Davallia hirta, from the New Hebrides in the South Pacific Ocean by Captain Hoskins, of the same dwarf habit as the type, but the frond apices and the pinnæ are bi- or tri-furcate.

DAVALLIA INTERMEDIA, *Hort.*

Gard. Chron. 1889, vol. xxv. p. 31 (Report of R.H.S. Floral Committee); Veitchs' Catlg. of Pl. 1889, p. 5.

A supposed hybrid between Davallia Mooreana (D. pallida) and D. decora, raised from prothallium produced in a pan in which the spores of the two species had been sown.

D. intermedia shows clearly the characteristics of the two supposed parents.

The plant of free growth, of elegant aspect, is very suitable for suspended baskets in the warm conservatory.

DAVALLIA PALLIDA, *Mett.*

Syns. *D. Mooreana*, Mast.

Gard. Chron. 1869, p. 964, with fig.; Veitchs' Catlg. of Pl. 1870, p. 20, fig. p. 9; Fl. and Pom. 1872, p. 21, figs.

A native of Borneo, whence it was introduced through Thomas Lobb, this Davallia, one of the most beautiful of the genus, is remarkable for a graceful habit, a large size combined with a small sub-division, pale green colour, a smooth surface and bullate sori.

DAVALLIA PARVULA, *Wall.*

Nich. Dict. Gard. vol. i. fig. 627; Veitchs' Catlg. of Pl. 1868, p. 7, fig.; Fl. and Pom. 1872, p. 108, fig.

Introduced from Borneo through Thomas Lobb, a dwarf-growing stove fern with finely divided leaves, averaging from 1 to 1½ in. in height, on creeping rhizomes of from 4 to 6 in. in length.

DAVALLIA PENTAPHYLLA, *Blume.*

Lowe's British and Exotic Ferns, vol. viii. pl. xviii.

A distinct dwarf fern, native of Java and the Malay Archipelago, whence it was introduced.

The fronds consist of a terminal segment and usually two pairs of lateral ones.

DAVALLIA REPENS, *Desv.*

Syns. *D. hemiptera*, Bory.

Veitchs' Catlg. of Pl. 1869, p. 12, fig. p. 4.

A beautiful stove fern, similar in appearance to some of the fine filmy species, with a dwarf compact habit, delicate cut fronds attaining some 4 or 6 in. in length.

DAVALLIA STRIGOSA, *Sw.*

Syns. *Microlepia strigosa*, Presl.

Hooker's Sp. Fil. i. 47; Gard. Chron. 1862, p. 399 (advt.); Veitchs' Catlg. of Pl. 1862, p. 13.

An elegant fern of moderate size introduced by the late John Gould Veitch from Nagasaki, Japan, with bright green hairy fronds about 2 ft. in height, ovate acuminate in form, bipinnate or tripinnate with roundish oblong or somewhat trapeziform pinnules more or less lobed or toothed on the margin.

DAVALLIA TENUIFOLIA, *Sw.*, var. BURKEI.

Gard. Chron. 1895, vol. xviii. p. 102 (Report of R.H.S. Floral Committee); Veitchs' Catlg. of Pl. 1896, p. 6, with fig.

A variety sent from New Guinea by David Burke.

As distinguished from the type the fronds are longer, quite pendulous, the pinnæ more distant and the ultimate segments narrower and more elongated.

DAVALLIA TENUIFOLIA, *Sw.*, var. VEITCHIANA, *Hort.*

Veitchs' Catlg. of Pl. 1885, p. 12, fig. p. 7; Fl. and Pom. 1882, p. 122; Gard. Chron. 1882, vol. xvii. p. 648.

Probably the most finely cut and graceful fern in cultivation, sent to us from the Straits Settlements by the late Dr. J. T. Veitch, whose name it bears. As distinguished from Davallia tenuifolia, the fronds are arching, almost drooping, the foliage more lace-like, the pinnæ longer, more slender, and the ultimate segments quite minute.

DICKSONIA BERTEROANA, *Hook.*

Hooker's Sp. Fil. i. 23a; Veitchs' Catlg. of Pl. 1880, p. 13, with fig.; Nich. Dict. of Gard. vol. i. same fig.

Introduced from Juan Fernandez through George Downton, and of special interest as one of the very few plants in cultivation from that remarkable island.

DICKSONIA DAVALLOIDES, *R. Br.*, var. YOUNGII.

Syns. *Dennstaedtia davalloides*, var. *Youngii.*

Veitchs' Catlg. of Pl. 1877, p. 22, fig. p. 10.

This handsome fern, a native of the New Hebrides, was introduced through J. R. Young, Esq., of Sydney, N.S.W., with whose name it is associated.

The large fronds have a bold graceful aspect, attain upwards of 10 ft. in length, and are furnished with finely cut pinnules of a light cheerful green.

It needs the temperature of a warm conservatory for successful culture.

DOODIA ASPERA, *R. Br.*, var. MULTIFIDA, *Hort.*

Veitchs' Catlg. of Pl. 1879, p. 72.

A pretty greenhouse fern of dwarf habit, sent to us from N.S.W. by Charles Moore, Esq., late Superintendent of the Botanic Gardens, Sydney.

It is of dwarfer habit than the type, with crested fronds of a pinkish colour when young.

DOODIA DURIUSCULA, *Moore.*

Syns. *D. media*, R. Br., var. *duriuscula.*

Gard. Chron. 1868, p. 1114.

A native of New Caledonia, introduced through the late John Gould Veitch, of neat tufted habit, with arching gracefully curved fronds which are practically evergreen.

GYMNOGRAMME FLEXUOSA, *Desv.*

Syns. *Cryptogramme retrofracta*, Hook.

Hooker's Sp. Fil. v. 129; Gard. Chron. 1865, p. 531.

Introduced from Tropical America, where it has a wide distribution, it is noteworthy for finely cut fronds and a flexuose rachis.

GYMNOGRAMME JAPONICA, *Hook.*

Gard. Chron. 1863, p. 628 (Notice of Exhibit).

A species closely allied to the Javanese Gymnogramme javanica, from Japan through the late John Gould Veitch.

GYMNOGRAMME PEARCEI, *Moore.*

Gard. Chron. 1864, p. 340.

One of the finest of the Silver Ferns with tall triangular bright green fronds finely cut into small narrow segments with a superficial resemblance to Asplenium ferulaceum. The lower part of the stipes as well as the caudex have a covering of white powder, confined to about 2 in. from the very base. The remaining part of the stipes, as well as the lamina of the frond, are smooth and shining. It is still a popular fern in collections largely grown for decorative purposes.

For its introduction we are indebted to Richard Pearce, who discovered it during his journey in Chili.

GYMNOGRAMME PEARCEI, *Moore,* var. ROBUSTA, *Hort.*

Veitchs' Catlg. of Pl. 1888, p. 11.

A variety of more vigorous habit than the type, of freer growth. Like the original species, the base of the stipes as the crown is covered with a white powder, the remainder of the frond green.

GYMNOGRAMME SCHIZOPHYLLA, *Baker.*

Veitchs' Catlg. of Pl. 1881, p. 14, fig. p. 7.

A native of the West Indies; a remarkable peculiarity of this Gymnogramme is the furcation of the rachis at about two-thirds of its length, where it is proliferous, each frond producing a young plant at the point.

GYMNOGRAMME VEITCHII, *Hort.*

Gard. Chron. 1896, vol. xix. p. 652 (Report of Show); Gard, Chron. 1884, vol. xvi. p. 446; Veitchs' Catlg. of Pl. 1895, p. 7.

A beautiful Silver Fern with finely-cut pinnæ raised from spores of Gymnogramme Pearcei robusta sown with spores of a variety of G. chrysophylla, and it may reasonably be assumed a hybrid between these two beautiful forms.

LINDSAYA RETUSA, *Mett.*

Syns. *Davallia retusa,* Cav.

Veitchs' Catlg. of Pl. 1886, p. 11.

An elegant stove fern of vigorous growth introduced from Sumatra through Curtis.

The fronds from 2 to 3 ft. long, have pale crimson stipes and rachids and light green pinnæ: of extremely light appearance, one of the best for basket culture.

LOMARIA BLECHNOIDES, *Bory.*

Gard. Chron. 1861, p. 499 (advt.).

An effective hardy greenhouse fern of moderate size, from Chili, with fronds, which grow in a spreading tuft from a short caudex, about 10 in. to 1 ft. in height, lance-shaped in form and deeply pinnatifid.

LOMARIA CILIATA, *Moore.*

Gard. Chron. 1866, p. 290; Veitchs' Catlg. of Pl. 1867, p. 8, fig.

An elegant dwarf-growing tree fern, from New Caledonia by the late John Gould Veitch.

The arborescent character of the stem, together with the lobate character of the pinnæ, the undulate surface, truncate lobes, and spinulate teeth, present the most striking distinction, and render this species one of the most interesting in the genus.

LOMARIA DISCOLOR, *Willd.*, var. BIPINNATIFIDA, *Hort.*

Veitchs' Catlg. of Pl. 1878, p. 68; Moore in Gard. Chron. 1877, vol. viii. p. 488.

This truly fine fern, one of the best for decorative purposes, was introduced from Melbourne through Peter C. M. Veitch.

It is a sub-arborescent species with numerous gracefully arching fronds, 18 to 24 in. in length, from a short robust stem; the pinnæ so closely set that they overlap, are cut to the rib, the sub-divisions being slightly crisped.

LOMARIA GERMAINII, *Hook.*

Syns. *L. crenulata*, Moore.

Gard. Chron. 1862, p. 399 (advt.); Hooker's Sp. Fil. iii. 152.

A hardy evergreen fern introduced by Richard Pearce from Chili. The plant forms a close tuft about 6 in. high; the sterile fronds narrow, lanceolate, almost pinnate, with small oblong acute crenulate divisions, and the fertile fronds taller on reddish stalks, linear and crenulate.

LOMARIA GIBBA, *Lab.*, var. BELLI, *Hort.*

Veitchs' Catlg. of Pl. 1870, p. 35.

A crested form of the typical Lomaria gibba, with tasselled fronds, of distinct and elegant appearance.

LOMARIA GIBBA, *Lab.*, var. CRISPA, *Moore.*

Moore in Gard. Chron. 1868, p. 682; Veitchs' Catlg. of Pl. 1870, p. 35.

A form of dwarfish habit, so densely leafy and wavy that the edges of the pinnæ have a crisped appearance.

LOMARIA LECHLERI, *Moore.*

Moore in Gard. Chron. 1866, p. 634.

An evergreen hardy greenhouse fern obtained from Chili through Richard Pearce.

The fronds, spread into a head of 18 to 20 in. in diameter, reach from 1 ft. to 18 in. in height. The aspect of the plant is as a very rigid form of Lomaria lanceolata, but the texture of the fronds as well as the size and form of the fertile ones show the distinction clearly.

LYGODIUM POLYSTACHYUM, *Wall.*

Moore in Gard. Chron. 1859, p. 671.

An ornamental species discovered in Moulmein by Thomas Lobb at an elevation of 1,000 ft., in general appearance and habit resembling several species of Gleichenia.

MARATTIA BURKEI, *Baker.*

Gard. Chron. 1897, vol. xxii. p. 315 (Report of Show); Baker in Gard. Chron. 1897, vol. xxii. p. 427, fig. 129, p. 435.

Discovered by David Burke, probably in Columbia, but as no locality was given, the native country cannot be known with certainty.

It is closely allied to Marattia alata.

NEPHRODIUM HOPEANUM, *Baker.*

Syns. *Lastrea Hopeana*, T. Moore.

Moore in Gard. Chron. 1882, vol. xviii. p. 744.

Imported from the South Sea Islands, and named by Mr. Baker from specimens previously collected in Fiji by Lieut. Hope.

The slender stipes and bipinnatifid fronds cut into narrow falcate segments make this plant light and graceful, and particularly suited for decorative purposes.

NEPHRODIUM MAXIMOWICZII, *Baker.*

Syns. *Lastrea Maximowiczii*, Moore.

Moore in Gard. Chron. 1881, vol. xv. p. 626; Baker, Syns. Fil. 499.

Introduced from Japan, a hardy evergreen fern of dwarf habit, dense glossy green fronds, striate along the pinnæ on the upper surface.

NEPHRODIUM OPACUM, *Hort.*

Syns. *Lastrea opaca*, Hook.

Veitchs' Catlg. of Pl. 1862, p. 12; Gard. Chron. 1862, p. 399 (advt.).

A hardy species introduced from Yokohama through the late John Gould Veitch, with tufts of opaque, dark green fronds, more or less olive-green when young, bipinnately divided.

NEPHRODIUM RICHARDSII, *Baker*, var. MULTIFIDUM, *Hort.*

Syns. *Lastrea Richardsii*, var. *multifida*, Moore.

Moore in Gard. Chron. 1882, vol. xv. p. 104; Veitchs' Catlg. of Pl. 1881, p. 15, fig. p. 9; Fl. des Serres, 1880, tt. 2401, 2402.

A beautiful crested fern sent by Charles Moore, Esq., late of the Botanic Gardens, Sydney, N.S.W.

It differs from the type in having the frond, tips and pinnæ cut into numerous narrow finger-like segments; distinct and ornamental.

NEPHROLEPIS DAVALLOIDES, *Kunze*, var. FURCANS, *Hort.*

Veitchs' Catlg. of Pl. p. 22, 1876, fig. p. 46, 1877; Fl. and Pom. 1877, p. 18, fig.

A crested form of the beautiful Java fern, Nephrolepis davalloides, sent by Messrs. J. Baptist & Sons, of Sydney, N.S.W.

The plant is of robust growth, produces numerous arching fronds from 3 to 4 ft. long, bifurcate at the apex, as well as at the apex of each pinna.

NEPHROLEPIS DUFFII, *T. Moore.*

Moore in Gard. Chron. 1878, vol. ix. p. 622, fig. 113, p. 623; Fl. and Pom. 1878, p. 171, fig.

Introduced from Australia, and named by Dr. Moore in honour of Mr. Duff, an *employé* in the Sydney Botanic Gardens, who discovered the species on Duke of York's Island.

The numerous fronds arch gracefully, and are further forked at the extremity of the tips.

NEPHROLEPIS PLUMA, *T. Moore.*

Moore in Gard. Chron. 1878, vol. ix. p. 588, fig. 108, p. 589.

The tubers, peculiar to the species, were discovered in the sterile fronds of a Platycerium imported from Madagascar.

A deciduous species, the fronds entirely die in the winter months, and new ones arise in spring from the tubers in small fascicles.

The plume-like fronds from 4 to 5 ft. in length, 4 in. broad, are pendulous.

NEPHROLEPIS RUFESCENS, *Presl.*, var. TRIPINNATIFIDA, *Baker.*

Syns. *N. exaltata*, Schott.

Gard. Chron. 1887, vol. i. p. 476, figs. 90, 91, pp. 477, 481; Veitchs' Catlg. of Pl. 1887, p. 10, fig. p. 3.

A native of the Fiji Islands, and one of the most beautiful of all forms of the widely distributed type species.

A stove fern with tufted sub-erect fronds, 2 to 3 ft. high, and pinnæ cut in segments of endless variety of length and form.

NOTHOCHLÆNA MOLLIS, *Kunze.*

Gard. Chron. 1861, p. 499 (advt.).

This fine hardy greenhouse fern, indigenous to the South of Chili, is of tufted habit, with long narrow lanceolate fronds, bipinnate with crowded deeply lobed pinnules.

OSMUNDA JAVANICA, *Blume.*

Gard. Chron. 1894, vol. xvi. p. 761 (Report of Show); Veitchs' Catlg. of Pl. 1884, p. 74; Fl. and Pom. 1882, p. 122.

A handsome evergreen fern from Java, with erect fronds 1 to 2 ft. high, with crimson stipes and lance-shaped pinnæ 4 to 6 in. long, of leathery texture, and of a grass-green colouring.

OSMUNDA REGALIS, *L.*, var. CORYMBIFERA, *Hort.*

Syns. *O. japonica*, Thunb., var. *corymbifera*, Mast.

Gard. Chron. 1883, vol. xix. p. 466; Veitchs' Catlg. of Pl. 1883, p. 16, fig. p. 8; Fl. and Pom. 1882, p. 122; *id.* 1883, p. 104, fig.

A very distinct dwarf-growing variety of the Royal Fern, with the tips of the fronds and pinnæ peculiarly crested.

The fronds crowded, and the stipes, about 4 in. in height, erect, of a distinct rosy pink.

It was introduced from Japan through Charles Maries, has proved exceedingly ornamental for pot culture, and interesting as one of those forms with multifid fronds which are reproduced in the same character when raised from spores.

PELLÆA BELLA, *Baker.*

Syns. *Platyloma bellum*, Moore.

Moore in Gard. Chron. 1873, p. 213; Fl. and Pom. 1873, p. 157, fig.

A pretty greenhouse fern from California, allied to Pellea brachyptera, with smaller pinnæ and of more slender growth.

PLATYCERIUM VEITCHII

FERNS

PELLÆA BRACHYPTERA, *Baker.*

Syns. *Platyloma brachypterum*, Moore.

Gard. Chron. 1873, p. 141; Fl. and Pom. 1873, p. 157, fig.

A dwarf-growing cool greenhouse fern with fronds 4 to 6 in. in length, introduced from California, remarkable for the decided blue-green colour of the fronds.

PELLÆA GLAUCA, *J. Sm.*

Syns. *Cheilanthes glauca*, var. *hirsuta*, Moore.

Gard. Chron. 1861, p. 499 (advt.).

A beautiful coldhouse fern of dwarf habit, from the Andes of Peru, with tripinnate or even quadripinnate fronds, stiff on rigid foot-stalks of a deep brown colour.

PLATYCERIUM ALCICORNE, *Desv.*, var. HILLII, *Moore.*

Syns. *P. Hillii*, Moore.

Gard. Chron. 1878, vol. x. p. 429, figs. 74, 75; Veitchs' Catlg. of Pl. 1883, p. 16, fig. p. 9.

This Elk's Horn Fern remarkable for the decided green colour of the fronds, the foliage of Platyceriums usually of a greyish hue from the abundant white stellate scales spread over their surface.

P. Hillii is a native of Queensland, Australia, named in compliment to Mr. Walter Hill, late Superintendent of the Botanic Garden at Brisbane, through whom it was introduced.

PLATYCERIUM ALCICORNE, *Desv.*, var. MAJUS.

Veitchs' Catlg. of Pl. 1873, p. 13, fig. p. 9.

A fine form of the Stag's Horn Fern, from Australia.

The sterile fronds attain a large size, and the growth is more free than that of the type.

PLATYCERIUM ALCICORNE, *Desv.*, var. VEITCHII, *Hort.*

Gard. Chron. 1896, vol. xix. p. 652 (Report of R.H.S. Floral Committee).

A form introduced from Australia, distinct in having unusually stout erect fertile fronds, of leathery substance, narrower than those of other species in cultivation, and of a dark green colour.

POLYPODIUM ALBO-SQUAMATUM, *Blume.*

Syns. *Pleopeltis albido-squamata*, Presl.

Hooker's Garden Ferns, t. 47; Gard. Chron. 1863, p. 1180.

Introduced from Borneo, but previously only known as a native of Java.

The fronds from 1 to 3 ft. in length, droop gracefully and the pinnæ covered on the upper surface along the margin with small white scales suggested the specific name.

POLYPODIUM FOSSUM, *Baker.*

Syns. *Pleopeltis fossa*, Moore.

Moore in Gard. Chron. 1882, vol. xviii. p. 586; Veitchs' Catlg. of Pl. 1883, p. 73.

A native of the Malayan Archipelago, sent to Chelsea from the Botanic Garden at Leyden.

The gracefully arching fronds spring from a creeping rhizome, and, with the comparatively dwarf habit of the plant render the species desirable for clothing rockeries or for ferneries.

POLYPODIUM KRAMERI, *Franch. & Sav.*

Moore in Gard. Chron. 1881, vol. xv. p. 136; Veitchs' Catlg. of Pl. 1884, p. 74.

A pretty little hardy fern in the way of Polypodium phegopteris, introduced from Japan. A dwarf-growing species with slender stipes 3 to 4 in. long, suitable for rockwork, and an interesting companion to the native P. Phegopteris and the North American P. hexagonopterum.

POLYPODIUM NERIIFOLIUM, *Schk.*, var. CRISTATUM, *Hort.*

Veitchs' Catlg. of Pl. 1897, p. 7, with fig.

A distinct form of the South American Polypodium neriifolium, obtained by sowing the spores of that species with spores of a crested form of the common P. vulgare.

The fronds are from 3 to 4 ft. long, the pinnæ markedly crested or tasselled.

POLYPODIUM SCHNEIDERIANUM, *Hort. Veitch.*

Veitchs' Catlg. of Pl. 1896, p. 8, with fig.; Gard. Chron. 1894, vol. xv. p. 665; The Garden, 1894, vol. xlv. p. 472.

Raised by our *employé*, George Schneider, from Polypodium aureum and P. vulgare elegantissimum from spores of these two species sown together.

The plants show a decided blending of the characters of both the parents, and there is little doubt as to a hybrid origin. The fronds attain a length of 2 to 3 ft., are about 18 in. wide, with closely set, narrowly oblong pinnæ, and gently undulated margins.

PTERIS LONGIFOLIA, *L.*, var. MARIESII, *Hort.*

Veitchs' Catlg. of Pl. 1895, p. 9, with fig.

A variety of the old Indian fern, Pteris longifolia, sent by Charles

Maries, at the time Superintendent to the Maharajah Scindia of Gwalior at the State Gardens.

It differs from the type in having shorter fronds and more narrow pinnæ.

PTERIS LUDENS, *Wall.*

Gard. Chron. 1894, vol. xv. p. 761, fig. 101, p. 783; Veitchs' Catlg. of Pl. 1895, p. 85, fig.

A remarkable and at the same time beautiful stove fern, native of Malaysia and the Philippines, with fronds of two quite distinct forms. The barren ones vary in shape from triangular to hastate, more or less lobed; the fertile are of variable shape, on long stipes, more or less pinnatifid.

PTERIS PALMATA, *Willd.*

Syns. *Litobrochia nobilis*, Moore.

Gard. Chron. 1862, p. 932; Hooker's Garden Ferns, t. 22.

Introduced to this country from Rio de Janeiro by Mr. J. Wicks, and placed by him in our hands for distribution.

In the younger stages the fronds are marked with a greyish band, lost as they reach maturity; about 15 in. in length, the breadth across the centre is of the same proportion.

The fern is of large size for the group of net-veined Pterids to which it belongs, and to which the specific name nobilis in the older designation is due.

PTERIS QUADRIAURITA, *Retz.*, var. ARGYRÆA, *Moore.*

Syns. *P. argyræa*, Moore.

Gard. Chron. 1859, p. 671; Fl. Mag. 1861, pl. 4; Lowe's New and Rare Ferns, pl. x.

This, the first well-marked variegated fern introduced to cultivation, is still noteworthy for its distinct and novel character, due to the presence of a well-defined conspicuous stripe of silvery grey along the centre of each pinnæ.

From Central India through Thomas Lobb.

PTERIS SERRULATA, *L. f.*, var. CRISTATA.

Moore in Proc. R.H.S. vol. iii. p. 289; Gard. Chron. 1863, p. 1180.

A form introduced from Japan through the late John Gould Veitch, with the general habit of the typical species, but with fronds more or less bipinnately divided, giving a crested appearance.

SCOLOPENDRIUM VULGARE, *Sm.*, var. SCALARIFORME, *Hort.*

Veitchs' Catlg. of Pl. 1895, p. 11, with fig.

A pleasing form of the favourite old Harts' Tongue Fern, with curious crisped and corrugated margins of the fronds.

SELAGINELLA ATROVIRIDIS, *Spring.*

Gard. Chron. 1859, p. 603 (advt.).

This fine species of dwarf habit introduced from Borneo through Thomas Lobb has very dark green leaves.

SELAGINELLA CANALICULATA, *Baker.*

Veitchs' Catlg. of Pl. 1884, p. 16.

A robust caulescent species from the East Indies, with erect stems, about the thickness of an ordinary writing pencil; clothed with numerous leafy scales. The apical portion bright green, towards the base pale crimson.

SELAGINELLA CAULESCENS, *Spring.*

Gard. Chron. 1861, p. 499 (advt.).

Introduced from Central India.

The fronds bright green, somewhat rigid, of from 8 to 12 in. in height, and of neat, compact habit.

SELAGINELLA GRANDIS, *Moore.*

Gard. Chron. 1882, vol. xviii. p. 40, figs. 7, 8; Fl. and Pom. 1882, p. 123; Veitchs' Catlg. of Pl. 1883, p. 18, fig. p. 11.

This bold, handsome species, one of the most beautiful in cultivation, is remarkable for a deep grass-green colour and great width of frond.

The stems erect, produce numerous branches which arch gracefully to one side at the apex, and are, when fringed with the tail-like fructifications, very elegant.

It was introduced from Borneo through Charles Curtis, and exhibited for the first time before the Royal Horticultural Society under the provisional name of Selaginella platyphylla.

SELAGINELLA GRIFFITHII, *Spring.*

Gard. Chron. 1861, p. 499 (advt.).

Introduced from Borneo through Thomas Lobb, this dwarf, elegant species has pale green fronds with a beautiful metallic lustre. The stems grow 8 or 10 in. high, branch pinnately, and droop gracefully at their extremities.

SELAGINELLA LOBBII, *Moore.*

Syns. *S. cognata*, Hort.

Gard. Chron. 1859, p. 603 (advt.).

Interesting for the fern-like branches and rich lustrous metallic hue, and introduced from Borneo through Thomas Lobb.

FERNS

SELAGINELLA POULTERI, *Hort.*

Veitchs' Catlg. of Pl. 1868, p. 17, fig. p. 7.

A miniature form of Selaginella denticulata which it somewhat resembles, but the habit of growth is more dense.

TODEA FRASERI, *H. & G.*, var. WILKESIANA.

Syns. *T. Wilkesiana*, Brack.

Gard. Chron. 1870, p. 795, with fig.; Veitchs' Catlg. of Pl. 1871, p. 13, fig.

This beautiful miniature tree fern was first discovered by the United States' Exploring Expedition in Ovolau, one of the Fiji Islands, and subsequently imported by the Veitchian people.

It has a slender stem 18 to 20 in. high, surmounted by ten to twelve spreading fronds of a broadly-lanceolate outline, 2 ft. or upwards in length, of thin texture and pellucid character, distinguishable in this from others of the Leptopteris group to which it belongs.

TODEA MOOREI, *Baker.*

Syns. *T. grandipinnula*, Moore.

Gard. Chron. 1886, vol. xxv. p. 752; Veitchs' Catlg. of Pl. 1887, p. 13.

This handsome fern originated in the Chelsea fernhouse under circumstances which suggested that it might be a hybrid between Todea Fraseri and T. hymenophylloides, the form, however, subsequently proved to be identical with a species found on Lord Howe's Island.

The obvious and characteristic peculiarity of the plant is the leafy aspect of the fronds, the pinnæ of which much overlap at the edges on account of the free growth, and from their unequal development give an irregular outline.

TODEA SUPERBA, *Col.*

Syns. *Leptopteris superba*, Presl.

Moore in Gard. Chron. 1861, p. 697; Veitchs' Catlg. of Pl. 1870, p. 35.

This beautiful fern from New Zealand requires a cool greenhouse temperature and a moist shady position for successful cultivation.

The fronds cut into minute crowded segments, have an upward manner of growth and a rich moss-like appearance; from 18 to 24 in. in length, they curve gracefully from the centre, and the dark green of the older are a perfect contrast to the brighter verdure of the younger.

TRICHOMANES PLUMA, *Hook.*

Lowe's New and Rare Ferns, pl. lxiii.—A.

An exceedingly beautiful and rare species, found near Sarawak in Borneo by Thomas Lobb.

The rhizome, creeping and covered with bright brown scales, has fronds 4 to 6 in. long, most delicately cut in segments, which, when covered with the dew-like moisture essential to the growth of Filmy Ferns, amply justify the specific name—" plume-like " or " feathery."

WOODSIA POLYSTICHOIDES, *Eat.*, var. VEITCHII, *Hook.*

Hooker's Garden Ferns, pl. 32; Gard. Chron. 1862, p. 399 (advt.).

Found in Japan by the late John Gould Veitch. The narrow, almost linear fronds, 6 or 8 in. or more in height, are pinnately divided, the pinnæ about 1 in. in length, linear-oblong, and distinctly auricled with a row of sori near each margin, the surface above and beneath covered with short close hair.

WOODWARDIA ORIENTALIS, *Swartz.*

Gard. Chron. 1862, p. 399 (advt.); Veitchs' Catlg. of Pl. 1862, p. 14.

Raised from spores sent from Japan by the late John Gould Veitch. The fronds have on the upper surface, more or less profusely, bulbiform plants opposite the sori, sometimes sufficiently numerous to cover the whole of the frond.

CONIFEROUS TREES

CONIFEROUS TREES

ABIES BRACTEATA, *Nutt.*

Gard. Chron. 1853, p. 435; Bot. Mag. t. 4740; Fl. des Serres, tom. ix. t. 109; Gard. Chron. 1889, vol. v. p. 242, fig. 44, p. 241; Man. Con. 1900, ed. 2, p. 493, figs. 127, 128, and a full-page illus.

Introduced through William Lobb from the Santa Lucia mountains, South California, in 1853, first seen by the Scot, David Douglas, travelling in South California for the Horticultural Society of London, 1830-1832, and shortly afterwards by Dr. Thomas Coulter, but no ripe seed obtained.

Theodore Hartweg, at that time in the service of the Horticultural Society of London, again made the attempt in 1846, but with no better success, and it was not till 1852 that William Lobb sent seed to Exeter, and from plants raised in this country of this consignment originated all the older specimens of Abies bracteata in Europe.

In describing the species in the Botanical Magazine, Sir Joseph Hooker remarks, "Perhaps the introduction of no Conifer, not even that of the Deodar, has excited a more lively interest in horticulture and arboriculture than that of the present species with its porcupine fruits."

ABIES CONCOLOR, *Lindl. & Gord.*

Syns. *Picea concolor*, A. Murr.

Gard. Chron. 1890, vol. viii. p. 748, figs. 147, 148; Man. Con. 1900, ed. 2, p. 502, fig. 129.

Abies concolor was introduced from the Sierra Nevada of California through William Lobb in 1851, and at the same time seed was received by the Scottish Oregon Association from Southern Oregon.

The plants raised from Lobb's consignment were distributed under the name of A. lasiocarpa, and those sent to the Scottish Oregon Association as A. grandis.

ABIES FIRMA, *Sieb. & Zucc.*

Gard. Chron. 1861, p. 265; *id.* 1879, vol. xii. p. 198; Man. Con. 1900, ed. 2, p. 506, figs. 130-132.

Abies firma was introduced in 1861 by the late John Gould Veitch, and again in 1878 by Charles Maries.

Isolated specimens may be seen in favoured localities in this country, but in general, though on the plains of Japan one of the finest and largest of the Japanese Abies, this fir proved disappointing.

ABIES GRANDIS, *Lindl.*

Syns. *Picea grandis*, Loudon.

Man. Con. 1900, ed. 2, p. 512, fig.

The tallest tree of the genus from the valleys of Western Oregon never reaching a greater elevation than 4,000 ft.; found also in Vancouver Island and British Columbia, this Conifer spreads southwards to Mendocino in North California.

Abies grandis was discovered by David Douglas during an excursion to the Columbia River in 1830, and he sent seed to the Horticultural Society of London, but few germinated.

No further consignment of the species was received for a quarter of a century till William Lobb in 1851 sent a small quantity to Exeter; about the same time John Jeffrey was able to despatch a limited supply to the Scottish Oregon Association.

ABIES MAGNIFICA, *A. Murr.*

Syns. *A. nobilis*, Lindl., var. *robusta*, Carr.

Gard. Chron. 1885, vol. xxiv. p. 652, figs.; Man. Con. 1900, ed. 2, p. 517.

From seed of this species by John Jeffrey in 1851 the Scottish Oregon Association distributed specimens to the members as the Abies amabilis of Douglas.

In the following year William Lobb also sent seed as A. amabilis, but on germination the plants were found to differ from the species in cultivation under that name; these were distributed as A. nobilis robusta, the name adopted by Carrière in the second edition of his Traité Général des Conifères.

ABIES MARIESII, *Mast.*

Masters in Gard. Chron. 1879, vol. xii. p. 788, figs.; Man. Con. 1900, ed. 2, p. 520.

Discovered by Charles Maries on Mount Hakkoda near Aomori, the northern seaport of the main island of Japan, in 1878. The following year seed sent to Coombe Wood gave poor results.

On Hakkoda it is common at 4,000-5,000 ft. mixed with deciduous trees; in Nikko the fir ascends higher, but occurs sparingly.

Abies Mariesii is an alpine tree with a comparatively restricted habitat, occupying a geographical position between that of A. Veitchii and A. sachalinensis, the nearest affinity A. homolepis.

ARAUCARIA IMBRICATA

STRETE RALEGH, DEVON

CONIFEROUS TREES

ABIES MICROSPERMA, *Lindl.*

Lindl. in Gard. Chron. 1860, p. 22; Man. Con. 1900, ed. 2, p. 425.

The specific name microsperma was given by Lindley to a Spruce Fir brought from Hakodate by the late John Gould Veitch, a weakly plant unsuitable for the climate of the British Isles.

ABIES SACHALINENSIS, *Mast.*

Syns. *A. Veitchii*, Lindl., var. *sachalinensis*, Schmidt.

Gard. Chron. 1879, vol. xii. p. 588, fig. 97; Man. Con. 1900, ed. 2, p. 538, fig. 138.

Originally discovered by Freidrich Schmidt, a German botanical traveller, in the island of Saghalien, in 1866, and described by him as a variety of Abies Veitchii.

Nothing further was heard till 1878, when, re-discovered by Maries in Yeso, it was introduced the following year: the species has proved too tender to be a success.

ABIES VEITCHII, *Lindl.*

Syns. *A. Eichleri*, Lauche; *Picea Veitchii*, Hort.

Gard. Chron. 1861, p. 23; *id.* 1880, vol. xiii. p. 275, fig.; Man. Con. 1900, ed. 2, p. 541, fig.

Discovered in 1860 on Fuji-yama, the "sacred" mountain of the Japanese, named by Dr. Lindley in honour of the late John Gould Veitch, and in 1879 introduced by Charles Maries, when young plants were widely distributed.

Mayr observed cones in great abundance every third year, and in intermediate years, but few—the tree apparently unable to support the continual exhaustion of a heavy annual crop.

The same observer distinguishes two forms; the type, in which the apical end of the cone-bract is exserted and bent downwards, and the Nikko form, a local variety with smaller cones, the cone-bracts of which do not protrude beyond the scale.

ARAUCARIA IMBRICATA, *Pavon.*

Syns. *Dombeya chilensis*, Lamarck.

Masters in Gard. Chron. 1890, vol. vii. p. 587, figs.; Man. Con. 1900, ed. 2, pp. 296-302, figs.

Araucaria imbricata was discovered in 1780 by Don Francisco Dendariarena, a Spaniard, and later by two others, Doctors Ruiz and Pavon, who had been sent to Peru to investigate the forests on behalf of the Spanish Government. These explorers sent the first dried specimen to Europe, to a Frenchman, Dombey, after whom it was named. In 1795 Captain Vancouver reached the coast of Chili, and with him the botanist Archibald Menzies.

Menzies procured cones seed and young plants which he succeeded in bringing alive to Europe. He gave them to Sir Joseph Banks, and some found their way to the Royal Gardens, Kew; one lived in an unhealthy condition till the autumn of 1892, when it incontinently died.

For many years after Menzies' introduction of A. imbricata the Conifer remained scarce till William Lobb sent a large supply of seed in 1844, and the tree became generally planted; to this consignment many of the oldest specimens in this country are traceable.

ARAUCARIA RULEI, *F. Müller.*

Syns. *A. Niepratschki*, Hort. Lemoine.

Gard. Chron. 1861, p. 868, figs.; Man. Con. 1900, ed. 2, p. 304.

Originally discovered by William Duncan, a plant collector in the employ of Mr. John Rule of Melbourne, Victoria, about the year 1860, the tree, a native of New Caledonia, was found on the lofty summit of an extinct volcano, and introduced to British Gardens in 1863.

A Certificate of Merit was awarded by the Royal Botanic Society in June 1879, as Araucaria Niepratschki, on being exhibited by Messrs. Veitch.

CEPHALOTAXUS OLIVERI, *Mast.*

Syns. *C. Griffithii*, Oliver.

Masters in Jour. Linn. Soc. vol. xxvi. p. 545; *id.* in Gard. Chron. 1903, vol. xxxiii. p. 227, fig. 93.

A new species from China from seed collected in the Province of Hupeh, noticeable for the regularly two-ranked manner in which the leaves are disposed, in close approximation, as the teeth of a comb.

Yet too early to predict the position this tree will occupy in British Arboreta, but as a handsome, desirable addition to Coniferous subjects it should ever hold a high position.

CRYPTOMERIA JAPONICA, *Don*, var. ELEGANS, *Mast.*

Syns. *C. elegans*, Carrière.

Masters in Jour. Linn. Soc. vol. xviii. p. 498; Gordon's Pinetum, ed. 2, p. 73; Man. Con. 1900, ed. 2, p. 264, figs. 80, 81.

Introduced by the late John Gould Veitch, a most fortunate and distinct addition to Pineta.

In the southern and western counties of England this Conifer produces such masses of foliage that the weight causes the apex to incline.

CRYPTOMERIA JAPONICA, *D. Don*, var. LOBBII, *Carrière.*

Syns. *C. Lobbii*, Hort.

Man. Con. 1900, ed. 2, p. 264, fig.

This form introduced through Thomas Lobb, from the Buitenzorg

Botanic Gardens, Java, had been sent twenty years earlier by Siebold from Japan.

Of more compact habit than the type, the branches shorter and more ramified, the leaves longer, of a darker green.

CUPRESSUS OBTUSA, *C. Koch.*

Syns. *Retinospora obtusa*, Sieb. & Zucc.

Gard. Chron. 1861, p. 265; Man. Con. 1900, ed. 2, p. 220, fig. 64.

Sent from Japan by the late John Gould Veitch, probably the first timber-producer of that country: a favourite subject for the mutilation and dwarfing in which the Japanese delight, and in which they so greatly excel. The species and several varietal forms are in general cultivation.

CUPRESSUS OBTUSA, *C. Koch*, var. FILICOIDES, *Kent.*

Syns. *Retinispora nobleana*, Hort.; *R. filicoides*, Gordon.

Man. Con. 1900, ed. 2, p. 221, fig. 65; Gordon's Pinetum, ed. 2, p. 364; Gard. Chron. 1876, vol. v. p. 235, fig.

A variety received with the type from Japan by the late John Gould Veitch, with fern-like foliage and corrugated coruscated branchlets.

CUPRESSUS OBTUSA, *C. Koch*, var. LYCOPODIOIDES.

Syns. *Retinospora lycopodioides*, Gord.

Gordon's Pinetum, ed. 2, p. 364; Man. Con. 1900, ed. 2, p. 222.

Introduced by the late John Gould Veitch, with the type species, in 1861, the foliage that of a Lycopode.

CUPRESSUS PISIFERA, *C. Koch.*

Syns. *Retinispora pisifera*, Sieb. & Zucc.

Gard. Chron. 1861, p. 265; *id.* 1876, vol. v. p. 235, fig.; Man. Con. 1900, ed. 2, p. 224, figs.

Cupressus pisifera and all varieties of Japanese origin in cultivation are due to the late John Gould Veitch.

Many as C. aurea, C. filifera, C. plumosa, and C. squarrosa are among the most frequently met with, and the most decorative of Conifer. Though mostly found with C. obtusa, they are in old age scarcely distinguishable, but long recognized as distinct species by the Japanese, who designate each by a different vernacular name; C. obtusa is Hi-no-ki and C. pisifera, Sa-wa-ra.

CUPRESSUS PISIFERA, *C. Koch*, var. SQUARROSA, *Mast.*

Syns. *Retinispora squarrosa*, Sieb. & Zucc.

Masters in Jour. R.H.S. vol. xiv. 297; Gordon's Pinetum, ed. 2, p. 371; Man. Con. 1900, ed. 2, p. 225, fig. 69.

So very a distinct variety of Cupressus pisifera Siebold and Maximowicz held it a distinct species, the real origin unknown. Beissener pointed out that it was a juvenile form of C. pisifera from cuttings with primordial leaves only. Introduced by the late John Gould Veitch in 1861.

FITZROYA PATAGONICA, *Hook. f.*

Bot. Mag. t. 4616; Paxt. Fl. Gdn. vol. ii. p. 147, pl. 387; Man. Con. 1900, ed. 2, p. 198; Lindl. in Jour. Hort. Soc. London, 1851, vol. vi. p. 264.

From Valdivia in 1849 by William Lobb to the Veitchian firm, this species, in cultivation since introduction, has proved hardy, though not altogether a satisfactory subject for British Pineta, as the same climatic conditions essential for Saxe-gothæa conspicua and Libocedrus tetragona lack in the British Isles.

Richard Pearce affirmed this Conifer to be the Fitzroya which supplies the valuable alerze timber of the Chilians, and not Libocedrus tetragona as stated by most travellers.

Among the largest specimens in the British Isles are those at Killerton, Exeter; at Upcott, near Barnstaple; at Penjerrick, Cornwall; at Fota Island, Cork, and one at Kilmacurragh, Co. Wicklow.

JUNIPERUS CALIFORNICA, *Carr.*

Syns. *J. occidentalis*, Hook., *J. pyriformis*, Lindl.

Gard. Chron. 1855, p. 420; Man. Con. 1900, ed. 2, p. 167; Sargent's Silva of North America, vol. x. t. 517.

Introduced in 1853, at the same time as Sequoia gigantea, by William Lobb, discovered on the mountains of San Barnardino in California, where it forms a low tree 10 to 12 ft. high, and distributed under Dr. Lindley's name, Juniperus pyriformis, on account of the pear-shaped fruit. The climate of this country is unsuited to the tree, and few are in cultivation.

JUNIPERUS CHINENSIS, *Linn.* vars. ALBO-VARIEGATA, *Hort.*, and AUREA, *Hort.*

Man. Con. 1900, ed. 2, p. 169.

The varieties *albo-variegata* and *aurea* were first sent to this country by Robert Fortune, and subsequently by the late John Gould Veitch. The type reached the Royal Gardens, Kew, through a young gardener, William Kerr, at that time in the employ of the Government in China.

JUNIPERUS RIGIDA, *Sieb. & Zucc.*

Gard. Chron. 1861, p. 23; Man. Con. 1900, ed. 2, p. 188.

Juniperus rigida, introduced to gardens by the late John Gould Veitch in 1861, was found growing luxuriantly at a great elevation on the Hakone ridges, attaining a height of from 15 to 20 ft.: as a decorative object this Juniper has proved superior to many others.

CONIFEROUS TREES

KETELEERIA DAVIDIANA, *Franch.*

Syns. *Pseudotsuga Davidiana*, C. E. Bertr.; *Abies Davidiana*, Franch.

Gard. Chron. 1903, vol. xxxiii. p. 85, figs. 37, 38; Man. Con. 1900, ed. 2, p. 486.

A Chinese species of unusual interest discovered by the French missionary Père David, after whom it was named. Seed sent by Wilson in 1901 has germinated freely.

KETELEERIA FORTUNEI, *Carr.*

Syns. *Abietia Fortunei*, Kent.

Masters in Gard. Chron. 1887, vol. ii. p. 440; Man. Con. 1900, ed. 2, p. 485.

This remarkable Conifer originally discovered by Fortune in 1844, near a temple at Foo-chow-foo—only a single tree, was rediscovered in 1873 in the same locality by Dr. Hance, and five years later by Charles Maries, who found it in quantity on the coast range of Fo-Kien (Fu-chau) associated with Pinus massoniana.

LARIX LEPTOLEPIS, *Endl.*

Syns. *Abies leptolepis*, Sieb. & Zucc.

Gard. Chron. 1861, p. 23, 1883, vol. xix. p. 88, fig. 13; Man. Con. 1900, ed. 2, p. 398, fig. 102.

This larch, first found wild by the late John Gould Veitch during an ascent of Fuji-yama in 1860, was introduced the following year. Suitable for the climatic conditions which prevail in Scotland, and frequently planted in that country in preference to Larix europea.

LIBOCEDRUS MACROLEPIS, *Benth. & Hook.*

Gard. Chron. 1901, vol. xxx. p. 467; The Garden, 1902, vol. lxii. p. 183; Man. Con. 1900, ed. 2, p. 255.

Raised from seed collected at Szemao by Wilson, in 1900.

In a young state a singularly handsome species, mature trees are still more beautiful objects. In Southern Yunnan commonly planted in the courtyards of temple grounds, in a wild state this Conifer chooses ravines usually associated with a water-course.

Logs frequently found in the forest strata in a semi-fossilized condition are in this state valued by the Chinese as coffins for the higher classes.

It will only be hardy in these islands in the warm corners of South-West Ireland and Cornwall.

LIBOCEDRUS TETRAGONA, *End.*

Syns. *Thuia tetragona*, Hort.

Gard. Chron. 1849, p. 563 (Notice of New Plants); *id.* 1861, p. 505; Paxt. Fl. Gdn. 1850, p. 46, with fig.; Man. Con. 1900, ed. 2, p. 256, fig. 74.

Introduced through William Lobb from Chili in 1849, but rare in this country, owing to absence of that excessive atmospheric humidity prevalent in the districts the plant inhabits.

In the Gardeners' Chronicle (*l.c. supra*) it is erroneously described as the Alerze timber of the Chilians, a timber almost indestructible by weather.

PICEA AJANENSIS, *Fischer.*

Syns. *Abies ajanensis*, Kent; *A. microsperma*, Lindl.; *A. Alcockiana*, Hort. (in part.)

Masters in Gard. Chron. 1880, vol. xiii. p. 115; *id.* vol. xiv. p. 427, figs.; Bot. Mag. t. 6743; Man. Con. 1900, ed. 2, p. 425.

Picea ajanensis introduced in 1861 by the late John Gould Veitch from Japan, was distributed under the name of Abies Alcockiana from the unfortunate circumstance that the seed of both species came home mixed.

One of the handsomest species, in May loaded with cones of the brightest crimson.

PICEA ALCOCKIANA, *Carr.*

Syns. *Abies Alcockiana*, Veitch; *Veitchia japonica*, Lindl.

Gard. Chron. 1880, vol. xiii. p. 212, fig. 43 (Picea); *id.* 1861, p. 23 (Abies); *id.* 1861, p. 265 (Veitchia); Man. Con. 1900, ed. 2, p. 429.

Named in honour of Sir Rutherford Alcock, one time British Minister at Tokio, to whom the late John Gould Veitch, the discoverer, was indebted on his travels in Japan.

On Mount Fuji-yama, at an altitude of 6000-7000 ft., seed was saved and sent home in 1860.

Rare in British Pineta, and at one time frequently confused with Picea ajanensis, the seed of the two species having been sent in one consignment.

The genus Veitchia, of Lindley, was founded on imperfect specimens of two mutilated cones and a few seeds of this fir, sent by the late John Gould Veitch, and in publishing a description of this supposed new genus Lindley says:—"We cannot do otherwise than associate with this extraordinary genus the name of John Gould Veitch, its active and intelligent discoverer, the introducer of so many fine trees previously unknown in this country."

PICEA POLITA, *Carr.*

Syns. *Abies polita*, Sieb. & Zucc.

Masters in Jour. Linn. Soc. vol. xviii. p. 507; Gard. Chron. 1880, vol. xiii. p. 233, fig. 44; Man. Con. 1900, ed. 2, p. 446, fig.

A native of Japan, found by the late John Gould Veitch in 1861. Rare in its native isles and only as isolated specimens of a miserable aspect, it is yet widely scattered over the mountainous districts from the extreme south to the 38th parallel of the north latitude.

PICEA POLITA

PENCARROW, CORNWALL

It thrives poorly in the too dry atmosphere of England; better in New Zealand, where it has been introduced, and it is cultivated by the Japanese for the decoration of gardens and temple enclosures.

The specific name polita, "polished or adorned," was selected in reference to the lustrous smoothness of the leaf and leaf-bud.

PINUS ARMANDI, *Franch.*

Masters in Gard. Chron. 1903, vol. xxxiii. p. 66, figs. 30, 31.

Raised at Coombe Wood from seed collected in the Province of Hupeh, Central China, by Wilson in 1900, this five-leaved pine of the Cembra section has a smooth bark, slender leaves, and oblong cones.

It had already been met with in China by the Pères David, Farges, and Delavay.

PINUS COULTERI, *Don.*

Syns. P. *macrocarpa*, Lindl.

Man. Con. 1900, ed. 2, p. 325; Lindl. Bot. Reg. 1840, vol. xxxi. misc. 61.

Though discovered by Dr. Thomas Coulter in 1832 on the west side of the Saint Lucia at 3,000-4,000 ft. elevation, it was not till David Douglas sent seed and specimens from the same locality to the Horticultural Society of London, under the name of Pinus Sabiniana, that this pine was known, and from this seed was raised the oldest specimens growing in Great Britain.

A further quantity subsequently collected was sent by William Lobb in 1851-1852.

The species is remarkable for the large size of the cones, sparingly produced.

PINUS DENSIFLORA, *Sieb. & Zucc.*

Gard. Chron. 1861, p. 265; Man. Con. 1900, ed. 2, p. 327.

Pinus densiflora, introduced to British gardens in 1861 by the late John Gould Veitch, had previously been grown in the horticultural establishment founded by Dr. Siebold at Leyden, but the pine was not in general cultivation.

PINUS KORAIENSIS, *Sieb. & Zucc.*

Gard. Chron. 1861, p. 1114; Gordon's Pinetum, ed. 2, p. 306; Man. Con. 1900, ed. 2, p. 334, fig. 94 (cone).

Introduced from Japan in 1861 by the late John Gould Veitch, but not endemic to the islands—its home the neighbouring peninsula of Corea.

In 1892 James H. Veitch met with several medium-sized specimens when crossing that peninsula, and in 1899 seed was gathered in Yuen-chiang, in South China, by Wilson.

PINUS PARVIFLORA, *Sieb. & Zucc.*

Gard. Chron. 1861, p. 265; *id.* 1878, vol. x. p. 624, fig. 103; Man. Con. 1900, ed. 2, p. 353.

Introduced to British Gardens by the late John Gould Veitch in 1861.

Cultivated everywhere in pots throughout Japan, dwarfed and distorted in every way, trained to every conceivable monstrosity, this pine when in the forest groves is a light and graceful object. The small size, well furnished trunk and light foliage are adaptable to small lawns.

PINUS PENTAPHYLLA, *Mayr.*

Man. Con. 1900, ed. 2, p. 356.

A species closely allied to Pinus parviflora, endemic to Japan.

Cones brought home by the late John Gould Veitch were supplemented by a few seeds from Charles Maries in 1879, from which plants raised at Coombe were subsequently distributed as P. parviflora, but their destination is unknown.

PINUS THUNBERGII, *Parlatore.*

Syns. *P. massoniana*, Sieb. & Zucc.

Man. Con. 1900, ed. 2, p. 384, fig.

Pinus Thunbergii, sent to Europe by Siebold in 1855, and to Great Britain with P. densiflora by the late John Gould Veitch, was unfortunately distributed as P. massoniana.

The curiously trained trees seen by Siebold, Maries, and others, probably belong to this species. One of the most remarkable may be seen in front of the Naniwaja tea-house, another in the village of Karasaki, and a third and most curious of all in the garden attached to the monastery at Kinkakuja, trained in the form of a Chinese junk.

A figure of this extraordinary tree, which has taken over three centuries of patient labour to produce, is given in the Manual of Coniferæ (*l.c. supra*) and as a photogravure plate in Traveller's Notes.

PODOCARPUS MACROPHYLLUS, *D. Don*, vars. ARGENTEO and AUREO-VARIEGATUS, *Kent.*

Man. Con. 1900, ed. 2, p. 151; Gordon's Pinetum, ed. 2, p. 331.

The type species first became known to science in the early part of the eighteenth century, through Kæmpfer, and the two variegated forms above named are due to Fortune: re-introduced by James H. Veitch in 1892.

PODOCARPUS NUBIGENA, *Lindl.*

Paxt. Fl. Gdn. 1851, p. 162, fig. 128; Jour. Hort. Soc. London, 1851, vol. vi. p. 264; Jour. R.H.S. xiv. p. 234; Gard. Chron. 1891, vol. x. p. 171, fig. 23; *id.* 1902, vol. xxxi. p. 113, fig.; Man. Con. 1900, ed. 2, p. 153.

Discovered in Southern Chili by William Lobb in 1846, and introduced the following year to Exeter.

A disappointing subject under cultivation, the climatic conditions of this country evidently unsuited to its requirements.

PRUMNOPITYS ELEGANS, *Philippi.*

Syns. *Podocarpus andina*, Pœpp.

Lindl. in Gard. Chron. 1863, p. 6; Man. Con. 1900, ed. 2, p. 155; Masters in Jour. R.H.S. vol. xiv. p. 244; Gard. Chron. 1902, vol. xxxi. p. 113, fig.

Introduced in 1860 through Richard Pearce, this Conifer has proved hardy over the greater part of Great Britain wherever planted, and in Ireland.

Specimens are growing at Eastnor Castle; Tortworth Castle; Menabilly, Cornwall; Kilmacurragh, Co. Wicklow; in Fota Island, and Lakelands, Co. Cork.

SAXEGOTHÆA CONSPICUA, *Lindl.*

Gard. Chron. 1887, vol. ii. p. 684; figs. 130, 131; *id.* 1889, vol. v. p. 782, fig.; Paxt. Fl. Gdn. vol. i. p. 111, figs.; Jour. Hort. Soc. London, 1851, vol. vi. p. 258; Man. Con. 1900, ed. 2, p. 158.

Discovered by William Lobb in 1846 in Southern Chili, this Conifer attracted great interest on introduction, but hopes entertained of a distinct addition to British Arboreta were not realized, and the Saxegothæa is now rarely seen.

SCIADOPITYS VERTICILLATA, *Sieb. & Zucc.*

Gard. Chron. 1861, p. 22; Fl. des Serres, 1861, tom. iv. p. 241; Nicholson in Woods and Forests, 1884, vol. i. p. 132, figs.; The Garden, 1890, vol. xxxviii. p. 499, fig.; Man. Con. 1900, ed. 2, p. 287, figs.; Bot. Mag. t. 8050.

Thomas Lobb sent the first living specimen of this remarkable tree in 1853, having met with it in the Botanic Gardens at Buitenzorg, Java; it arrived at Exeter in feeble health, and shortly died. In 1861 the late John Gould Veitch brought home seed, and to this source most of the older specimens growing in this country may be traced. It was seen by James H. Veitch in the Province of Mino on the Nakasendo below Nakatsu-gawa, growing by the roadside, but that the trees were growing naturally is improbable. Except in the neighbourhood of Nakatsu-gawa, the Sciadopitys is not regarded as a garden plant in Japan, and is not often seen in old gardens, but usually in the neighbourhood of temples.

Wherever rhododendrons thrive, this fine Conifer may be planted, as the conditions for successful cultivation of the former are suited to the requirements of the latter, and good specimens, once the tree becomes established, are soon formed.

SEQUOIA GIGANTEA, *Torr.*

Syns. *S. Wellingtonia*, Seem.; *Wellingtonia gigantea*, Lindl.; *Washingtonia californica*, Winsl.

Lindl. in Gard. Chron. 1853, p. 823; Gard. Chron. 1853, p. 819; *id.* 1854, p. 118 (Note on wood and cone); *id.* 1855, p. 70 (a Lecture on Wellingtonia gigantea); Fl. des Serres, tom. ix. p. 93; *id.* p. 121; Man. Con. ed. 2, p. 275, figs. 84, 85.

This, the Wellingtonia, or Mammoth Tree, giant in the forest primeval, the largest of all coniferous subjects, unsurpassed by any of any other Natural Order, with the possible exception of the Eucalypti of Western Australia, was probably first discovered by John Bidwill in 1841, but nothing definitely known till in 1852 it was again met with by the hunter Dowd.

First introduced to Europe through William Lobb, who sent seed and a living specimen to Exeter in 1853.

A full account of the introduction of this tree to this country is given in Veitchs' Manual of Coniferæ (*l.c. supra*).

THUIA DOLABRATA, *Linn.*

Syns. *Thujopsis dolabrata*, Sieb. & Zucc.

Gard. Chron. 1861, p. 23 (Thujopsis); *id.* 1882, vol. xviii. p. 556, figs.; Man. Con. 1900, ed. 2, p. 236.

Thuia dolabrata, first known to Europeans through the Swedish botanist Thunberg, who gathered specimens, communicated to Linnæus, in Japan in 1776. These specimens became the property of the Linnæan Society of London, and descriptions were published by David Don in Lambert's Genus Pinus in 1828.

The first living plant to reach England was sent by Thomas Lobb in 1853 from the Botanic Garden at Buitenzorg in Java; arriving in an exhausted condition, all attempts to save it proved fruitless. Shortly afterwards a plant received by Captain Fortescue, planted in Devonshire, had better fortune, but it was not till the late John Gould Veitch in 1861 sent seed to Chelsea, and Fortune some to Ascot, that T. dolabrata became generally distributed, and could take that high rank as an ornamental tree it has won in this country. In the Japanese hill districts bordering the shores of Lake Yumoto, it is the forest carpet.

THUIA DOLABRATA, *Linn.*, var. LÆTEVIRENS, *Mast.*

Syns. *Thujopsis lætivirens*, Lindl.

Lindl. in Gard Chron. 1862, p. 428; Gordon's Pinetum, ed. 2, 399 (as T. dolabrata nana); Masters in Jour. Linn. Soc. vol. xviii. 486; Man. Con. 1900, ed. 2, p. 237.

Sent from Japan by the late John Gould Veitch, a dwarf-spreading

THUYOPSIS DOLOBRATA

KILLERTON, DEVON

shrub with more slender branches, more divided than the type, and smaller leaves of a brighter green.

THUIA GIGANTEA, *Nutt.*

Syns. *T. Lobbii*, Hort.

Nicholson in Woods and Forests, 1884, p. 190, with fig.; Man. Con. 1900, ed. 2, p. 239, fig. 72, and full-page illus.

Thuia gigantea was known prior to 1807, in which year James Don mentions it in Hortus Cambridgiensis under T. plicata. It again appeared under the same name in Lambert's Genus Pinus, published in 1828, and in 1834 as T. gigantea in Michaux' work on North American Trees, in a description by Thomas Nuttall. David Douglas, when travelling for the Horticultural Society of London, 1825-1827, named it T. Menziesii, after his countryman, and later, the tree received the name of T. excelsa from Borgord, who described it from specimens collected by Mertens on the island of Sitka.

From the first consignment of seed received at Exeter in 1853, plants were distributed under the name of T. Lobbii, Hort.

TORREYA CALIFORNICA, *Torr.*

Syns. *T. Myristica*, Hook..

Gard. Chron. 1854, p. 519; *id.* 1884, vol. xxii. p. 681, fig. 116; Bot. Mag. t. 4780; Fl. des Serres, tom. ix. p. 175; Man. Con. 1900, ed. 2, p. 117.

A native of the elevated regions in the Sierra Nevada, California, by William Lobb rediscovered, and seed and specimens sent in 1851.

The credit of the first information is to David Douglas, who met with specimens while travelling for the Horticultural Society of London, but he failed at introduction. One of the best may be seen at Tortworth Court, Gloucestershire.

TSUGA DIVERSIFOLIA, *Maxim.*

Syns. *Abies Tsuga*, Sieb. & Zucc.

Gard. Chron. 1861, p. 23; Man. Con. 1900, ed. 2, p. 467.

Raised from cones from Japan by the late John Gould Veitch in 1861, unfortunately mixed with those of Tsuga Sieboldii, and both species were cultivated under the names of Abies Tsuga and A. T. nana. The introducer, unknown to himself, was the discoverer.

TREES AND SHRUBS DECIDUOUS AND CLIMBING PLANTS

TREES AND SHRUBS—DECIDUOUS AND CLIMBING PLANTS

ABUTILON VITIFOLIUM, *DC.*

Syns. *Sida vitifolia*, Cav.

Lindl. Bot. Reg. vol. xxx. t. 57; Bot. Mag. t. 4227 and t. 7328; The Garden, 1883, vol. xxiii. p. 224, pl. 369; *id.* 1897, vol. li. p. 334, pl. 1117; Gard. Chron. 1889, vol. vi. p. 156, fig. 21.

A handsome deciduous shrub or small tree from seed collected in Chili by William Lobb in 1844, first flowered under glass in 1845, and said to have been brought to Europe by Captain Cottingham, in 1836, and cultivated in his garden in Dublin, where it proved to be quite hardy.

In England it needs the protection of a greenhouse or wall except in favoured localities. In colour the flowers are variable, sometimes white and occasionally blue, a peculiarity that has caused some confusion in the nomenclature. Both forms are figured in the Botanical Magazine (*l.c. supra*), on separate plates, and also in The Garden, 1897 (*l.c. supra*).

ACER ARGUTUM, *Maxim.*

Nicholson in Gard. Chron. 1881, vol. xv. p. 725, fig. 132; Sargent's Trees and Shrubs, 1903, vol. i. p. 131, pl. lxvi.

A very fine maple introduced through Charles Maries from Japan, with leaves acutely five-pointed, and the wings of the samara widely apart.

It is hardy in this country and in the United States of America. The light green graceful foliage in summer and the purple branches in winter render it an attractive addition to hardy ornamental trees.

ACER CARPINIFOLIUM, *Sieb. & Zucc.*

Nicholson in Gard. Chron. 1881, vol. xv. p. 565, fig. 105.

A very distinct maple, in the leaf-form and veining closely resembling the Hornbeam (Carpinus), from which is derived the specific name.

A rare tree in Japan, whence it was introduced through Charles Maries, probably the only really respectable specimen is at Coombe Wood.

ACER CRATÆGIFOLIUM, *Sieb. & Zucc.*

Nicholson in Gard. Chron. 1881, vol. xvi. p. 75 ; Flora Japonica, t. 147.

A slender tree with unequally lobed bright green leaves about 3 in. long, not unlike those of the Hawthorn, introduced from Japan through Charles Maries.

ACER CRATÆGIFOLIUM, *Sieb. & Zucc.*, var. VEITCHII, *Nichols.*

Nicholson in Gard. Chron. 1881, vol. xvi. p. 75.

A vigorous form of the type with leaves finely marked white and rose on a dark green ground, from Japan in the same consignment.

ACER DAVIDII, *Franch.*

Jour. R.H.S. 1904, vol. xxix. p. 348, figs. 86, 90; Sargent's Trees and Shrubs, pt. iv. pl. lxxxiii.

Introduced to Coombe Wood through Maries from North China and distributed as Acer sp. for many years. From seed sown in January 1902, collected in Central China by Wilson, this species was again raised, and found to be identical with Maries' unnamed species. Herbarium specimens having also been sent by the last-named collector, made it possible to identify the species as A. Davidii of Franchet.

A variable plant, the leaves sometimes attain a length of 8 in. with a breadth of 5 in., on the first appearance of a reddish-bronze tint, they become when mature rich shining green with yellowish-green veins.

ACER DIABOLICUM, *K. Koch.*

Syns. *A. pulchrum*, Hort.

Nicholson in Gard. Chron. 1881, vol. xv. p. 532, fig. 100; Sargent's Trees and Shrubs, 1903, vol. i. p. 133, pl. lxvii.

A noble maple somewhat resembling Acer platanoides, from Japan through Charles Maries. This large-leaved species grows rapidly in this country, apparently quite at home, and attracts attention by the very large and bold foliage.

The specific name is said to have been suggested by the two horn-like processes between the wings of the " keys."

ACER DISTYLUM, *Sieb. & Zucc.*

Nicholson in Gard. Chron. 1881, vol. xv. p. 499, fig.

A distinct maple with simple leaves which attain a large size in this country, introduced from Japan through Charles Maries; it succeeds admirably, and is superior to many species and varieties in ordinary cultivation.

The "keys" in erect racemes form a feature during the early summer months when the light green colour is relieved by the dark green of the foliage.

ACER FRANCHETI, *Pax.*

Jour. R.H.S. 1904, vol. xxix. p. 353, fig. 88; Sargent's Trees and Shrubs, pt. iv. pl. lxxxvii.

A species allied to the Himalayan Acer villosum, from which it may be distinguished by trilobed leaves, the small teeth on the margin and the simple inflorescence. A common tree in Hupeh, Central China, raised from seed collected in that locality by Wilson.

ACER GRISEUM, *Pax.*

Gard. Chron. 1903, vol. xxxiii. p. 100; Jour. R.H.S. 1904, vol. xxix. p. 354.

A handsome species, the Chinese representative of a maple found in Japan, Acer nikoense.

Attaining in Central China a height of from 15 to 40 ft., the young foliage is beautifully coloured in the early spring, and the bark peels as in the common Silver Birch.

Discovered by Dr. Henry, and introduced to cultivation through Wilson, who sent seed in 1901.

ACER HENRYI, *Pax.*

Hooker's Ic. Pl. t. 1896; Gard. Chron. 1903, vol. xxxiii. p. 100; Sargent's Trees and Shrubs, pt. iv. p. 181.

A Chinese species of the Negundo section of Acer, specimens of which were first collected in the Province of Hupeh by Dr. Henry.

The species is remarkable for the great length of the samara in a young state of a bright red colour. The petioles of the long leaves have three ovate acuminate leaflets.

Young plants at Coombe were raised from seed collected in Central China in 1903.

ACER LÆTUM, *C. A. Mey.*, var. CULTRATUM, *Pax.*

Jour. R.H.S. 1904, vol. xxix. p. 354, fig. 101.

This form with five-lobed leaves of a lively shining green colour, borne on a rather dwarf-growing plant, was raised at Coombe Wood in 1902 from seed collected in Central China by Wilson.

ACER LÆTUM, *C. A. Mey.*, var. TRICAUDATUM, *Rehd.*

Jour. R.H.S. 1904, vol. xxix. figs. 100, 102.

A new form discovered in Central China named by Mr. A. Rehder of the Arnold Arboretum, Mass., U.S.A.

The leaves are three to five lobed, the lower pair often obsolete; the apices acutely pointed, and the petiole, bright rose-pink, contrasts strongly with the dark green of the blade.

It was raised at Coombe Wood from seed sown in February 1902.

ACER LÆVIGATUM, *Wall.*, var. FARGESII, *Rehd.*

Jour. R.H.S. 1904, vol. xxix. p. 353, fig. 91.

A form of the Indian Acer lævigatum of dwarfer habit and with smaller leaves, discovered on mountains south of the Yangtsze in the Province of Hupeh, Central China, and raised at Coombe in 1902.

The leaves leathery in texture, entire or slightly notched, are when first produced of a bright red and very ornamental colour.

Native of mountainous regions, this variety will probably prove to possess a hardier constitution than the Indian type, which will not withstand English winters.

ACER NIKOENSE, *Maxim.*

Syns. *Negundo nikoense*, Nichols.

Nicholson in Gard. Chron. 1881, vol. xvi. p. 815; Garden and Forest, 1893, p. 155, fig. 26.

This remarkable species, named after the district in Japan where it was first discovered, has a wide distribution, and is in all probability of Chinese origin, as it has lately been met with in the mountainous region of Central China, undoubtedly wild. The trifoliate leaves, densely hairy on the under surface, are in autumn a rich vinous red.

The samara are large with spreading wings, and the cells like the other parts of the plant thickly covered with hair.

It is quite hardy at Coombe Wood, where a specimen, one of the first raised, is now over 20 ft. high and 25 ft. through.

ACER PALMATUM, *Thunb.*

Syns. *A. polymorphum*, Sieb. & Zucc.

Veitchs' Catlg. of Pl. 1877, p. 80, figs.

Acer palmatum, known in gardens as A. polymorphum, was re-introduced by the late John Gould Veitch from Japan (though it had formerly been introduced through Dr. Siebold) as well as many and various forms. The varieties known as *atropurpureum dissectum, palmatifidum, roseo-marginatum,* and *sanguineum* were distributed in 1877.

ACER PICTUM, *Thunb.*, var. CONNIVENS, *Nichols.*

Nicholson in Gard. Chron. 1881, vol. xvi. p. 375.

A form of Acer pictum, from Japan, lobes of the leaves cut to the base of the blade, and the nerves on the under surface pubescent, bearded in the axils; the wings of the fruit erect with converging tips.

ACER PICTUM, *Thunb.*, var. MONO, *Pax.*

Jour. R.H.S. 1905, vol. xxix. pts. i., ii., and iii.

A form of the type introduced from the Province of Hupeh, Central China, through Wilson.

The leaves three to five lobed, with attenuated apices; the upper surface shining green and the under covered with a soft velvety pubescence.

From seed sown in February 1901.

ACER RUFINERVE, *Sieb. & Zucc.*

Nicholson in Gard. Chron. 1881, vol. xv. p. 42; Flora Japonica, vol. ii. t. 158.

Introduced from the mountains of the Central Island of Japan through Charles Maries.

A large tree with leaves variable in size and outline, resembling in habit the common Sycamore. The specific name refers to the reddish hairs which clothe the veins on the under surface of the leaf, and to the red colour of the peduncles and petioles.

ACER SINENSE, *Pax.*, var. CONCOLOR, *Pax.*

Jour. R.H.S. 1904, vol. xxix. p. 348, fig. 92.

A new form of the type with leaves of a large size coloured green on both surfaces, detected by Wilson in South Wushan on the sides of streams, raised at Coombe from seed sown in April 1901.

The leaves are somewhat as those of Acer pictum, five-lobed with acuminate apices, reddish in colour when first produced, when mature bright green.

ACER SUTCHUENENSE, *Franch.*

Jour. R.H.S. 1904, vol. xxix. p. 353, figs. 93, 96.

Discovered by Dr. Augustine Henry in the Province of Szechuan, close to the border of Hupeh, and introduced to cultivation through Wilson.

It is closely allied to Acer Henryi, a trifoliate species, but differs in the inflorescence.

ACER TETRAMERUM, *Pax.*, var. LOBULATUM, *Rehd.*

Jour. R.H.S. 1904, vol. xxix. p. 353, fig. 94.

A graceful variety with leaves resembling those of a Birch, coarsely serrated along the margin; discovered in Central China, and raised at Coombe from seed sown in April 1901.

HORTUS VEITCHII

ACTINIDIA CHINENSIS, *Planch.*

Maries in The Garden, 1882, vol. xxi. p. 101; Gard. Chron. 1903, vol. xxxiii. p. 248; *id.* vol. xxvi. p. 211; Jour. R.H.S. vol. xxviii. pt. i.; Veitchs' List of Novelties, 1904, p. 3, figs.

A handsome hardy climber introduced from Central China of the very first order.

It was known for some time prior to introduction, having been found by Fortune when travelling on behalf of the Royal Horticultural Society, and from his specimens Planchon's description is written in Hooker's London Journal of Botany, 1847, vol. vi. p. 303. Maries also detected it in the North Island of Japan, but failed to introduce it.

A rapid grower, valuable for very handsome foliage, covered with bright red hairs in a young state. The flowers, not yet seen in cultivation, are bright yellow, very handsome, and followed by edible fruits about the size of walnuts with a flavour resembling ripe gooseberries.

ACTINIDIA KOLOMIKTA, *Ruprecht.*

Masters in Gard. Chron. 1880, vol. xiv. p. 262; Rev. Hort. 1872, p. 395, fig. 43.

An interesting semi-scandent shrub, by Maries, who met with it in Yesso, Japan, in the year 1878, though it had previously been discovered on several occasions, and had appeared in France a few years earlier.

Now seldom met with in gardens, it deserves a place for the brilliant crimson and white autumn tints.

ÆSCULUS CALIFORNICA, *Nutt.*

Bot. Mag. t. 5077; Gard. Chron. 1902, vol. xxxi. p. 187, fig. of fruit; Fl. des Serres, 1858, tom. iii. 2nd ser. p. 39.

This, one of the rarest of the horse-chestnuts, a native of the western slopes of the mountains of California, where it is known as the Californian Buck's Eye, was raised from seed sent to Exeter by William Lobb, and produced fine thyrsi in July 1858.

Unfortunately, though hardy in this country, the absence of sufficient sun prevents the tree freely flowering.

A specimen in the Bath Botanic Garden produced both flowers and fruit in 1901, and from this material the figure in the Gardeners' Chronicle (*l.c. supra*) was prepared.

According to Professor Sargent, who describes the species as one of the most beautiful in the genus, the flowers are white tinted with rose, but in this country they are wholly white.

ÆSCULUS TURBINATA, *Blume.*

Rev. Hort. 1888, p. 496, figs. 120-124; Gordon in Gard. Mag. Sept. 21st, 1901, p. 614; Gard. Chron. 1902, vol. xxxi. p. 187, fig.

This Japanese Horse-chestnut closely resembles our own (Æsculus

Hippocastanum), but is readily distinguished by the greyish under surface of the leaves. Though often confused with Æ. chinensis, it is probable that that species is not in cultivation in this country.

One of the first specimens to be reared is growing at Coombe Wood, where it has attained a height of 16 ft., with a head 12 ft. in diameter and a circumference of stem at 3 ft. from the ground of 12 in.

It flowered in 1901.

ARISTOLOCHIA HETEROPHYLLA, *Hemsl.*

Hemsley in Jour. Linn. Soc. vol. xxvi. p. 361.

A quick-growing climber with variable leaves, raised from seed collected in Central China.

The species was first detected by Dr. A. Henry, and from this material Mr. Hemsley described it (*l.c. supra*):—"In foliage this resembles A. Kæmpferi, Willd., presenting similar variations; but the narrower and somewhat smaller perianth differs in the very sharply bent tube, and the equally 3-lobed limb similar to that of the North American A. Sipho, Ait."

BETULA ALNOIDES, *Ham.*, var. PYRIFOLIA, *Franch.*

The type species is found in Southern and Western China, and occurs also in Northern India and Upper Burmah.

The variety, as the name indicates, has leaves resembling a species of Pyrus, and was introduced to cultivation through Wilson, from seed collected in the Province of Hupeh, Central China.

BETULA MAXIMOWICZII, *Regel.*

Veitchs' Catlg. of Trees and Shrubs, 1888-1889.

The finest of the Japanese Birches, introduced to cultivation through James H. Veitch, who met with it in the neighbourhood of Hokkaido.

A tree 80 to 90 ft. high, in its native habitat it is easily distinguished from other birches by a pale smooth orange-coloured bark.

BRANDISIA RACEMOSA, *Hemsl.*

Hooker's Ic. Pl. t. 2383.

Introduced through Wilson from Central China, but so far of little use in cultivation.

It is an unusually handsome plant, as a reference to the plate (*l.c. supra*) will at once show, and it forms a dwarf shrub producing masses of densely flowered racemes of bright scarlet blossom.

At Coombe Wood it flowered imperfectly for the first time in 1904.

BUDDLEIA ALBIFLORA, *Hemsl.*

Gard. Chron. 1902, vol. xxxii. pp. 118, 139; Veitchs' List of Novelties, 1904, p. 5, fig.

A species named by Mr. Hemsley from specimens collected in Central China by Dr. Henry, whose notes state "Flowers white," and on this account the specific name albiflora was chosen.

The flowers, however, are not white, but pale mauve with an orange-yellow throat.

BUDDLEIA NIVEA, *Duthie.*

Gard. Chron. 1905, vol. xxxviii. p. 275, fig. 102.

A new species from Central China and of doubtful promise.

The flowers not so striking as those of some of the species recently introduced, but this defect is compensated for by the great beauty of the foliage, the whole under surface of which is, together with the young wood and leaves, covered with a dense white woolly tomentum.

The flowers in tail-like panicles at the ends of the branch are rose-purple in colour, individually small, but in a mass conspicuous.

BUDDLEIA VARIABILIS, *Hemsl.*, var. MAGNIFICA, *Hort.*

Gard. Chron. 1905, vol. xxxvii. p. 115 (Report of R.H.S. Floral Committee); *id.* p. 157 (Report of R.H.S. Floral Committee); Flora and Sylva, 1905, vol. iii. p. 339.

A very striking form of the well-known type introduced from the Province of Hupeh, Central China, through Wilson.

It differs from the brilliant *Veitchiana*, which it equals in size of flower-spike, in profusion of bloom, in a more constricted thyrsus, and in having flowers of a deeper richer shade of violet-purple.

BUDDLEIA VARIABILIS, *Hemsl.*, var. VEITCHIANA, *Hort.*

Veitchs' List of Novelties, 1903, p. 4, fig.

A superor form of the variable typical species, introduced from the Province of Hupeh, Central China.

The plant is more robust than the type, the flower spike larger, and more richly coloured;—the spikes in some instances measure over 2 ft. in length, and form continuous masses of bloom, not divided into globose axillary clusters as in the type.

Perfectly hardy in this country, it has been extensively planted since 1901.

CÆSALPINIA JAPONICA, *Sieb. & Zucc.*

Nicholson in Gard. Chron. 1888, vol. iv. p. 513, fig.; Gard. Mag. 1888, p. 445, fig.; The Garden, 1891, vol. xl. p. 588, pl. 837.

A beautiful shrub from Japan, and the only representative of the genus known to be hardy in Great Britain.

Of spreading, semi-scandent habit, it attains but a moderate height, and needs for support a wall or wooden trellis. The stems and branches, armed with numerous hard, curved prickles, are furnished with cut foliage of a soft fern-like aspect.

The raceme bears from twenty to thirty bright canary-yellow flowers, each 1 in. in diameter, with which the red filaments and anthers form a contrast. Flowers produced for the first time in this country in 1887 on specimens which had stood many years uninjured by the frost at Coombe Wood.

CARPINUS CORDATA, *Blume.*

Veitchs' Catlg. of Trees and Shrubs, 1888, p. 37.

This, one of the largest of the Hornbeams and certainly one of the finest, forms a conspicuous ingredient in the forests of the northern island of Japan, where it was met with by James H. Veitch during his journeyings in the Far East. Previously found by Maries in the same district, and through him first introduced to cultivation.

The leaves, large and striking in appearance, are 6 to 7 in. long and from 3 to 4 in. broad, with the characteristic venation of the genus; the long, hop-like catkins of fruit attain a length of 6 in or more, and are also very striking.

CELASTRUS HYPOGLAUCUS, *Hemsl.*

This fine species, which takes its specific name from the glaucous hue of the under surface of the large handsome leaves, was first detected by Dr. A. Henry in the Province of Hupeh, Central China, and subsequently introduced to cultivation from the same locality.

The leaves, 6 in. long by 2½ in. broad, are of a deep pea-green on the upper surface, glaucous beneath; the young wood purple, covered with waxy bloom.

CELASTRUS LATIFOLIUS, *Hemsl.*

Hooker's Ic. Pl. t. 2206.

This fine species of Celastrus, first made known to botanists through Dr. A. Henry, who collected specimens in the Province of Hupeh, Central China, was introduced to cultivation from the same locality from seed collected in 1900.

The leaves large, 6 in. broad by 8½ in. long, are broadly ovate, or nearly orbicular, acuminate with a cordate base, dark green on the upper surface, paler beneath, strongly veined, and serrated along the margin.

It is a common shrub in the neighbourhood of Ichang, known as Nan-shan-yeh. The root and leaves, powdered and mixed with flour, are scattered over growing crops of cabbage, turnip, and others of a similar nature, to kill obnoxious insects.

CHAMÆBATIA FOLIOLOSA, *Benth.*

Bot. Mag. t. 5171; Gard. Chron. 1859, p. 652, Bentham in Plantæ Hartwegianæ, p. 308.

A half-hardy shrub with flowers resembling those of a Potentilla, and finely-dissected foliage not unlike that of some species of Milfoil (Achillea). The shrub grows to a height of 2 or 3 ft., and has an agreeable balsamic odour.

The branches, when mature, are covered with smooth reddish-brown bark; when young, densely clothed with gland-tipped hairs.

It was introduced to Exeter in a living state from California through William Lobb in 1859.

CLEMATIS GOURIANA, *Roxb.*

A species introduced to cultivation through Wilson, who sent seed from Hupeh in 1901.

The plant, a hardy climber, has compound leaves composed of five petioled leaflets, irregularly serrate along the margin towards the apex; the petioles and young wood of a reddish-purple.

The flowers, produced in September, individually small, cream-white in colour, are showy from a great profusion, and are also very fragrant.

CLEMATIS MONTANA, *Wall.*, var. GRANDIFLORA.

Bot. Mag. t. 4061.

Both this variety and the type, natives of Northern India, are worthy of a place in every garden on account of the profusion of large, white, delicately fragrant flowers usually produced in early summer and in the autumn months.

The variety *grandiflora* flowered for the first time at Exeter in 1844.

CLEMATIS MONTANA, *Wall.*, var. RUBENS, *O. Kuntze.*

Flora and Sylva, 1905, vol. iii. p. 252, col. pl.

A magnificent form of the well-known type introduced to Coombe Wood from the Province of Hupeh, Central China, where it occurs on scrub-clad mountain-sides at elevations of 5,000-9,000 ft.

In foliage and habit this form closely resembles the typical species, but the stems and leaf-stalks are of a reddish-purple hue, as are the young leaves.

The rosy-red flowers differ in a marked degree from those of the type, are as large as those of the variety *grandiflora*, and are also produced twice during a season, in early May and in September.

Plants flowered for the first time at Coombe Wood in September 1903.

CLEMATIS MONTANA RUBENS

CLEMATIS PATENS, *Morr. & Decne*, var. JOHN GOULD VEITCH.

Fl. Mag. 1867, pl. 394; Fl. des Serres, 1869, tom. xviii. p. 85; Veitchs' Catlg. of Pl. 1868, fig. p. 12.

A double blue Clematis with numerous narrow perianth segments, at the time of introduction, 1865, the finest of its kind; from Japan by the late John Gould Veitch.

CORNUS PAUCINERVIS, *Hance*.

A small-growing, free-flowering species somewhat resembling the American Cornus stricta, inhabiting Central and Western China, where it is met with as a fluviatile shrub at 1,000-3,600 ft.

CORNUS STOLONIFERA, *Michx.*

Gard. Chron. 1875, vol. iv. p. 678, fig. 138, p. 679.

Raised from seed sent by a correspondent from the Rocky Mountains, where it is known as "The Red Osier Dogwood."

By misadventure it received the name of Cornus capitata, and specimens under that name unfortunately distributed.

C. stolonifera is a hardy deciduous shrub, of rapid growth in moist situations, forming large clumps 5 to 6 ft. high, multiplying freely by prostrate or subterranean suckers. Ornamental in summer when it blossoms freely, in autumn when laden with white berries, or in winter, when stripped of its leafy honours, the bright red bark of the annual shoots are conspicuous.

CORYLOPSIS HIMALAYANA, *Griff.*

Bot. Mag. t. 6779.

This delicate graceful shrub, the rarest species of the genus in cultivation, produces spikes of creamy white flowers with a primrose-like smell early in the year, preceding the foliage by some three to four months. Discovered by Griffith in Bhotan, north of the Assam valley, at elevations of 5,000-8,000 ft., seed was sent to the Royal Gardens, Kew, by Dr. King, in 1879, and to the Veitchian firm about the same time.

CORYLOPSIS PAUCIFLORA, *Sieb. & Zucc.*

Gard. Chron. 1893, vol. xiii. p. 335 (Report of R.H.S. Show); Bot. Mag. t. 7736.

As with the other species of Corylopsis hardy in gardens, the primrose-yellow flowers appear in early spring before the leaves. It may be distinguished from the better-known C. spicata by a more slender habit, fewer flowered spikes, and smaller leaves, longer in proportion to their breadth than those of C. spicata.

It was introduced from Japan, where it occurs in various localities.

CORYLOPSIS SINENSIS, *Hemsl.*

Hemsley in Gard. Chron. 1906, vol. xxxix. p. 18, fig. 12.

Introduced through Wilson from Central China, this species flowered for the first time at Coombe in the spring of 1905.

As a species closely allied to Corylopsis spicata, of Japan, the chief distinctions are as given by Mr. Hemsley (*l.c. supra*) :—

"Corylopsis sinensis differs from C. spicata, to which it is closely allied, in the stipules of the flowering branches being broader than long; in the leaves being broadest above the middle; in the orbicular petals being suddenly narrowed into a distinct claw; in the yellow anthers and white seed."

CORYLOPSIS SPICATA, *Sieb. & Zucc.*

Bot. Mag. t. 5458; Fl. des Serres, tom. xx. t. 2135; Gard. Chron. 1865, p. 172, fig.; *id.* 1881, vol. xv. p. 510, fig.; Nicholson in Woods and Forests, 1884, p. 333, fig.

Introduced by Fortune and by the late John Gould Veitch in the year 1864.

As the generic name implies, this shrub has the general appearance of the common Nut-bush (Corylus), its habit deciduous and its leaves long-stalked, heart-shaped, and feather-veined. The great value to the garden is with the flowers, in themselves small, yellow in colour, but produced profusely and in conspicuous racemes before any leaves appear.

The flowers, the colour and odour of the cowslip, are singly borne in the axils of greenish bracts, some eight to twelve in a raceme.

CORYLUS TIBETICA, *Bat.*

A species remarkable in having the young wood covered with rough ferruginous hairs, and the involucre surrounding the nuts spiny.

The leaves, large and handsome, are broadly ovate in outline, cordate at the base, with biserrate margins; it was introduced from Central China.

CRATÆGUS PINNATIFIDA, *Burge*, var. MAJOR, *N. E. Brown.*

Syns. *C. tartarica*, Hort. Veitch.

N. E. Brown in Gard. Chron. 1886, vol. xxvi. p. 621, fig. 121, p. 620.

Introduced by us from Tartary, and valuable for the berries, of large size and of a fine bright red colour.

DAVIDIA INVOLUCRATA, *Baillon.*

Plantæ Davidianæ, pt. ii. p. 60, t. 10; Veitchs' List of Novelties, 1903, p. 4, fig.; Gard. Chron. 1903, vol. xxxiii. p. 236, fig. 98.

This truly remarkable plant was first made known to science from specimens collected by Abbé David on the mountains of the Mu-Pin, west of Szechuan, Central China, in 1871.

Seeds were later sent to M. Maurice de Vilmorin, in Paris in 1897, and a few plants raised. For its introduction to gardens in this country we are indebted to Wilson, who succeeded in introducing a quantity of fertile seed from which plants were raised in 1903. Davidia involucrata is a deciduous tree, with handsome cordate leaves, resembling those of the lime. The feature of the plant is the presence of two large white bracts subtending a cluster of red stamens, freely produced from dwarf shoots. So far the tree has proved perfectly hardy in the county of Surrey and of most vigorous growth, but the plants are yet too young to flower.

DENDROMECON RIGIDUM, *Benth.*

Bot. Mag. t. 5134; Jour. of Hort. July 31st, 1902, p. 102, with fig.; Fl. des Serres, tom. iv. 2nd ser. 1861, p. 43; The Garden, 1896, vol. l. p. 292, pl. 1087.

This half-hardy shrubby plant with yellow poppy-like flowers, meriting the name Dendromecon or Tree-Poppy, was first discovered by David Douglas in California, in the dry rocky ranges from San Diego to Clear Lake, and it is also found more abundantly south of Point Conception and on Santa Rosa Island. Long after discovery it was only known from herbarium specimens until raised from seed sent home by William Lobb. Still rare in gardens, it needs the protection of a wall, except in favoured localities, but is undoubtedly a handsome plant, the length of time during which the bright yellow flowers are produced not the least valuable of many qualities.

DEUTZIA DISCOLOR, *Hemsl.*

Gard. Chron. 1904, vol. xxxv. p. 371; Veitchs' List of Novelties, 1905, p. 2, fig.

Several forms of this variable species have been introduced to Europe, principally through Jesuit missionaries who transmitted the seeds collected by natives in China.

The variety *purpurascens* is perhaps the best known, and has helped largely in the production of hybrid forms raised by M. Lemoine. The type, a dwarf shrub producing a wealth of small white or pinkish flowers, was introduced through Wilson, among numerous forms, from the Province of Hupeh, Central China.

DEUTZIA GLOBOSA, *Duthie.*

A species introduced from Western Hupeh, Central China, first flowered at Coombe Wood during the summer of 1905.

Characterized by dense globose panicles of creamy-white, medium-sized flowers, the filaments of the stamens are petaloid but not lobed at the apex.

DEUTZIA GRACILIS, *Sieb. & Zucc.*

Gard. Chron. 1851, p. 513 (advt.); Fl. des Serres, 1850-1851, tom. vi. t. 611.

A hardy Japanese shrub grown in almost every garden for forcing purposes, with white flowers useful in winter or early spring.

Introduced from Japan through Thomas Lobb, it was first exhibited on May 3rd 1851, at the Horticultural Society's Show, Chiswick, and again on May 14th of that year at the Royal Botanic Society's Show in Regent's Park, on both occasions a Medal awarded.

DEUTZIA MOLLIS, *Duthie.*

A distinct species from Central China.

The leaves ovate or elliptic lanceolate with the characteristic stellate hairs on the upper surface clothed beneath with a soft pubescence of a velvety appearance.

The white flowers in a flat corymbose inflorescence.

DEUTZIA PLANIFLORA, *Duthie.*

A medium-sized shrub with flowers of the purest white, raised from seed collected in Central China.

The leaves somewhat membraneous, lanceolate acuminate, with serrate margins: the flowers in thyrsoid panicles; the obtuse petals spread flat and the broad filaments prominently lobed.

DEUTZIA REFLEXA, *Duthie.*

A species collected in Central China by Wilson, a somewhat slender-growing shrub with greyish wood and narrow lanceolate leaves serrate along the margin. The flowers on slender pedicels in corymbose panicles are pure white in colour, the lateral margins of each petal folded back have a distinct and curious appearance.

DEUTZIA WILSONI, *Duthie.*

A vigorous-growing species with coriaceous elliptic-lanceolate leaves, serrate along the margin, rough to the touch.

The flowers, individually the largest known in the genus, are in erect thyrsoid panicles.

It was raised from seed collected in Central China by Wilson, after whom it is named, and first flowered at Coombe Wood during the summer of 1905.

TREES AND SHRUBS—DECIDUOUS

DIERVILLA SESSILIFOLIA, *Buckl.*

Gard. Chron. 1897, vol. xxii. p. 17, fig. 3, p. 14.

A North American species with honey-yellow flowers, in dense heads on short cymes, imported from North Carolina in 1889, and first exhibited in flower in July 1897.

DIPELTA FLORIBUNDA, *Maxim.*

A handsome deciduous shrub allied to the Weigela, with a great—and probably undeserved—reputation, introduced to cultivation through Wilson, who sent living roots in 1902, and in 1904 a supply of seed.

The plant inhabits almost inaccessible cliffs, and seldom produces fertile seed. The flowers are tubular, white and pink in colour, freely produced; plants growing at Coombe Wood have proved hardy, but not yet shown signs of flower.

'DIPTERONIA SINENSIS, *Oliver.*

Gard. Chron. 1903, vol. xxxiii. p. 22; Jour. R.H.S. 1903, vol. xxvii. p. 60, figs. 18, 19.

A tree closely allied to the Maple with pinnate leaves of from five to seven pairs of ovate-lanceolate leaflets, with serrate margins.

The small white polygamous flowers, in terminal panicles, are followed by numerous winged fruits which, composed of two connate carpels surrounded by a continuous membraneous wing, resemble those of the Wych elm.

It was raised from seed from the Province of Hupeh, and is an interesting shrub.

ELEUTHEROCOCCUS HENRYI, *Oliver.*

Hooker's Ic. Pl. t. 1711; Gard. Chron. 1905, vol. xxxviii. p. 403, fig. 151.

An Araliaceous shrub discovered in the Province of Hupeh, Central China, and named in compliment to Dr. A. Henry.

Wilson introduced it to cultivation from the same locality, plants raised from seed flowering at Coombe in August 1905.

The plant clothed with five-foliate leaves, has spiny branches terminated by globular heads of flower, resembling the inflorescences of the Ivy. These are succeeded by jet-black fruits deeply coloured.

The root bark is used as a drug by the natives.

ELEUTHEROCOCCUS LEUCORRHIZUS, *Oliver.*

Hooker's Ic. Pl. sub t. 1711; Gard. Chron. 1905, vol. xxxviii. p. 404, fig. 153.

Like the preceding species, originally discovered by Dr. A. Henry and subsequently introduced to cultivation through Wilson.

It is distinguished by the Chinese by the white bark of the root, famous as a drug, that of Eleutherococcus Henryi being red.

EUPTELIA DAVIDIANA, *Baill.*

Hooker's Ic. Pl. t. 2787.

An extremely interesting tree, widely distributed in Central and Western China, where specimens have been obtained by many travellers, the first Père David, after whom it is named. The species is variable, and different forms have at various times been given specific rank, but Mr. Hemsley (*l.c. supra*) does not find sufficient distinctive characters to form more than one species.

Seed collected in the Province of Hupeh in 1900 germinated, and plants raised at Coombe Wood have proved perfectly hardy.

The plant forms a shrub or small tree with neat, nearly orbicular leaves terminated by a thick mucro-like apex, and colours well in autumn; the wood resembles that of the Hazel.

EVODIA RUTÆCARPA, *Benth.*

An interesting member of the Rue family, known in the Himalaya and Japan, introduced to cultivation from the Province of Hupeh, Central China.

The pinnate leaves consist of four to six pairs of lanceolate leaflets 1 in. or more in length; the petioles red and the under surface with a silky pubescence.

The flowers, in terminal much-branched corymbs, are greenish-white in colour, with the powerful odour characteristic of the Rue family.

FRAXINUS BRACTEATA, *Hemsl.*

Hemsley in Jour. Linn. Soc. vol. xxvi. p. 84.

Specimens of this handsome Ash were first collected by Dr. A. Henry in the neighbourhood of Ichang, Central China, from which locality it was subsequently introduced.

The leaves, very light and handsome, are composed of eight to eleven pairs of sub-opposite leaflets, glossy deep green above, bright green beneath, with entire margins, and the tree attains a height of some 40 ft.

Mr. Hemsley (*l.c. supra*) remarks :—"It differs from F. retusa, Champ., in having entire leaflets, less capillary pedicels, and obtuse fruits; and from F. Griffithii, Clarke, in the very differently shaped fruit."

FRAXINUS MARIESII, *Hook.*

Bot. Mag. t. 6678.

An ally of the South European Manna Ash, which it resembles in a profusion of small white flower in dense broad panicles during the early spring months.

A native of China, discovered in the province of Kiu-kiang by Charles Maries, who sent home seed from which plants were raised and flowered for the first time at Coombe in May 1882.

FREMONTIA CALIFORNICA, *Torr.*

Bot. Mag. t. 5591; Fl. des Serres, 1877, p. 175; The Garden, 1873, vol. iii. p. 54, fig.; *id.* 1882, vol. xxii. p. 115, fig.; *id.* 1886, vol. xxix. p. 8, pl. 525; Flora and Sylva, vol. ii. p. 279, fig.

This singular and beautiful hardy shrub was first raised in this country from seed sent by Mr. Robert Wrench from California to the Gardens of the Horticultural Society, where a plant flowered for the first time in April 1854. The specimen remained unique for a number of years, as all attempts at propagation failed, but the plant died on removal when the arboretum at Chiswick was given up.

Seed subsequently by William Lobb, from California, a number of plants were raised and afterwards distributed.

Fremontia californica, the only species in the genus, is found on dry hills from Pitt River to San Diego, abundant in the foothills of the Southern Sierra Nevada. It was first discovered during Colonel Fremont's adventurous expedition to the Rocky Mountains in 1846, and bears the name of that distinguished officer.

HAMAMELIS ARBOREA, *Mast.*

Syns. *H. japonica*, Hook. f.

Gard. Chron. 1881, vol. xv. p. 205, fig. 38; The Garden, 1891, vol. xxxix. p. 546, pl. 809; Bot. Mag. t. 6659.

A handsome species, originally introduced from Japan, a near ally of the American Hamamelis virginica, but with larger and more showy flowers.

It is a very desirable hardy shrub, not only for the structural interest the flowers possess for the botanist, but, from a horticultural standpoint, of considerable beauty, and, moreover, furnishing an additional link between the flora of the Eastern United States and that of Japan.

The long strap-shaped crinkled petals of the flowers are of a bright yellow colour, the centre rich purple, borne in early spring while the plant is still leafless, and very conspicuous.

HAMAMELIS MOLLIS, *Oliv.*

Bot. Mag. t. 7884; Flora and Sylva, vol. ii. p. 104.

This, the rarest and largest-flowered of all the Witch-Hazels, found in Kiang-su in the district of Kiu-kiang, China, by Charles Maries, and sent by him to Coombe, was for twenty years overlooked till Mr. George Nicholson, late Curator of the Royal Gardens, Kew, brought the plant to notice.

HYDRANGEA HORTENSIA, *DC.*, var. MARIESII, *Hort.*

The Garden, 1898, vol. xliv. p. 390, pl. 1196.

This most curious of the many varieties of the common Hydrangea was introduced from Japan through Charles Maries.

The inflorescence is remarkable in having sterile flowers confined to the outer edge of the corymb, of large size, each measuring from 3 to 3½ in. across, of a pink or delicate mauve-pink hue.

HYDRANGEA HORTENSIA, *DC.*, var. ROSEA.

Rev. Hort. 1904, p. 544, col. pl.

A beautiful form introduced from Japan, producing in a normal state flowers of a beautiful deep rose remarkable in that they assume a peculiar porcelain blue tint, for which a ferruginous soil is essential.

HYDRANGEA LONGIPES, *Hemsl.* (*non Franch.*).

Hemsley in Jour. Linn. Soc. vol. xxiii. p. 274.

A new species first sent by Dr. Henry from the Province of Hupeh, Central China, and subsequently introduced to cultivation from the same locality.

A shrub 2 to 3 ft. in height with long-stalked ovate-rotund leaves resembling those of Hydrangea scandens, and terminal cymes surrounded by white sterile blooms on stalks at some distance from the fertile flowers.

LINDERA TZUMU, *Hemsl.*

Hemsley in Jour. Linn. Soc. vol. xxvi. p. 392.

A tree 20 to 50 ft. high with variable foliage introduced from the Province of Hupeh, Central China.

According to Bretschneider, this is the t'ze tree of the Chinese Classics, valued by the ancient Chinese for its timber.

Of rapid growth in this country, seedling trees at Coombe Wood having attained a height of 8 ft. in four years; the leaves broadly ovate, or rhomboid, cuneate at the base, are sometimes trilobed; the under side glaucous, the petioles and principal veins of a reddish tinge; the young wood brightly spotted with purple.

LIRIODENDRON CHINENSE, *Sarg.*

Sargent's Trees and Shrubs, 1903, vol. i. p. 103, t. 51; Gard. Chron. 1903, vol. xxxiv. p. 370; Hooker's Ic. Pl. 1905, pl. 2785; Flora and Sylva, 1905, vol. iii. p. 202.

This species was formerly regarded as a geographical form of the variable North American Tulip Tree, Liriodendron tulipifera, but more complete specimens have shown sufficient characters to render it a distinct species. It has been collected by several travellers, including Maries, Dr. A. Henry, and Wilson, and through the last named was introduced to cultivation.

The flowers green, are smaller than those of L. tulipifera and more spreading; when fully open they do not present the tulip-like appearance, there is a difference in the construction of the pistil, and the leaves are more glaucous.

In the native habitat it forms a tree 15 to 20 ft. high, though occasionally isolated examples are met which greatly exceed this height.

LONICERA GYNOCHLAMYDÆA, *Hemsl.*

Hemsley in Jour. Linn. Soc. vol. xxiii. p. 362.

A dwarf-growing shrub first from the Patung district, Central China, and subsequently introduced to cultivation from the Province of Hupeh.

The flowers, remarkable for the cap-like production of the calyx over the connate bracteoles, are rose-coloured, produced on two-flowered peduncles in the axils of the leaves.

LONICERA KŒHNEANA, *Rehd.*

Rehder in Sargent's Trees and Shrubs, 1902, pt. i. pl. xxi.

A Chinese species first sent as dried specimens from Central China by Dr. A. Henry, and "dedicated to Professor E. Kœhne, the distinguished German botanist, whose arrangement of the cultivated species of Lonicera in his Deutsche Dendrologie is the best and most natural hitherto published."

A strong-growing shrub with ovate acuminate leaves, in the axils of which the two-flowered peduncles are produced, flowered at Coombe during the summer of 1905 from seed collected in Central China by Wilson.

LONICERA TRAGOPHYLLA, *Hemsl.*

Bot. Mag. t. 8064; Hemsley in Jour. Linn Soc. vol. xxiii. p. 367; Jour. R.H.S. 1903, vol. xxviii. p. 63, fig. 24; Gard. Chron. 1904, vol. xxxvi. p. 151, suppl. illus.; Veitchs' List of Novelties, 1905, p. 3, figs.; Sargent's Trees and Shrubs, pl. xlvi.

A Chinese Honeysuckle flowered in June 1904 at Coombe, figured in the Botanical Magazine (*l.c. supra*) in March 1906, the month of the publication of *Hortus Veitchii.*

The flowers in dense heads, each composed of from twelve to fifteen yellow trumpet-shaped blooms, have exserted stamens. The blooms are the largest produced by any species of the sub-genus (Periclymenum), and the species is the only representative in Central and Eastern Asia of the almost exclusively Mediterranean sub-section Eucaprifolium, Spach.

LOROPETALUM CHINENSE, *R. Br.*

Gard. Chron. 1880, vol. xiv. p. 620; *id.* 1883, vol. xix. p. 152, fig.; Bot. Mag. t. 7979.

A native of the mountains of China, introduced through Charles Maries in 1880, a member of the family to which the Sweet Gums (Liquidamber) and Witch Hazels (Hamamelis) belong, and originally described under the name of H. chinensis by Robert Brown in Abel's Narrative of a Journey in the interior of China, p. 375.

A free blooming shrub with flowers clustered in small heads terminating the branches, the calyx pale green, the long linear strap-shaped petals pure white.

As with many not perfectly hardy plants, it is not generally met with in gardens, but it blooms profusely in winter and early spring, in a small state, and is well adapted for conservatories and the winter garden during the early months of the year.

MAGNOLIA SALICIFOLIA.

Veitchs' Catlg. of Trees and Shrubs, 1902, p. 45.

Of this Japanese species, not yet flowered in Europe, seed was collected on Mount Hakkoda by James H. Veitch, but failed to germinate, and was later successfully introduced through the Arnold Arboretum, U.S.A.

A small tree with slender branches, willow-like leaves 5 to 6 in. long, light green above, silvery white beneath.

MAGNOLIA SOULANGEANA, *Hort.*, var. NIGRA, *Hort.*

Nicholson in The Garden, 1884, vol. xxv. p. 276, pl. 434.

Mr. Nicholson, late Curator of Kew, the first authority of his day on cultivated trees and shrubs, is of opinion that this is of hybrid origin, the possible parents the purple-flowered Magnolia (M. obovata) and the Yulan (M. conspicua), and this is probable, as both the species suggested have been cultivated in the gardens of the Japanese and Chinese from time immemorial.

The original plant from Japan through the late John Gould Veitch.

MAGNOLIA STELLATA, *Maxim.*

Syns. *M. Halleana*, S. B. Parsons; *Burgeria stellata*, Sieb. & Zucc.

The Garden, 1878, vol. xiii. pl. cxxxii.; Fl. Mag. 1878, n.s. pl. 309; Bot. Mag. t. 6370.

A delightful hardy shrub long known to science as Burgeria stellata, a name founded on an erroneous observation as to the nature of the fruit,

MAGNOLIA STELLATA

THE ROYAL GARDENS, KEW

and subsequently as Magnolia Halleana, given by Mr. S. B. Parsons of Flushing, U.S.A., in compliment to Mr. G. R. Hall, who in 1862 introduced the plant to America.

Found wild on Fuji-yama, it had long been cultivated, and those first sent home were obtained from gardens in Nagasaki by Oldham in 1862.

Plants introduced by Messrs. Veitch from Japan, flowered for the first time in this country at Coombe in March 1878.

The flowers star-shaped, pure white in colour, are produced early in the year before the leaves appear.

MELIOSMA MYRIANTHA, *Sieb. & Zucc.*

Gard. Chron. 1902, vol. xxxi. p. 30, fig.

An uncommon shrub, the only species of the genus, with small greenish-yellow flowers densely borne in branched panicles, resembling some species of Spiræa. The plant, though not perfectly hardy in all parts of England, flowers and seeds freely at Coombe Wood.

It was introduced from Japan, but has by no means a wide distribution.

NEILLIA SINENSIS, *Oliver.*

Hooker's Ic. Pl. t. 1540; Jour. R.H.S. 1903, vol. xxviii. p. 61.

A beautiful hardy shrub, from Central China, which attains a height of from 4 to 5 ft., furnished with an elegant currant-like leafage, and produces in great abundance drooping racemes of rose-pink flowers.

Perfectly hardy, and flowered for the first time in this country in the summer of 1904.

NYSSA SINENSIS, *Oliver.*

Hooker's Ic. Pl. t. 1964.

A deciduous shrub or small tree discovered in the Province of Hupeh Central China, by Dr. A. Henry, and subsequently introduced to cultivation from the same locality.

The discovery of this genus in China and in the Himalaya is interesting as it was formérly supposed to be restricted to the Eastern United States.

PERTYA SINENSIS, *Oliver.*

Hooker's Ic. Pl. t. 2214.

A genus of the Mutisia group of the Compositeæ, first made known in China from specimens collected in the Province of Hupeh by Dr. A. Henry.

A bush some 6 ft. high of thin graceful reddish-purple stems clothed with glabrous oblong-lanceolate leaves in the axils of which the peduncled capitula of flowers are formed.

Plants growing at Coombe Wood were raised from seed from Central China.

POPULUS LASIOCARPA, *Oliver.*

Jour. R.H.S. 1903, vol. xxviii. p. 65, fig. 27.

Discovered by Dr. A. Henry as a tree 20 to 40 ft. high, in the Province of Hupeh, Central China, and introduced to cultivation from the same locality.

The leaves are large, probably the largest of any species in the genus, the plant perfectly hardy.

PRUNUS (CERASUS) PSEUDO-CERASUS, *Lindl.*, var. JAMES H. VEITCH.

Gard. Chron. 1896, vol. xix. pp. 466, 517, fig. 79.

The finest of all forms of the Japanese Cherry, cultivated in gardens for the beauty of their flowers. When travelling in that country, James H. Veitch heard by chance in Tokio of a nurseryman who specialized cherries in a district he was unable to visit, and he wrote for the most distinct forms the man grew. Subsequent trial in England proved all to be valueless save this variety.

In this country it forms a small tree or bush-like shrub, with blossoms in early spring, later than in Japan.

The deep rose-pink double flowers are some 2 in. in diameter, the young leaves serrated, of a bright bronze tint.

PYRUS DELAVAYI, *Franch.*

Plantæ Delavayanæ, 1890, p. 227.

This species, one of the most interesting and ornamental of the whole family, attains the dimensions of an ordinary orchard apple tree, and in flower presents a similar appearance.

The foliage of two forms, the young leaves pinnatifid and hairy, give the tree the appearance of a species of Cratægus; the mature leaves obovate, entire, and nearly glabrous, present quite a different aspect. The interesting fruits ovoid in shape are about the size of those of Pyrus Maulei.

Intermediate in structure between those of the quince and apple, having seeds arranged in rows as in the quince, but numerically the same as in the apple.

A native of the high plateaux of Yunnan, first discovered by Père Delavay, and subsequently introduced to cultivation.

RHODODENDRON CALENDULACEUM, *Torr.*

Syns. *Azalea calendulacea*, Michux.

Watson in Gard. Chron. 1892, vol. xii. p. 742.

A North American species, said to cover wide stretches of country in Virginia, introduced to Exeter in 1837, and very fragrant.

The gorgeous flowers in June are of every shade of lemon-yellow to brilliant fire-red. Many forms appeared, raised in gardens by selection and cross-fertilization, to which varietal names have been given.

RHODODENDRON DILATATUM, *Miq.*

Syns. *Azalea dilatata*, Hort.

Bot. Mag. t. 7681.

A native of Japan, of the Azalea section of the genus, with deciduous foliage and flowers with only five stamens, first made known through Siebold, and introduced by us in 1883.

RIBES GLACIALE, *Franch.*

A species from Central China with leaves resembling those of the gooseberry and the wood as the scaly bark of the currant.

The flowers inconspicuous, greenish, diœcious, followed by small red fruit.

RIBES LOBBII, *Asa Gray.*

Syns. *R. subvestitum*, Hook.

Gard. Chron. 1883, vol. xix. p. 11, fig.; Bot. Mag. t. 4931.

A native of California introduced to Exeter through William Lobb, this hardy Currant is now seldom seen, although an early flowering character should commend it to the lover of hardy shrubs.

The flowers, small and fuchsia-like, are dull purple in colour, the sepals tipped with green.

ROSA SERICEA, *Lindl.*, var. PTERACANTHA.

Gard. Chron. 1905, vol. xxxviii. p. 260, figs. 98, 99.

A form of the type with red spines dilated into wings at the base in an extraordinary degree, frequently the entire length of an internode; the foliage neat and attractive, the flowers white, and the fruits yellow.

Found by several travellers in various localities in China, and introduced to cultivation through Wilson, who met with it in the Province of Hupeh.

ROSA SORBIFLORA, *Focke.*

Gard. Chron. 1905, vol. xxxvii. p. 227, fig. 96.

A species closely allied to the Banksian Rose with numerous small white flowers, clustered in a corymbose inflorescence as in the Sorbus. From the heights in West Hupeh, it has not as yet flowered in this country.

ROSE QUEEN ALEXANDRA.

Gard. Chron. 1901, vol. xxx. p. 30, fig.

Raised by Seden from Crimson Rambler, fertilized with Rosa multiflora simplex, it was the Gold Medal Rose of the year 1901. Of climbing habit, clothed with large trusses of rich rose-coloured flowers in great profusion.

RUBUS BAMBUSARUM, *Focke.*

Hooker's Ic. Pl. sub t. 1952.

This handsome Rubus, first discovered by Dr. A. Henry, and introduced to cultivation through Wilson, who collected seed in the Province of Hupeh, Central China, in 1900. The leaves resemble those of R. Henryi, but being trilobed are so divided as to be trifoliate, and in exceptionally vigorous growths often five-foliate.

The leaflets are narrow, lanceolate, resembling those of Bamboos or Willows, dark green above, covered on the under surface with a cream-white or dun-coloured indumentum. In some districts the leaves are dried and used as tea.

The flowers, not showy, develop black edible fruit.

This bramble is a handsome pillar plant of a vigorous constitution, the long trailing branches attaining a length of from 10 to 12 ft. in a summer's growth.

RUBUS BIFLORUS, *Buch.*

Syns. *R. leucodermis*, Hort. (non Douglas).

Bot. Mag. t. 4678.

A handsome Bramble originally from Nepal, cultivated under the name of Rubus leucodermis, a name well-deserved from the pure white stem, but the name had already been given by Douglas to a North American species.

R. biflorus is a curious ornamental plant, quite hardy, a striking object in the shrubbery from the white-washed appearance of the stems, which when examined are found to be covered by a minute, perfectly white, pulverulent substance.

The pure white flowers are followed by orange-coloured fruit, edible and of pleasant flavour.

TREES AND SHRUBS—DECIDUOUS

RUBUS CHROÖSEPALUS, *Focke.*

Hooker's Ic. Pl. t. 1952.

A species discovered by Dr. A. Henry, and later introduced to cultivation from Central China in 1900.

It is a simple-leaved species, the leaves aptly compared by Dr. W. O. Focke, the author of the specific name, to those of Tilia alba.

The flowers in a lax panicle are remarkable in having the inner surface of the reflexed sepals coloured—compensation for the want of petals.

RUBUS CONDUPLICATA, *Duthie.*

A new species from the Province of Hupeh, Central China, through Wilson, who sent seed in 1900.

Simple-leaved with trilobate leaves, dark green on the upper surface, white beneath with pinkish veins: the wood covered with a thick waxy white substance gives the plant a distinctly ornamental appearance.

The white flowers are not sufficiently showy to be of value from a horticultural standpoint.

RUBUS COREANUS, *Miq.*

Moore in Jour. of Bot. 1875, p. 230; Jour. Linn. Soc. vol. xxiii. (In. Fl. S.), p. 230.

An erect-growing bush, native of the Corean Archipelago and Central China, introduced to cultivation from Hupeh.

The stems are covered with a conspicuous waxy white covering; the leaves pinnate, with 2 to 3 pairs of pinnæ and a terminal rhomboid leaflet, dark metallic green in colour, pale green beneath.

The fruit varies in colour from red to yellow and black.

RUBUS FLAGELLIFLORUS, *Focke.*

A handsome Rubus found in the mountain woods of Central China: a strong-growing plant with long whip-like growths 6 to 8 ft. in length, bearing white flowers, which suggested the specific name.

The leaves handsome, cordate, acuminate, have an irregularly serrate margin. In the shade the leaves are beautifully marbled on the upper surface, as in certain forms of Rex Begonia; the under surface covered with a thick cream- or dun-coloured tomentum, as is the wood of the young growth.

RUBUS HYPARGYRUS, *Focke.*

This handsome Chinese bramble has trifoliate leaves of a peculiar metallic lustre on the upper surface, covered beneath with a white felt-like tomentum.

The young wood and spines are red, the small flowers pink, followed by woolly fruit.

Raised from seed collected in Central China, this fine species flowered at Coombe Wood in 1905.

RUBUS ICHANGENSIS, *Hemsl. & O. Kuntze.*

Hemsley in Jour. Linn. Soc. vol. xxiii. p. 231.

Introduced through Wilson, who collected seeds in 1901, in the neighbourhood of Ichang, in which locality it had previously been met with by Dr. Henry, from whose material it was described.

Mr. Hemsley states (*l.c. supra*), "The elongated, cordiform, distantly toothed, glabrous, or early glabrescent leaves, associated with small flowers and few ovaries are especially characteristic of this species."

RUBUS INNOMINATUS, *S. Moore.*

Gard. Chron. 1905, vol. xxxviii. p. 291, fig. 112.

A handsome bramble introduced to cultivation from the Province of Hupeh, with leaves usually trifoliate, green above, white beneath; the stems covered with a curiously soft pubescence which gives a velvety impression to the touch.

The feature is the bright orange-scarlet colour of the fruit.

Throughout the month of September, when the large panicles of these are at their best, the plant presents a remarkably striking and ornamental picture, as decorative as any spring-flowering shrub. The fruits edible, resemble in flavour those of the common blackberry, but differ in being of a larger size and more flat in general shape.

RUBUS IRENÆUS, *Focke.*

A trailing species with heart-shaped or occasionally slightly trilobed leaves, dark green above, dun-coloured beneath with serrate margins, suitable for covering low banks or for the margins of woods and plantations where the long shoots can ramble unrestricted.

RUBUS LAMBERTIANUS, *Ser.*, var. GLABER, *Hemsl.*

A simple-leaved species with remarkable young shoots glandular and sticky to the touch, with the stipules of the leaves divided into about five unequal bristle-like segments.

The leaves cordate at the base, slightly trilobed, acute, with dentate margins bright green and glossy on the upper surface, are pale green beneath.

The inflorescence consists of a terminal panicle of small white flower followed by red fruit.

RUBUS LASIOSTYLUS, *Focke.*

Bot. Mag. t. 7426; Gard. Chron. 1904, vol. xxxiv. p. 170; Veitchs' List of Novelties, 1905, p. 6, figs.

This handsome Chinese bramble was introduced to the Botanic Gardens of this country through the discoverer, Dr. Augustine Henry, in 1899, from the Province of Hupeh, and later sent home by Wilson, who found it in the same locality.

Remarkable for white stems, closely set with prickles, and a very woolly fruit of sub-globose shape, resembling a raspberry as it also does in flavour.

RUBUS LEUCOCARPUS, *Benth.*

A species raised from seed collected in Central China. The leaves pinnate, consist of several pairs of pinnæ; the leaflets small with serrate margins.

The flowers in small sessile clusters in the axils of those of the current season's growth are of a pink colour, individually small, followed by white or red fruit.

RUBUS NIVEUS, *Wall.*

A strong-growing species often reaching a height of 20 ft., clothed with three to five foliate leaves, covered with a soft pubescence which gives a satiny impression to the touch.

The lateral leaflets oblique, the terminal triangular in outline with coarsely toothed margins, the under surface white.

The small white flowers are in axillary and terminal corymbs followed by purplish fruit.

RUBUS PALMATUS, *Thunb.*

Bot. Mag. t. 7801.

A white-flowered species with elegantly lobed palmate leaves, not perfectly hardy in other than favoured localities in this country, and a suitable subject for clothing pillars in the cool greenhouses or the winter garden.

A native of the central mountains of Japan, Korea, and China, it was from the first-named country introduced to cultivation.

RUBUS PARKERI, *Hance.*

A species introduced through Wilson from the Province of Hupeh, Central China, where it is commonly met with in the wooded slopes of glens and gorges.

The leaves ovate acute, cordate at the base, with an irregularly toothed margin have a soft pubescence on the under surface.

The flowers small, in large panicles, are followed by dull red fruit of an indifferent flavour.

RUBUS PLAYFAIRII, *Hemsl.*

Hemsley in Jour. Linn. Soc. vol. xxiii. p. 235.

A species named in compliment to Mr. G. M. H. Playfair, by whom it was discovered at Pakhoi, in the Province of Kwang-tung, South-Eastern China, in 1889, and introduced to cultivation from seed collected in 1901.

Mr. Hemsley states (*l.c. supra*), "This is the only shrubby bramble with pedately divided leaves known to occur within our limits; and it is a very well-marked species readily distinguished from the few other species of this group inhabiting Eastern Asia."

RUBUS WILSONI, *Duthie.*

A species raised from seed collected in Central China in 1901 by Wilson, after whom it is named.

The leaves pinnate with two pairs of ovate-lanceolate pinnæ and a terminal cordate leaflet, dark green above, paler beneath, have serrate margins. The wood of the stem is very prickly.

SAPINDUS MUKOROSSI, *Gærtn.*

This Chinese representative of the genus makes a handsome tree 20 to 25 ft. high with pinnate leaves 1 to 1½ ft. in length.

The flowers are small, white in colour, in very large panicles at the end of the branches. The fruits round in form, about ¾ in. in diameter, have a horny coat which, saponaceous in water, is used as soap by the Chinese. The round very hard seeds are threaded and used as rosaries by the Buddhist priests.

Plants are growing at Coombe Wood.

SCHIZOPHRAGMA INTEGRIFOLIA, *Oliver.*

Jour. R.H.S. 1903, vol. xxviii. p. 62.

A climbing or trailing plant remarkable in having an inflorescence surrounded by large white bracts, similar to those produced by Mussænda frondosa.

Plants introduced from Central China by Wilson have proved hardy at Coombe Wood, though at present too young to flower.

TREES AND SHRUBS—DECIDUOUS

SPIRÆA HENRYI, *Hemsl.*

Jour. R.H.S. 1903, vol. xxviii. p. 61, fig. 20.

A Chinese species, first detected by Dr. A. Henry, a compact shrub with neat foliage and white flowers in corymbs in the axils of the leaves along the whole length of the previous season's growth.

Raised from seed collected in Central China, the shrub has proved perfectly hardy and flowered with unusual freedom at Coombe Wood during the summer of 1904.

SPIRÆA VEITCHII, *Hemsl.*

A new species discovered in Central China by E. H. Wilson and through him introduced to cultivation.

The plant forms a neat compact shrub, with thin growths 6 to 8 ft. long, of a reddish-brown clothed with small glaucous-green, oblong-lanceolate leaves serrate along their apical portion.

The flowers on the short side growths along the whole length of the previous year's shoots are in terminal corymbs pure white, very showy in a mass.

SPIRÆA WILSONI, *Duthie.*

A new species, in some respects intermediate between Spiræa Henryi and S. Veitchii.

The flowers white, are in large flat corymbs composed of several smaller corymbs, and the peduncle and pedicels are covered with silky hair.

A native of the scrub-clad mountains of Western Hupeh, 7,000-8,500 ft. elevation, it was first flowered during the summer of 1905 at Coombe Wood.

STEPHENANDRA FLEXUOSA, *Sieb. & Zucc.*

Veitchs' Catlg. of Trees and Shrubs, 1882-1883, p. 30.

A graceful Rosaceous shrub of Spiræa-like habit, introduced through Charles Maries from Japan.

The slender stems arch gracefully, are covered with crimson bark and furnished with trilobed deeply-cut leaves about 2 in. long, the flowers white, in small erect clusters.

STRANSVÆSIA UNDULATA, *Decne.*

A handsome Photinia-like shrub with entire lance-shaped, coriaceous, shining leaves, entire along the margin with an undulate surface, collected in bunches at the end of the growths.

The small white flowers are in flat corymbs, followed by brilliantly coloured hawthorn-like fruits, only equalled by the exquisite colouring of the foliage in the autumn.

A form introduced with the type known as *fructo-luteo* has yellow fruit.

STUARTIA MONADELPHIA, *Sieb. & Zucc.*

This species inhabits the forests of Central China at elevations of 5,000-7,500 ft., forming a tree some 25 to 30 ft. in height, with a slender trunk; the thin bark continually peels in flakes.

The elliptic leaves are acuminate with a serrate margin, clothed with silky hairs on the principal veins of the under surface.

The fragrant white flowers the same size as those of Stuartia Pseudo-camellia, but not so cup-shaped.

STUARTIA PSEUDO-CAMELLIA, *Maxim.*

Syns. *S. grandiflora*, Briot.

Gard. Chron. 1888, vol. iv. p. 187, fig. 22; Bot. Mag. t. 7045; The Garden, 1893, vol. xliii. pl. 899; Rev. Hort. 1879, p. 430.

A species closely allied to the old North American garden plants—Stuartia pentagyna and S. virginica, the three constituting the whole of this interesting genus in cultivation in this country prior to the introduction of S. monadelphia from China.

S. Pseudo-camellia was introduced from Japan, but appears to have been cultivated in the United States of America for several years; and in France in the nursery of M.M. Thibaut and Keteleer as early as 1868.

A handsome shrub, with creamy-white flowers resembling those of a single-flowered Camellia, and foliage at all times beautiful, in some seasons the brilliant crimson with gold autumnal tints render it exceptionally attractive.

STYRAX OBASSIA, *Sieb. & Zucc.*

Gard. Chron. 1888, vol. iv. p. 131, fig. 12; Jour. of Hort. 1888, p. 513, fig. 73; Bot. Mag. t. 7039.

One of the most interesting of all hardy shrubs introduced from Japan, and first made known by Siebold and Zuccarini in Flora Japonica, 1835, vol. i. t. 56, where a description and figure are given.

The difference in the size and form of the leaves is noticeable, the larger attaining 10 in. in diameter, sometimes alternate, some larger than others, and usually one of great size at the apex of the branch; the bases of the petioles are sheaths, and entirely conceal the bud, as in the London Plane. The flowers pure white, with yellow stamens, are in racemes 4 to 7 in. long, with a hyacinth scent.

Styrax obassia was introduced to this country through Charles Maries, and flowered for the first time at Coombe Wood in June 1888.

TETRACENTRON SINENSE, *Oliver*.

Hooker's Ic. Pl. t. 1892.

The representative of a new genus of the Magnoliaceæ, first discovered by Dr. A. Henry in Hupeh, and subsequently introduced to cultivation from the same locality through Wilson: in the native habitat a tree 20 to 50 ft. high, with alternate ovate-elliptic leaves, serrate along the margin.

The flowers minute, on drooping spikes 4 to 6 in. in length, of singular botanical interest, are of little value from a horticultural standpoint.

TILIA HENRYANA *Szy*.

Jour. R.H.S. 1903, vol. xxvii. p. 66.

A species named in compliment to Dr. A. Henry, somewhat resembles Tilia Tuan, but has broader leaves and a different inflorescence.

Raised from seed collected in Central China, the Lime has proved quite hardy.

TILIA MIQUELIANA, *Maxim*.

Bean in Gard. Chron. 1895, vol. xviii. p. 766; Gard. and For. 1893, p. 113, fig. 19.

Introduced through Charles Maries from Japan, where it forms a common forest tree on the hill slopes in the north island, often attaining a height of 100 ft.

TILIA MIQUELIANA, *Maxim*., var. CHINENSIS, *Szy*.

Hooker's Ic. Pl. t. 1927.

A form of the type first met with by Dr. A. Henry in the Province of Hupeh, Central China, and later introduced to cultivation from the same locality.

TILIA OLIVERI, *Szy*.

Hooker's Ic. Pl. sub t. 1927.

Discovered in Central China, and seed received at the same time by Messrs. Veitch in 1900.

The leaves heart-shaped, unequal at the base, shortly acuminate at the apex, have unequally serrate margins; glabrous on the upper surface, covered beneath with a white tomentum.

The plants growing at Coombe Wood are at present too young to show the true character.

TILIA TUAN, *Szy.*

Jour. R.H.S. 1903, vol. xxviii. p. 66.

A handsome Lime tree with obliquely-ovate leaves semi-cordate at the base, dark green on the upper surface, covered beneath with a stellate white tomentum: raised at Coombe Wood from seed collected in Central China.

VIBURNUM CEANOTHOIDES, *Wright.*

As the specific name implies, this Viburnum closely resembles a Ceanothus in general appearance, forming a bush 4 to 6 ft. in height with leaves cuneate and toothed.

The flowers white, in corymbs, are succeeded by very numerous bright red fruit.

A native of the plateaux of Yunnan, first discovered by W. Hancock Esq., of the Chinese Imperial Maritime Customs, and described from material sent by him to Kew.

VIBURNUM CORYLIFOLIUM, *Hook. f. & Thoms.*

A species from Central China through Wilson, with dark green leaves, hairy on both surfaces and strongly veined, in outline and general appearance not unlike a Corylus. The wood in a young state is very tomentose.

VIBURNUM DILATATUM, *Thunb.*

Syns. *V. Mariesii*, Hort.

Bot. Mag. t. 6215.

A handsome hardy shrub, with apparently a wide range in Japan, having been collected in various localities throughout the whole length of the archipelago.

Introduced through Charles Maries, it flowered at Coombe Wood for the first time in England in June 1875.

VITIS ACONITIFOLIA, *Hance.*

Jour. R.H.S. vol. xxviii. 1904, p. 392, fig. 87.

A Chinese species with leaves of variable form, but more or less resembling those of the Monkshood, introduced by the Veitchian firm, and exhibited for the first time in September 1903, before the Royal Horticultural Society.

TREES AND SHRUBS—DECIDUOUS

VITIS ARMATA, *Diels & Gilg.*

Jour. R.H.S. 1904, vol. xxviii. p. 392, figs. 83, 88, 104; Veitchs' List of Novelties, 1904, p. 4, fig.

A peculiar vine with stems and leaf-stalks beset with fleshy processes resembling prickles.

It is a native of Central China, from seed collected by Wilson, and plants raised proved hardy and very ornamental.

The leaves heart-shaped, 7 to 8 in. broad and 9 to 10 in. long, are three-angled, glabrous in all parts; the autumnal colour rich and brilliant.

VITIS ARMATA, *Diels & Gilg.*, var. VEITCHII, *Hort.*

Jour. R.H.S. 1904, vol. xxviii. p. 393, fig. 89, and col. pl. p. 389.

A more vigorous form of the type, superior in every way, and probably the finest of the Chinese vines: the autumn coloration of a solid richness quite unexampled.

VITIS DELAVAYI, *Franch.*

Jour. R.H.S. 1904, vol. xxviii. p. 393, fig. 102.

A Chinese species of great promise with trifoliate leaves, aptly named in compliment to Père Delavay, a French missionary, to whom is due much for his labour in making known the Flora of the Chinese Empire.

VITIS FLEXUOSA, *Thunb.*, var. CHINENSIS.

Jour. R.H.S. 1904, vol. xxviii. p. 393, fig. 107.

The leaves of this form vary to a great extent, are glossy when young, the upper surface shining green, dull when mature, with traces of purple on the under surface as in the type.

It was introduced to cultivation from the Province of Hupeh, Central China, in 1900.

VITIS FLEXUOSA, *Thunb.*, var. WILSONI.

Jour. R.H.S. 1904, vol. xxviii. p. 394, fig. 90; Veitchs' List of Novelties, 1905, p. 5, figs.

A peculiarly attractive form of the type, from China through Wilson, after whom it is named. The leaves small, are remarkable for a deep bronzy-green hue, a shining metallic lustre, the under surface bright purple: first exhibited in September 1903.

VITIS HENRYANA, *Hemsl.*

Jour. R.H.S. 1904, vol. xxviii. p. 394, fig. 92; Veitchs' List of Novelties, 1905, p. 4, figs.; Gard. Chron. 1905, vol. xxviii. p. 309, fig. 122.

This Chinese species was first collected by Dr. A. Henry, in whose honour the vine is named.

A trailing subject with prettily variegated leaves of five lance-shaped leaflets and serrate margins: the variegation silvery white lines margined with pink along the principal veins, the interspaces of the darkest green.

It was introduced by Wilson.

VITIS INCONSTANS, *Miq.*

Syns. *Ampelopsis tricuspidata*, Sieb. & Zucc.; *A. Veitchii*, Hort.

Masters in Gard. Chron. 1869, p. 838; Lynch in Gard. Chron. 1880, vol. xiv. p. 664, fig.; La Belg. Hort. 1877, vol. xxvii. p. 224.

Introduced to Europe from Japan by the late John Gould Veitch, and distributed under the provisional name of Ampelopsis Veitchii.

It is believed to have been previously sent by Oldham to the Royal Gardens, Kew, and is stated to have been known in the United States of America prior to cultivation in Great Britain.

Scarcely any other climber has attained so great a popularity, and no climber requires so little attention when once planted. It withstands the hot, dry, smoky atmosphere of towns with impunity, and in autumn the foliage assumes a brilliant tint of crimson quite unequalled in the plant kingdom.

VITIS LEEOIDES, *Maxim.*

Jour. R.H.S. 1904, vol. xxviii. p. 395, figs. 95, 96.

A pinnate-leaved vine, the leaves in five leaflets resembling those of the allied genus, Leea, introduced from Central China.

A handsome species, the under surface of the foliage coloured bright claret-purple, a striking contrast to the glaucous-green hue of the upper half.

VITIS MEGALOPHYLLA, *Diels & Gilg.*

Jour. R.H.S. 1904, vol. xxviii. p. 395, figs. 86, 97.

The most remarkable of all Chinese species of Vitis, with pinnate or bipinnate leaves 2 to 3 ft. across, resembling individually those produced by some Araliads or Kœlreuteria paniculata, dark green on the upper surface, pale glaucous-green beneath, glabrous in all parts. The stems of the same glaucous-green hue as the under surface of the leaves, and of rapid growth, attain a height of 8 to 10 ft. in a season.

One of the most handsome climbing plants, of light and graceful appearance, a valuable addition to pergola and pillar plants, hardy in the British Isles.

VITIS OBTECTA, *Wall.*

Jour. R.H.S. 1904, vol. xxviii. p. 395.

A compound-leaved species from China through Wilson with digitate leaves resembling in appearance those of a miniature horse-chestnut; the terminal leaflet, the largest, measures from 4 to 5 in. in length, the smallest about 2½ in., oblanceolate in outline terminated by an acuminate point.

VITIS REPENS, *W. & A.*

A native of India, first made known to science from specimens collected in that country. Introduced from Central China through Wilson, who sent seed in 1901, and of little interest.

The leaves of the species are dark green, sometimes slightly trilobed with bidentate margins and red petioles.

The specific name repens appears to have been given in allusion to the rambling habit of growth characterizing the species.

VITIS SINENSIS, *Diels & Gilg.*

Jour. R.H.S. 1904, vol. xxviii. p. 396, fig. 99.

A species with very variable leaves, those at the base usually simple, broadly ovate in outline, more or less trifid; later more lobed, they become at about 5 ft. from the ground deeply cut into three or five distinct leaflets; the figure in the Journal of the Royal Horticultural Society (*l.c. supra*) shows the two extreme forms of leaf-variation.

It was exhibited for the first time before the Royal Horticultural Society on September 1st, 1903.

VITIS THOMSONI, *M. Laws.*

The Garden, 1903, vol. lxiv. p. 203, fig.; Jour. R.H.S. 1904, vol. xxviii. p. 396, figs. 85, 100; Veitchs' List of Novelties, 1904, p. 4, fig.

A graceful vine with purple leaves and stems, the former digitately compound, composed of five small leaflets, the terminal, measuring 3½ in. in length by 1 in. in breadth, the largest.

When first produced of a bright claret-purple, ultimately suffused on the upper surface with green as the plant matures, but on the under the purple hue retained all through the summer, glistens with a metallic lustre.

As an ornamental species perhaps one of the best of the Chinese vines and certain to become popular.

VITIS TOMENTOSA, *Planch.*

A Chinese species with cordate leaves, sometimes slightly trilobed, resembling those of the common vine (Vitis vinifera).

The young wood, petioles, the upper and under surface of the leaf are covered with soft hair.

ZELKOVA ACUMINATA, *Planch.*

Syns. *Z. Keaki*, Maxim.; *Planera acuminata*, Lindl.

Nicholson in Woods and Forests, 1884, p. 176, figs.; Lindl. in Gard. Chron. 1862, p. 428.

A handsome deciduous tree discovered near Yeddo, Japan, by the late John Gould Veitch, through whom it was introduced to this country.

In the native habitat a tree 90 to 100 ft. in height, with a straight stem, not unlike an elm.

The very valuable timber used by the Japanese in the construction of houses, ships, and the best kind of cabinet work, is hard and takes a fine polish; the gnarled stems and roots are used for the best kinds of lacquer work. So far in this country the tree has not become common, and specimens are of comparatively small dimensions, but the light graceful foliage, most pronounced in early spring, should commend the species to a wide class of planters.

TREES AND SHRUBS
EVERGREEN AND CLIMBING
PLANTS

TREES AND SHRUBS—EVERGREEN AND CLIMBING PLANTS

ABELIA SPATHULATA, *Sieb. & Zucc.*

Bot. Mag. t. 6601.

A beautiful free-flowering hardy shrub introduced from Japan through Charles Maries.

The flowers in pairs, white with a yellow throat, are subtended by a rosy calyx of four or five spreading lobes.

ARUNDINARIA NITIDA, *Mitford.*

Syns. *A. Khasiana*, Hort. (non Munro).

Gard. Chron. 1898, vol. xxiv. p. 211, fig.; Mitford in The Bamboo Garden, p. 73, fig.

Raised at Coombe from seed received in 1889 from Dr. Regel, at that time Director of the Botanic Gardens, St. Petersburg.

It is described in The Bamboo Garden (*l.c. supra*) as "By far the daintiest and most attractive of all its genus," and the tale is:—

"The story of this lovely species is somewhat curious. When the Bamboo Garden was being formed at Kew, Mr. Bean came across it in Messrs. Veitchs' nursery at Coombe Wood, where it was then named Bambusa nigra. . . . At that time the only Arundinaria known to have black stems was the Himalayan A. Khasiana, and with this species the plant now under notice was conjecturally identified. As A. Khasiana, accordingly, it was described by Mr. Bean in the Gardeners' Chronicle and by myself in The Garden. Attention, however, was called to the subject by Mr. Gamble's monograph of the Bambuseæ of British India, from which it is clear that this Arundinaria agrees only in its purple-black stems with A. Khasiana, and, moreover, that there is not among the Bambuseæ of the Himalayas any known plant corresponding with it."

Further inquiry showed that the seed received from Dr. Regel had been collected in North Szêchuan by the Russian, Potanin, and a new name required, nitida was chosen as appropriate to its brilliancy and beauty; an unusually graceful species.

ARUNDINARIA VEITCHII, *N. E. Brown.*

Syns. *Bambusa Veitchii*, Carrière; *B. albo-marginata*, Hort.

N. E. Brown in Gard. Chron. 1889, vol. v. p. 521; Mitford in The Bamboo Garden, p. 77, fig.

A dwarf-growing species from Japan, some 2 ft. high, with leaves

7 in. long by $2\frac{1}{4}$ in. broad, green on the upper surface, glaucous beneath.

The edges wither and turn white in autumn, the plant assuming a half-dead appearance during the winter months. The thick new foliage of the spring is robust, and for ousting weeds and noxious rubbish this Bamboo is very valuable.

AZARA MICROPHYLLA, *Hook. f.*

Gard. Chron. 1874, vol. i. p. 81, fig.; Veitchs' Catlg. of Pl. 1874, p. 3, fig.

A neat evergreen shrub, a native of Valdivia, introduced through Richard Pearce.

From the distichous arrangement of the shoots, the branches spread in one plane; the small, dark, shining green leaves in pairs, one of each pair darker than the other and slightly different in form.

The flowers small, inconspicuous, numerous, with a vanilla-like fragrance, are succeeded by bright red ornamental fruit.

In the West of England Azara microphylla forms a small tree 20 to 30 ft. high, with neat shining evergreen foliage.

BALBISIA VERTICILLATA, *Cav.*

Syns. *Cruckshanksia cistiflora*, Hook.

Bot. Mag. t. 6170.

Balbisia verticillata is a half-hardy evergreen shrub, with yellow flowers resembling a Cistus or an Ænothera, introduced to this country for the first time by the Horticultural Society in 1825, and subsequently lost; it was later sent to the Veitchian house, and flowered in September 1875.

Like other plants that love a cool, dry atmosphere, B. verticillata, difficult to cultivate, is liable to damp in a long wet winter.

It has a wide range in Chili and Peru, and is known to the natives of the country as the Flore de San José.

BERBERIDOPSIS CORALLINA, *Hook. f.*

Bot. Mag. t. 5343.

This charming evergreen shrub, a native of the Valdivian forests, introduced to this country through Pearce, is not only a beautiful plant, but botanically one of the most interesting, as it connects the two natural orders Berberideæ and Lardizabalæ united by Bentham and Hooker in the Genera Plantarum previous to its discovery—a link which proved the theory correct.

It is not a little remarkable that such a striking plant should have remained so long unknown.

The deep coral-red flowers, from which the plant takes its name, are on slender deep-red pedicels, usually in pairs, in the axils of the upper leaves arranged in drooping terminal racemes, of a thick consistence with a shining surface.

Unfortunately not perfectly hardy, there are in the south-west corner of England fine examples growing in the open.

BERBERIS ACUMINATA, *Franch.*

Plantæ Delavayanæ, t. 38.

This fine evergreen species was first made known through Père Delavay, who collected it in Central China in 1882.

In 1900 seeds were sent from the same locality, and plants raised flowered at Coombe Wood in 1904.

The foliage is distinct and handsome, 5 to 6 in. long, narrow, lanceolate, acuminate, with spiny margins: the young wood bright red.

The flowers in the axils of the uppermost leaves are large for a Berberis, creamy yellow in colour, on slender peduncles.

BERBERIS CONGESTIFLORA, *Gay*, var. HAKEOIDES, *Hook.*

Bot. Mag. t. 6770; Gard. Chron. 1901, vol. xxix. p. 295, fig.

This striking plant is quite unlike any other Barberry in cultivation.

It forms a large bush of decurved branches loaded with globose masses of yellow flower, sessile in the axils of the leaves and along the leafless terminations of the branches.

It was introduced from the Cordillera of Chile through Richard Pearce in 1861, and flowers annually at Coombe Wood.

BERBERIS DARWINII, *Hook.*

Moore in Gard. Mag. Bot. 1851, p. 129, fig.; Paxt. Fl. Gdn. 1851, t. 46; Bot. Mag. t. 4590; Fl. des Serres, tom. vii. p. 47; Gard. Chron. 1851, p. 167; *id.* 1884, vol. xxi. p. 452.

First discovered by the celebrated Charles Darwin, and named in compliment to him.

Probably the best known, if not the most beautiful, of all the species of Berberis at present in cultivation, introduced by William Lobb in 1849 from Chiloe, an island off the south coast of Chili.

The neat glossy foliage and rich golden-yellow flowers beautifully tinged with red are borne in dense profusion in early spring. A writer in the Gardeners' Chronicle (*l.c. supra*) states:—"If Messrs. Veitch had done nothing else towards beautifying our gardens, the introduction of this single species would be enough to earn the gratitude of the whole gardening world."

It is further interesting as one of the parents of B. × stenophylla, a very widely grown garden hybrid.

BERBERIS WALLICHIANA, *DC.*

Syns. *B. Jamesoni*, Hort. (non Lindl.).

Bot. Mag. t. 4656; Paxt. Fl. Gdn. vol. i. pp. 12 and 79, fig.

Originally detected in Nepal by Dr. Wallich, after whom it is named, and sent for cultivation through Thomas Lobb, and shortly afterwards by Dr. Hooker, from the Eastern Himalaya.

CAMELLIA GRIJSII, *Hance.*

A Camellia closely allied to the attractive little Japanese species, C. Sasanqua, inhabiting the glens and gorges of the Yangtsze and its chief tributaries at elevations of 1,000-2,000 ft.

An attractive shrub, seldom exceeding 5 ft. in height, with neat lanceolate-acuminate shining leaves and ivory-white flowers 1 to 1½ in. in diameter, produced abundantly in the early spring, and at intervals till late in autumn.

Introduced to cultivation through Wilson from Central China in 1901.

CAMELLIA SASANQUA, *Thunb.*

The Garden, 1893, vol. xliv. p. 328, pl. 930; Bot. Mag. t. 5152.

There are several forms of this beautiful Camellia in a wild state, and numerous garden varieties.

Long known in gardens, having been introduced by the East India Company in the early part of the 18th century, it appears to have been subsequently lost to British horticulture till re-introduced from Japan through Charles Maries.

CARYOPTERIS MASTACANTHUS, *Schauer.*

Syns. *Nepeta incana*, Thunb.

The Garden, 1883, vol. xxiv. p. 523, with fig.; Bot. Mag. t. 6799; Gard. Chron. 1884, vol. xxi. p. 148, fig. 30; Lindl. Bot. Reg. 1846, t. 2; La Belg. Hort. 1893, p. 273, col. pl.

Caryopteris Mastacanthus was first sent to this country by Fortune who found it wild near Canton.

First cultivated as a greenhouse plant, it was discarded for more showy occupants, and not being perfectly hardy was ultimately lost to cultivation.

For its re-introduction we are indebted to Charles Maries, who sent seed from China.

TREES AND SHRUBS—EVERGREEN

CASTANOPSIS CHRYSOPHYLLA, *A. DC.*

Syns. *Castanea chrysophylla*, Hook.

Gard. Chron. 1897, vol. xxii. p. 411, fig. 120; *id.* 1904, vol. xxxvi. p. 152, fig. 59; Bot. Mag. t. 4953.

A representative of a most remarkable genus, intermediate in character between the oak and the chestnut, the "Golden-Leaved Chestnut." The name in allusion to the bright golden-yellow colour of the under surface of the leaves, the upper bright lustrous green; as the branches move to the wind, the contrast presented by the two surfaces is singularly conspicuous.

A native of Oregon, a small tree, of a shrubby nature, found at high elevations on the Californian coast range south of the Bay of San Francisco. It attains full size and beauty in the humid climate of the coast valleys of Northern California, and is one of the noblest of the forest inhabitants, reaching a height of 100 to 150 ft., frequently free of branches up to 80 ft. above ground. Seed sent from California by William Lobb, probably about the year 1853, from which plants were raised. The finest specimen in this country is at Tortworth Court, Gloucestershire, the seat of Earl Ducie, which bears fruit freely every year, and from which plants have been raised.

There is also a good specimen in the gardens at Pencarrow, Cornwall.

CEANOTHUS FLORIBUNDUS, *Hook.*

Bot. Mag. t. 4806; Watson, Gray and Brewer, Botany of California, vol. i. p. 104.

A handsome and interesting hardy species raised from seed sent from California by William Lobb.

In Botany of California (*l.c. supra*) is written: "This species is as yet known only from the figure and original description drawn from cultivated specimens. But for the peculiar inflorescence it might be a form of Ceanothus dentatus."

The flowers are of the richest mazarine-blue that can be imagined, in numerous capitate globose corymbs, crowded at the extremity of numerous short stubby branches.

CEANOTHUS LOBBIANUS, *Hook.*

Bot. Mag. t. 4810; Fl. des Serres, 1854-1855, tom. x. p. 125.

This Californian species, introduced to Exeter by William Lobb, has many points in common with Ceanothus thyrsiflorus, and is possibly a hybrid form of that species.

CEANOTHUS OREGANUS, *Nutt. MSS.*

Bot. Mag. t. 5177.

This species was first detected by Douglas in the woods of Oregon, frequent from the Blue Mountains to the sea, and was also found by Nuttall and Tolmie; introduced to this country through William Lobb from the same locality; the flowers white, in copious lateral panicles during the month of May. The plant is now rarely seen.

CEANOTHUS PAPILLOSUS, *Torr. & Gray.*

Bot. Mag. t. 4815.

One of the discoveries of David Douglas, for the Horticultural Society of London, though not introduced to this country until William Lobb sent seed to Exeter, from which plants raised flowered in an open border for the first time in June 1854.

CEANOTHUS VEITCHIANUS, *Hook.*

Bot. Mag. t. 5127; Fl. des Serres, tom. iii. 2nd ser. 1858, p. 171; Sargent, Silva of North America, vol. ii. p. 43.

A magnificent hardy shrub introduced from California through William Lobb, and generally the most successful of the genus.

Though closely allied to Ceanothus floribundus, C. papillosus, and C. Lobbianus, it is distinguished by its foliage, and surpasses these in an abundance of bright mazarine-blue flowers, and a glossy, almost varnished, leaf-surface.

C. thyrsiflorus shows a tendency to cross with other species and produce natural hybrids; several had been suspected, and Dr. Parry, who long studied the Californian Ceanothus in the field, reached the conclusion that C. Lobbianus and C. Veitchianus are hybrids of this species.

CEANOTHUS VELUTINUS, *Douglas.*

Bot. Mag. t. 5165.

This species was first found on the Rocky Mountains, at considerable elevations, by David Douglas, but not introduced to cultivation until sent home by William Lobb.

A white-flowered shrub with singularly dark-green leaves, glossy from an aromatic resin exuded in hot weather, the under side velvety with a whitish down.

CEANOTHUS VEITCHIANUS

BICTON, DEVON

TREES AND SHRUBS—EVERGREEN

CLEMATIS ARMANDI, *Franch.*

Jour. R.H.S. 1903, vol. xxviii. p. 58, fig. 14; Gard. Chron. 1905, vol. xxxviii. p. 30, suppl. illus.

An evergreen species of Clematis allied to C. Meyeniana, and by some botanists considered only a form.

A native of Central and Western China, frequently met with at altitudes above 3,000 ft. on open scrub-clad mountains.

The flowers pure white, often rosy-pink at the back, 2 in. in diam., are in dense axillary clusters, the trifoliate evergreen leaves of leathery texture. Plants, from seed sent from Central China, flowered for the first time in this country at Coombe Wood in April 1905.

CLEYERA FORTUNEI, *Hook. f.*

Syns. *Eurya latifolia*, Hort., var. *variegata.*

Bot. Mag. t. 7434; Gard. Chron. 1862, p. 398 (advt.).

A handsome Japanese shrub, introduced to this country through the late John Gould Veitch, and about the same time by Fortune.

It is half-hardy except in favoured localities, remarkable for the great beauty of the bright green leaves, variegated with golden-yellow, scarlet towards the margin.

The plant long in cultivation under the name of Eurya, till specimens ultimately flowered which proved it to be a species of Cleyera.

COTONEASTER APPLANATA, *Duthie.*

An important addition to a genus of useful garden shrubs, raised from seed collected in Central China in 1900. The branches have a tendency to grow at right angles to the erect main stem in one plane only, and the plant is naturally adapted for growing on trellises or against a wall, and in such situations the neat dark-green foliage, and in autumn the brilliant scarlet berries, are particularly effective. The leaves small, heart-shaped, 1 in. long by ½ in. broad, dark green above, covered beneath with a white tomentum.

COTONEASTER BULLATA, *Duthie.*

This fine Cotoneaster, a new species raised from seed collected in Central China in 1900, has leaves dark green on the upper surface, paler beneath, slightly tomentose along the midrib and principal veins; they measure 2 in. long by 1¼ in. broad, the margins entire, the surface bullate, suggesting the specific name.

COTONEASTER HUMIFUSA, *Duthie.*

A dense-growing creeping species, introduced from Central China through Wilson, the leaves dark green above, pale and slightly tomentose on the under surface, 1 in. long by ½ in. broad.

The long shoots run over the ground as a carpet of bright green foliage, in autumn studded with vivid scarlet fruit.

The habit of this plant is admirably adapted for covering banks or rockeries.

DESFONTAINEA SPINOSA, *Ruiz & Pav.*

Gard. Chron. 1849, p. 564 (Notice of New Plants); *id.* 1854, p. 287; Bot. Mag. t. 4781; Fl. des Serres, tom. ix. p. 207.

An evergreen shrub with glossy holly-like leaves and tubular flowers often 2 in. long, bright scarlet, tipped with yellow.

Unfortunately not hardy in all localities, it is worth growing as a pot plant, if protected during winter and plunged in borders to flower in the summer months.

The plant was long a puzzle to systematic botanists, who differed in their opinion as to its affinities and the exact position in the Natural System, but it was finally placed in Loganiaceæ, a family of which the Buddleia is a familiar example.

Desfontainea spinosa was introduced to gardens through William Lobb, who sent plants to Exeter, where it flowered for the first time in August 1853.

ELÆAGNUS MACROPHYLLA, *Thunb.*

Bot. Mag. t. 7638.

A handsome species, the largest-leaved in the genus, with the young wood, leaves, flowers, and fruits, the latter bright red in colour, and very ornamental, covered with the characteristic lepidote scales.

It was introduced from Japan to Coombe Wood by Charles Maries, in 1879, and is perfectly hardy.

EMBOTHRIUM COCCINEUM, *Forst.*

Gard. Chron. 1849, p. 564 (Notice of New Plants); Bot. Mag. t. 4856; Fl. des Serres, 1858, tom. iii. 2nd ser. p. 37; The Florist, 1858, vol. xi. n.s. pl. 135; The Garden, 1876, vol. x. p. 566; pl. li.

A handsome evergreen shrub with racemes of the richest scarlet in profusion at the end of April and the beginning of May, justifying the popular appellation of "Flame Bush."

Like other South American shrubs not perfectly hardy in all parts of this country, it succeeds admirably in South Devon, Cornwall, and Ireland.

Interesting as a member of the order Proteaceæ, an order representative of Australian plants, formerly much cultivated in glasshouses in this country, although seldom seen in beds or borders.

Embothrium coccineum is a native of the Straits of Magellan and Tierra del Fuego, introduced through William Lobb, and later by Pearce. There are two forms, one having much brighter-coloured flowers than the other.

It flowered for the first time in this country at Exeter in May 1853.

ENKIANTHUS CAMPANULATUS, *Nichols.*

Nich. Dict. Gard. vol. i. p. 510; Bot. Mag. t. 7059.

A native of the northern part of Japan, frequently collected, though in the more southern provinces only known in cultivation, a pretty, quite hardy ericaceous shrub bearing in abundance in the spring of the year small bell-shaped ochreous-red flowers.

Charles Maries sent it to English gardens.

ESCALLONIA × LANGLEYENSIS, *Hort. Veitch.*

Gard. Chron. 1897, vol. xxii. p. 17, fig. p. 15; *id.* 1898, vol. xxiii. p. 11, fig. 4.

A valuable hybrid raised by Seden at Langley by crossing Escallonia Philippiana with the pollen of the dark variety of E. macrantha known in gardens as *sanguinea.* Perfectly hardy, of semi-scandent habit, in many respects intermediate between the two parents; the slender elongated branches furnished with small oval leaves of a dark lustrous green, produce along their whole length numerous erect branchlets, each with four to seven bright rose-carmine flowers.

ESCALLONIA MACRANTHA, *Hook. & Arn.*

Bot. Mag. t. 4473; Gard. Chron. 1849, p. 371 (Report of Exhibition); Fl. des Serres, 1850, p. 305.

First made known by Mr. Cuming, and shortly afterwards introduced in a living state through William Lobb, from Chiloe, it flowered in 1848, and, exhibited at the Garden Exhibition of the Horticultural Society held in July of that year, is now one of the most generally planted of the genus.

As a subject for sea-side planting, few shrubs are more charming or useful, combining beautiful glossy foliage with a profusion of rose-carmine flowers in early spring, and a dense habit of growth; a valuable subject for hedge-work in warmer spots.

Along the Cornish coast it may be seen in quantity, and the flowers are sold in the streets of watering-places.

ESCALLONIA MONTANA, *Philippi.*

Gard. Chron. 1873, p. 947.

This plant introduced from the mountains of Valdivia through Richard Pearce, first flowered in 1873.

The species closely allied to Escallonia rubra, has reddish blossoms and is not quite hardy, except in favoured localities.

ESCALLONIA ORGANENSIS, *Gardn.*

Bot. Mag. t. 4274.

This species first detected by Gardner, William Lobb sent seed about the same time from the Organ Mountains to Exeter, from which plants raised flowered during the summer of 1846.

The blooms of a deep rose colour, the midrib and margins of the leaves tinted red.

ESCALLONIA PHILIPPIANA, *Mast.*

Masters in Gard. Chron. 1873, p. 947; *id.* 1878, vol. x. p. 109, fig.; Bean in Gard. Chron. 1893, vol. xiv. p. 60.

This, one of the most distinct and the hardiest of all Escallonias in cultivation, was introduced from Valdivia through Richard Pearce, and first flowered in July 1873.

The pure white flowers, in dense profusion during the summer months, give the plant the appearance of some of the Spiræas.

It is further interesting as one of the parents of E. × langleyensis, a really valuable cross.

ESCALLONIA PTEROCLADON, *Hook.*

Gard. Chron. 1855, p. 36, with fig.; Bot. Mag. t. 4827.

A small hardy shrub, with leaves like a small-leaved myrtle, and pretty Epacris-like flowers, white tinged with red, which grows to a height of some 4 to 5 ft.; an abundant, fragrant bloomer.

The specific name is in allusion to the wings which clothe the young stems, but which in a measure peel off and disappear as the plant ages.

It is a native of Western Patagonia, first detected by William Lobb, who sent plants to Exeter which flowered in July 1854, on which occasion it was exhibited at one of the Horticultural Society's exhibitions held at Chiswick.

EUCRYPHIA CORDIFOLIA, *Cav.*

Nicholson in Gard. Chron. 1897, vol. xxii. pp. 246, 247, fig.

This species flowered, probably for the first time in Europe, at Coombe

Wood during the summer of 1897, and from material supplied from that source the figure in the Gardeners' Chronicle (*l.c. supra*) was prepared.

A native of Chili, attaining a height of 30 ft. or more, the dark persistent foliage sets off to advantage the snowy white flowers: not so hardy as the beautiful Eucryphia pinnatifolia, except in southern or south-western counties a sheltered position is required.

EUCRYPHIA PINNATIFOLIA, *Gay.*

Gard. Chron. 1880, vol. i. p. 337; The Garden, vol. xii. p. 544, col. pl.; Bot. Mag. t. 7067; Flora and Sylva, vol. i. p. 73, fig. p. 41.

An exceptionally fine and most interesting shrub or small tree from Chili, very locally distributed, confined, as far as at present known, to the Cordillera of Concepcion, where it forms a small tree about 10 ft. high, called "Nirhe" by the people.

The flowers in August, produced in immense quantities, are pure white, and in shape and set of the stamens similar to the ordinary St. John's Wort or a Stuartia.

The stamens are numerous, with long filaments tipped with golden-yellow anthers, red when the flower first expands, a contrast to the pure white perianth; they persist after the petals have fallen, and in themselves are very ornamental.

It will always remain a very choice shrub in this country on account of the difficulty of propagation and an intense objection to transplanting, but when once established no further attention is required.

EUONYMUS RADICANS, *Sieb. & Zucc.*, FOLIIS VARIEGATIS.

Gard. Chron. 1862, p. 398 (advt.).

A very pretty little shrub from Japan, of neat habit, small ovate leaves of a bright green, blotched and margined with silvery white.

FAGUS OBLIQUA, *Mirb.*

Gard. Chron. 1849, p. 563 (Notice of Novelties for the year 1849).

A handsome evergreen beech of great beauty growing from 30 to 40 ft. in height, and inhabiting Chili and Patagonia, whence it was introduced by William Lobb.

It is one of the Robles of the Chilenos, and is of value for a very solid, close-grained, heavy wood.

According to Sir Joseph Hooker, it occupies only the lower elevations of the mountains, and so cannot be perfectly hardy.

GAULTHERIA FERRUGINEA, *Cham. & Schlecht.*

Syns. *Andromeda hirsuta*, Arrab.

Gard. Chron. 1844, p. 38 (Report of Horticultural Society's Show); Bot. Mag. t. 4697.

A half-hardy evergreen shrub from the Organ Mountains, Brazil, through William Lobb, exhibited in flower for the first time before the Horticultural Society in January 1844.

The figure in the Botanical Magazine (*l.c. supra*) was prepared from plants raised at the Comely Bank Nursery, Edinburgh, in 1852, from seed collected by Gardner in the same locality as Lobb.

It is a handsome plant with urn-shaped drooping rose-pink flowers in short racemes from the axil of the leaf.

HAPLOPAPUS SPINULOSUS, *Hook. f.*

Bot. Mag. t. 6302.

This low bushy sub-shrub, with pinnatifid spiny leaves and numerous bright yellow composite flowers which open late in the year, native of the prairies which border the Rocky Mountains, from the boundary of the British possessions as far south as New Mexico.

It was raised from seed and first flowered in August 1874.

HYMENANTHERA CRASSIFOLIA, *Hook. f.*

Gard. Chron. 1875, vol. iii. p. 237; *id.* 1892, vol. xii. p. 411, fig. reproduced; Veitchs' Catlg. of Pl. 1877, p. 14, fig.; The Garden, 1877, vol. xi. p. 145, fig.

A hardy evergreen shrub of the violet family, with small inconspicuous blossoms, succeeded by pure white berries; an interesting subject during the autumn months, raised from seed sent by a correspondent from New Zealand, still rare in cultivation.

There is an especially good specimen on a wall in the Botanic Gardens at Cambridge and one at Coombe.

HYPERICUM HOOKERIANUM, *Wight & Arn.*

Syns. *H. oblongifolium*, Hook.

Bot. Mag. t. 4949.

A native of Northern India, Nepal, and the Himalayas, at elevations of 6,000-12,000 ft., found on the hills about Mufflong, Assam, by Thomas Lobb, through whom it was introduced.

It forms a neat bush with evergreen leaves and large rich yellow flowers, unfortunately not perfectly hardy in all localities.

ILEX PERNYI, *Franch.*

Jour. R.H.S. 1903, vol. xxviii. p. 59.

A species named after Père Paul Perny, a French Jesuit missionary; a discovery during his travels in China between the years 1850 and 1860.

A dense-growing species with small spiny leaves and red berries, probably allied to Ilex cornuta.

Raised from seeds collected in China by Wilson, it has proved perfectly hardy at Coombe Wood, and of very dwarf compact habit.

JASMINUM PRIMULINUM, *Hemsl.*

Hooker's Ic. Pl. t. 2384; Gard. Chron. 1903, vol. xxxiii. p. 197, fig. 83; Bot. Mag. t. 7981; Flora and Sylva, 1904, vol. ii. p. 168, col. pl.

A beautiful shrub first discovered by W. Hancock Esq., at Mengtse in Yunnan, and later by Dr. A. Henry and Wilson in the same locality. By the last-named living plants were sent to Coombe Wood, where they flowered in October 1901 for the first time.

In general appearance the plant resembles the well-known Jasminum nudiflorum, but the flowers and leaves are much larger and the plants when grown in the open almost evergreen.

In Dr. Henry's view, in an article in Flora and Sylva (*l.c. supra*), it is only a form of J. nudiflorum that has escaped from cultivation. In support of this theory he mentions the facts that it never sets seed, the flowers are often semi-double, and the shrubs were always seen growing in gardens or in hedges in the vicinity of villages, and never in woods and forests, but a greater knowledge may lead to a different opinion.

Plants of a high order, grown by Leopold de Rothschild Esq., Gunnersbury, were exhibited in London in January 1906, cultivated as a cold greenhouse subject.

LARDIZABALA BITERNATA, *Ruiz & Pav.*

Bot. Mag. t. 4501.

This singular evergreen climber introduced through G. T. Davy Esq., who found it in the Province of Concepcion and gave instructions for plants to be sent to him at Valparaiso, ultimately brought to Exeter by William Lobb; it is now in general cultivation.

It flowered for the first time in March 1849.

LEUCOTHOË DAVISIÆ, *Torrey.*

Syns. *L. Lobbii*, Hort.

Bot. Mag. t. 6247.

A handsome evergreen shrub with neat elegant racemes of white flower resembling those of the lily-of-the-valley.

Discovered by William Lobb in 1853 on the Sierra Nevada Mountains of California, at an elevation of 5,000 ft., introduced and distributed by the Veitchian firm under the name of Leucothoe Lobbii, and subsequently again gathered in the same locality by Miss N. J. Davis, after whom Dr. Torrey named it, a name adopted by Dr. Asa Gray in the Proceedings of the American Academy and in his work on the Botany of California.

Remarkable as the only species of the genus known on the western side of the American Continent.

LIGUSTRUM HENRYI, *Hemsl.*

Hemsley in Jour. Linn. Soc. vol. xxvi. p. 90.

An evergreen species of neat appearance first detected in the immediate neighbourhood of Ichang by Dr. A. Henry, after whom it is named, and from this locality afterwards introduced to cultivation.

A neat and attractive shrub with glossy dark green leaves somewhat variable in form, from 1½ to 2 in. long by 1¼ to 1½ in. broad, in outline from rotund-ovate to ovate-lanceolate.

Mr. Hemsley states (*l.c. supra*), "This is very near Ligustrum Tschonskii, differing markedly in the shape of the leaves, which in the present species vary much in size and shape on the same branch."

The flowers white, fragrant, and the fruits black.

LONICERA PILEATA, *Oliver.*

Gard. Chron. 1904, vol. xxxv. p. 243, figs.; Hooker's Ic. Pl. t. 1585; The Garden, 1904, vol. lxv. p. 235; Bot. Mag. t. 8060.

A hardy Chinese honeysuckle, with neat evergreen foliage, introduced from the Province of Hupeh.

It is a dwarf-spreading shrub with dark green foliage somewhat resembling that of the common privet. The flowers freely produced in the axils of the uppermost leaves are about ½ in. in length, greenish-white in colour, not showy, but delightfully fragrant.

It flowered for the first time in this country at the Royal Gardens, Kew, during April 1904.

MAGNOLIA DELAVAYI, *Franch.*

Plantæ Delavayanæ, p. 33, tt. 9, 10; Flora and Sylva, 1903, vol. i. p. 18.

An evergreen species named in compliment to Père Delavay who discovered it in Central China. It was afterwards collected by Dr. A. Henry and Wilson, and by the last-named traveller introduced to cultivation at Coombe Wood.

The plants are at present too small to flower, but from information derived from those who have seen it in the native habitat this Magnolia promises to be a great addition to gardens.

The flowers are said to be pure white, egg-shaped, of great substance and very massive.

TREES AND SHRUBS—EVERGREEN

MYRTUS CHEKEN, *Spreng.*

Syns. *Eugenia Cheken*, DC.

Bot. Mag. t. 5644.

A pretty evergreen Chilian plant introduced from Chili through William Lobb, suitable for walls in the southern and western parts of England, with pure white flowers plentifully produced during the summer months.

MYRTUS LUMA, *Barn.*

Syns. *Eugenia Luma*, Berg.

Bot. Mag. t. 5040.

A charming evergreen, quite equal in beauty to the common myrtle, a native of the colder parts of Chili from Concepcion to the island of Chiloe and Valdivia, sent to this country through William Lobb.

Blossoming in the summer months, the branches literally loaded with white flowers, almost concealing the foliage.

MYRTUS UGNI, *Molina.*

Syns. *Eugenia Ugni*, Hook. & Arn.

Gard. Chron. 1857, p. 267 (Note on the fruit); Bot. Mag. t. 4626; Fl. des Serres, tom. vii. p. 215.

A half-hardy evergreen shrub, the "Myrtilla" of the Chilians, allied to the common myrtle, and requiring the protection of a wall for successful cultivation out-of-doors, except in most favoured localities.

It is a native of Chili, wild on the hills near Valparaiso, introduced to this country through William Lobb.

The fruit a jet black, delicate juicy berry, large as a black currant, of an agreeable flavour and aroma, is cultivated in the private gardens of Valparaiso as a dessert. Hopes entertained, but not fulfilled, that it might be a useful addition to English hardy fruit, led to trials being made, C. Wentworth Dilke Esq. offering prizes for competition at the Grand Autumn Fruit Show of the London Horticultural Society in 1857.

NOTOSPARTIUM CARMICHÆLIÆ, *Hook. f.*

Bot. Mag. t. 6741; Gard. Chron. 1883, vol. xx. p. 169, fig. 26; Veitchs' Catlg. of Trees and Shrubs, 1883, p. 20.

Known to the inhabitants of New Zealand as the "Pink Broom," and introduced through P. C. M. Veitch.

In its native habitat a large shrub or small tree with green rush-like branches devoid of leaves, on which small clusters of rosy pink pea-shaped flowers are produced during the summer months.

Endemic to the archipelago, but chiefly confined to the Middle Island.

OLEARIA HAASTII, *Hook. f.*

Masters in Gard. Chron. 1872, p. 1195, fig.; Bot. Mag. t. 6592; Veitchs' Catlg. of Pl. 1874, p. 12, fig.; Gard. Chron. 1896, vol. xx. p. 532, fig.

Raised from seed sent by a correspondent from New Zealand to Exeter in 1858, Olearia Haastii, popularly recognized as the "Daisy Bush," is now well known, and a favourite subject for planting.

Perfectly hardy, it withstands the dry atmosphere and heat of summer with impunity, and produces in profusion white daisy-like flowers during the months of August and September.

OLEARIA MACRODONTA, *Baker.*

Syns. *O. dentata*, Hook. f.; *Eurybia dentata*, Hook.

Bot. Mag. t. 7065.

A hardy shrub from New Zealand, with evergreen holly-like leaves and numerous small daisy-like flowers, in broad corymbs terminating each branchlet.

Originally described by Sir Joseph Hooker in his flora of New Zealand as Eurybia dentata, this plant, in his later work, became united to Olearia, the specific name retained. This having, however, been already given to an Australian species, Mr. Baker changed it to O. macrodonta.

A handsome shrub, especially in the south and west of England, and in July has many large corymbs of white flower.

OSMANTHUS AQUIFOLIUM, *Siebold*, var. ILICIFOLIA, *Dippel.*

Syns. *Olea ilicifolia*, Haask.

Gard. Chron. 1858, p. 419 (advt.).

A native of China and Japan, introduced from the latter country through Thomas Lobb in 1856, a holly-like evergreen shrub with clusters of white flower in the axils of the leaves.

PERNETTYA CILIARIS, *Don.*

Syns. *Gaultheria ciliaris*, Cham. & Schlecht.

Gard. Chron. 1878, vol. x. p. 89, fig. 12; Jour. Hort. Soc. London, vol. vi. p. 268.

A pretty evergreen, producing in the early spring months racemes of lily-of-the-valley-like flowers in dense profusion.

A native of Mexico whence it was introduced, now almost certainly lost to cultivation.

PHILESIA BUXIFOLIA, *Lam.*

Gard. Chron. 1882, vol. xviii. p. 105, fig.; Bot. Mag. t. 4738; The Florist, 1854, n.s. vol. iv. pl. 85; Fl. des Serres, tom. ix. p. 41; The Garden, 1883, vol. xxiii. p. 380, col. pl.

A very curious plant, a native of Valdivia, with flowers similar in shape

OLEARIA HAASTII

THE ROYAL GARDENS, KEW

and appearance to those of the well-known Lapageria rosea, and a native of Valdivia, where it inhabits marshy places under trees. Long a desideratum in gardens till William Lobb sent plants to Exeter, which flowered in June 1853, for the first time.

Hardy in Devon and Cornwall, and in some parts of Ireland, it is best grown under glass in less favoured localities, treated as the Lapageria.

An interesting hybrid, intermediate in general character between the two, known as Philageria Veitchii, has been raised by crossing Lapageria with Philesia buxifolia.

QUERCUS ACUTA, *Thunberg.*

Syns. *Q. Buergeri*, Blume.

Woods and Forests, vol. i. 1884, p. 85, fig.; The Garden, 1881, vol. xix. p. 285, same fig.

A handsome evergreen oak sent from Japan by Charles Maries.

It forms a small tree in this country with handsome leathery leaves, resembling those of the common laurel in shape, but of a darker green. Seedling plants vary much in habit, and some forms selected have received the varietal names *pyramidalis* and *robusta*, differ from the type in habit as the names indicate.

The Oak is perfectly hardy in this country, and at Coombe Wood on cold, clayey soil, in an exposed situation there is a very noble symmetrical specimen 20 ft. in height.

QUERCUS GLABRA, *Thunb.*, var. LATIFOLIA, *Hort.*

Veitchs' Catlg. of Trees and Shrubs, 1881-1882, p. 22.

A large-leaved form of Quercus glabra, the finest evergreen oak of Japan. The type species has long been in gardens, but has not proved generally hardy; the leaves smaller than those of the form *latifolia*, are very fine, of a bright fulvous green.

RAPHIOLEPIS JAPONICA, *Sieb. & Zucc.*

Syns. *R. japonica*, var. *integerrima*, Hook.; *R. ovata*, Hort.

Gard. Chron. 1863, p. 694 (advt.); Fl. Mag. t. 299; Bot. Mag. t. 5510; The Garden, 1876, vol. ix. p. 597.

A pretty plant with thick evergreen shining leaves and spikes of large fragrant white flower.

A native of Japan, Bonin, and the Corea, introduced from the first-named country through the late John Gould Veitch.

RHAPHITHAMNUS CYANOCARPUS, *Miers.*

Syns. *Pœppigia cyanocarpa*, Bertero.

Gard. Chron. 1848, p. 474 (Report of R.H.S. Show); Bot. Mag. t. 6849.

A half-hardy evergreen shrub or small tree from Chili, where it was first detected by Mr. Miers, with neat foliage, and small tubular pale blue flowers, succeeded by globose bright blue fruit.

RHODODENDRON AURICULATUM, *Hemsl.*

Jour. R.H.S. 1903, vol. xxviii. p. 64, fig. 25.

A handsome species introduced through Wilson from Central China, common on precipitous cliffs north of the Yangtsze at elevations of more than 5,000 ft.

The flowers, not yet seen in England, are large, funnel-shaped, 3 in. in depth, 4 in. wide at the mouth, pure white or rosy pink in colour, and very striking.

The leaves some 8 in. in length by 2¾ in. broad, are prolonged into two ear-like processes at the point of insertion of the petiole with the blade.

RHODODENDRON CALIFORNICUM, *Hook.*

Gard. Chron. 1855, p. 391 (Note on Exhibit); Bot. Mag. t. 4863.

Discovered by William Lobb during his first collecting mission to California, and introduced to this country, where the plant flowered for the first time during June in the year 1855.

It produces compact trusses of flowers, deep rose-tinted, and is now rarely met with.

RHODODENDRON × EARLY GEM.

Gard. Chron. 1878, vol. ix. p. 335, fig. 57, p. 336.

A hybrid raised at Coombe Wood from Rhododendron præcox and R. dahauricum, the former itself the offspring of R. dahauricum and R. ciliatum.

Early Gem is a dwarf compact plant with small dark green persistent foliage and fairly large pinkish lilac flowers in dense profusion in March and early April. Always an admirable subject for forcing, and in favourable springs one of the earliest to flower in the open.

RHODODENDRON INDICUM, *Sweet*, var. OBTUSUM.

Syns. *R. obtusum*, Planch; *Azalea obtusa*, Lindl.

Lindl. Bot. Reg. vol. xxxii. t. 37; Nicholson in Gard. Chron. 1886, vol. xxv. p. 585, fig.

This charming little plant with vivid orange-scarlet flowers was originally introduced from Shanghai to the gardens of the Horticultural Society by Fortune, in 1844, but probably lost to cultivation until re-introduced from Japan by the Veitchian firm.

RHODODENDRON × MANGLESII, *Hort.*

Veitchs' Catlg. of Trees and Shrubs, 1885, p. 11; Gard. Chron. 1885, vol. xxiv. p. 48, fig.; The Garden, 1890, vol. xxxviii. p. 225, fig.

A beautiful hybrid obtained at Coombe from the Himalayan Rhododendron Aucklandii and a garden form of R. catawbiense known as *album grandiflorum.*

The flowers of the hybrid fully 4 in. in diameter, are of the purest white, with the exception of a few red spots on the upper segment of the corolla. The truss large and full-flowered, composed of from ten to fifteen blooms, resembles in this respect the American species, whilst in the foliage and calyx, and the size and quality of the flower the influence of R. Aucklandi may be traced.

It was dedicated to the memory of J. H. Mangles, Esq., of Haslemere under whose care this genus was so very greatly improved.

RHODODENDRON MICRANTHUM, *Turcz.*

Jour. R.H.S. 1903, vol. xxviii. p. 64.

This Rhododendron, a native of Central China, on cliffs north of the Yangtsze at elevations of more than 5,000 ft., forms a bush from 4 to 20 ft. high, and produces in great abundance small white flowers in erect racemes. The leaves small, are covered on the under surface with ferruginous scales. Plants raised from seed collected by Wilson are growing at Coombe Wood, and the species first flowered in this country at Caerhays Castle, Cornwall, in the spring of 1905.

RHODODENDRON OCCIDENTALE, *A. Gray.*

Syns. *Azalea occidentalis*, Torr. & Gr.

Bot. Mag. t. 5005.

Raised from seed sent from California by William Lobb, and much resembling the Azalea calendulacea of Eastern America, but differing in flowers white striped with pink on the exterior of the perianth, with a yellow blotch on the standard petal. In A. calendulacea the corolla is from yellow to orange changing to flame colour.

RHODODENDRON RACEMOSUM, *Franch.*

Gard. Chron. 1892, vol. xii. p. 62; The Garden, 1892, vol. xlii. p. 320, col. pl.; Bot. Mag. t. 7301.

A charming dwarf-growing species of Rhododendron raised from seed sent from the Jardin des Plantes, Paris, collected in Yunnan, Western China, by Père Delavay.

The plant quite hardy, produces in the spring months dense clusters of soft rose-pink flowers at the end of the short branches.

RHODODENDRON SCHLIPPENBACHII, *Max.*

Gard. Chron. 1894, vol. xv. p. 462, fig. 58; Bot. Mag. t. 7373; The Garden, 1894, vol. xlvi. p. 80, col. pl.

A Corean and Manchurian species discovered by Baron Alexandra von Schlippenbach on the shores of Possjet Sound, Manchuria. From the texture of the leaves, deciduous in this country, it obviously belongs to the Azalea section of the genus, of which it is the largest leaved.

The flowers of a delicate rosy lilac, are spotted at the base with darkish brown.

Rhododendron Schlippenbachii was introduced by James H. Veitch, who sent it to Chelsea in 1893 from Japan, where it was found cultivated, but it is said to have been first collected by a young Kewite, Richard Oldham, in 1863, collecting at that time in China; it was not introduced nor was it named till Maximowicz wrote a description from Baron Schlippenbach's specimens seven years later.

RUBUS JAPONICUS, *Veitch*, var. TRICOLOR, *Hort.*

Gard. Chron. 1894, vol. xvi. p. 95, fig. 15.

A slender-growing trailing plant from Japan, with rose-coloured stems, the petioles and leaves with large patches of white amongst the green, in a young state pinkish-white.

SARCOCOCCA PRUNIFORMIS, *Lindl.*

A neat dwarf evergreen shrub of Holly-like appearance, with shining bright green pointed leaves, in the axils small clusters of whitish flower, open in the winter, followed in the spring by bright blue fruit.

It is recorded from Afghanistan, Northern India, and southward from Ceylon and Sumatra, and was introduced from the Province of Hupeh, Central China.

SCHIZANDRA HENRYI, *Clarke.*

Gard. Chron. 1905, vol. xxxviii. p. 162, fig. 55.

A hardy climbing shrub from the Province of Hupeh, Central China, a member of a genus little known in gardens, forming with Kadsura a distinct tribe of the Magnoliaceæ. Botanically the flowers are interesting, but not showy; the leaves bright shining green have showy red petioles; the young stems winged.

After flowering, the receptacle becomes fleshy, enclosing mucilaginous berries, and the fruit is eaten by the local peasantry.

TREES AND SHRUBS—EVERGREEN

THIBAUDIA MICROPHYLLA, *Lindl.*

Lindl. in Gard. Chron. 1848, p. 23, fig.

An evergreen shrub with scarlet flowers, from seed collected in Peru by William Lobb.

The figure and description (*l.c. supra*) were prepared from Lobb's herbarium specimens, only seedling plants, not large enough to flower, at that time in cultivation.

Apparently the plant not hardy was never distributed; and it is doubtful if a member of this genus, as the flowers solitary and axillary are not in a racemose inflorescence.

TROCHODENDRON ARALIOIDES, *Sieb. & Zucc.*

Gard. Chron. 1894, vol. xv. p. 716, fig.; Bot. Mag. t. 7375.

A very singular hardy evergreen shrub, native of Japan, introduced to Coombe Wood; flowered in April 1894 for the first time.

With Euptelia a distinct tribe of the Magnoliaceæ, known as Trochodendræ, distinguished by an entire absence of sepals and petals.

The stamens radiate from the centre of the flower as the spokes of a wheel, suggesting the generic name.

VACCINIUM LEUCOSTOMUM, *Lindl.*

Lindl. in Gard. Chron. 1848, p. 7, fig.

Raised from seed sent by William Lobb from Peru, collected at Veto, 8,000 ft. above sea-level.

The flowers are described as "scarlet tipped with white"; the plant probably not hardy, was soon lost, and does not appear to be now in cultivation.

VALDIVIA GAYANA, *Remy.*

Gard. Chron. 1863, p. 366 (Report of R.H.S. Show); The Florist, 1863, p. 87.

This singular, small, half-hardy evergreen shrub, with a short pyramidal panicle of pretty rose-coloured flowers from a tuft of long, serrated rugose leaves, was exhibited in flower in April 1863, and is now, in all probability, also lost.

VIBURNUM BUDDLEIFOLIUM, *C. H. Wright.*

Gard. Chron. 1903, vol. xxxiii. p. 257.

A new species introduced to Coombe Wood from Central China through Wilson.

The leaves about 3½ in. in length, hairy on the upper surface, densely floccose beneath, resemble those of Viburnum rhytidophyllum, but are smaller; the flowers white, in terminal corymbose panicles.

The plant has not yet flowered.

VIBURNUM PROPINQUUM, *Hemsl.*

First discovered in the neighbourhood of Ichang, on the Yangtsze, by Dr. Henry, this species was afterwards introduced to cultivation through Wilson, who sent seed collected in 1901.

As a species it closely resembles the Western Viburnum Tinus, but differs in the nervation of the leaf and the small greenish-white flowers.

VIBURNUM RHYTIDOPHYLLUM, *Hemsl.*

Jour. R.H.S. 1903, vol. xxviii. p. 63, fig. 23.

A very striking shrub of an unusually promising nature with large, broadly-lanceolate leaves, strongly nerved on the upper surface, covered beneath with dense woolly tomentum. The branches terminate in corymbs of yellowish-white flowers 7 to 8 in. across.

Plants raised at Coombe Wood from seed collected in China have proved perfectly hardy.

VIBURNUM UTILE, *Hemsl.*

Hemsley in Jour. Linn. Soc. vol. xxiii. p. 257.

This neat, attractive Viburnum, a native of Central China, was first detected by Dr. A. Henry, and subsequently introduced to cultivation.

The leaves small, lanceolate-oblong, dark shining green above, are covered beneath with a thick coat of white stellate hair.

Mr. Hemsley states (*l.c. supra*), "This species is readily distinguished by its thick, entire leaves, glabrous and shining above and felted beneath. The branches are used for making pipe stems, according to Dr. Henry."

VIBURNUM VEITCHII, *C. H. Wright.*

Gard. Chron. 1903, vol. xxxiii. p. 257; Jour. R.H.S. 1903, vol. xxviii. pts. i. and ii. p. 63; Flora and Sylva, vol. ii. p. 209.

A species introduced from Central China through Wilson. The leaves about 5 in. in length, cordate at the base, with margins coarsely dentate, have an under surface densely tomentose.

The species with flowers white, in terminal corymbose panicles, is closely allied to two others, Viburnum rhytidophyllum and V. buddleifolia.

HERBACEOUS PLANTS

HERBACEOUS PLANTS

ACONITUM HEMSLEYANUM, *Pritz.*

A climbing Aconite, a hardy herbaceous perennial, with curious bulbils in the axils of the leaves and bracts, as in many lilies.

The flowers are of a rich deep blue colour, large and showy.

Living roots from Central China in 1901 failed to grow, but a better fate awaited a second consignment collected in 1903, and plants flowered at Coombe in the summer of 1905.

ACONITUM SCAPOSUM, *Franch.*, var. PYRAMIDALE.

A strong-growing hardy herbaceous perennial with foliage typical of the genus and pyramidal spikes of dark blue flowers crowded on the upper two-thirds of a scape 2 to 2½ ft. in height.

The flowers consist of a long blue spur with little or no hood and the small petals are whitish tipped with green.

It was raised from seed collected in Central China in 1901, and first flowered at Coombe Wood in August 1904.

ADENOPHORA CAPILLARIS, *Hemsl.*

Hemsley in Jour. Linn. Soc. vol. xxvi. p. 10.

A hardy herbaceous perennial about 2 to 2½ ft. high, with loose, graceful panicles of blue campanula-like flower. Mr. Hemsley (*l.c. supra*) states, "A very distinct and beautiful species remarkable for the exceeding slenderness of the pedicels, the narrow somewhat inflated corolla and the very thick tubular disk."

Discovered in the Patung district, Central China, introduced to cultivation through Wilson from the same locality and first flowered in the summer of 1905.

ADENOPHORA POLYMORPHA, *Ledeb.*

Gard. Chron. 1903, vol. xxxiv. p. 187.

This hardy herbaceous plant of the Campanula family of a height of 4 to 5 ft., has numerous dark blue bell-shaped blossoms, introduced from China, and first flowered in the Royal Gardens, Kew, in September 1903.

AMARANTHUS MELANCHOLICUS, *Linn.*, var. RUBER.

Gard. Chron. 1862, p. 398 (advt.).

A plant with striking blood-red foliage, introduced by the late John Gould Veitch from Japan, selected from amongst several varieties as the one most valuable for garden flower decoration, summer bedding, and general ornamental purposes.

AMARANTHUS SALICIFOLIUS, *Hort. Veitch.*

Gard. Chron. 1871, p. 1550, with fig.; Fl. Mag. 1871, pl. 557; Fl. des Serres, 1873, vol. xix. p. 3.

Introduced from the Philippines by the late John Gould Veitch.

This annual is of pyramidal form, 2 to 3 ft. high; the leaves, from 5 to 7 in. long by ¼ in. wide, of a bright bronzy-green, change as the plant matures to an orange-red.

ANDROSACE HENRYI, *Oliver.*

Hooker's Ic. Pl. t. 1973.

Detected by the Rev. E. Faber on Mount Omei and by Dr. A. Henry in the Province of Hupeh, from whose specimens it was originally described.

A somewhat large species for the genus with reniform crenate leaves on hairy petioles 4 to 6 in. in length, and umbels of pure white flower freely produced.

It was raised from seed collected in high ranges at elevations of 6,000-8,500 ft.

APERA ARUNDINACEA, *Hook. f.*

Gard. Chron. 1897, vol. xxii. pp. 282, 283, fig. 84.

A New Zealand grass raised from seed at Coombe Wood in January 1896, and flowered in October 1897; an unusually graceful plant.

ARTEMISIA LACTIFLORA, *Wall.*

A strong-growing hardy herbaceous perennial from 4 to 5 ft. with elegant lobed leafage and terminal panicles of milk-white flower, small individually, showy in a mass.

Sent from China by Wilson, flowered for the first time in 1903 at Coombe Wood; an unusually valuable plant.

ARUM PALÆSTINUM, *Boiss.*

Bot. Mag. t. 5509; N. E. Brown in Gard. Chron. 1882, vol. xvii. p. 428.

A singular Aroid discovered by the distinguished botanist and traveller M. Boissier, near Jerusalem, and from this locality plants secured flowered for the first time in 1865.

ARTEMESIA LACTIFLORA

The spathe is pale green suffused with purplish dots and blotches on the outside, rich velvety black with a yellowish-white base on the inner, quite free from odour.

ASTILBE DAVIDII, *Henry.*

Syns. *A. chinensis*, var. *Davidii*, Franch.

Henry in Gard. Chron. 1902, p. 95, fig. p. 103; Bot. Mag. t. 7880; The Garden, vol. lxii. p. 179, fig.

A beautiful hardy perennial, the best of its class, from the Province of Hupeh, through Wilson—a herb of a high order.

The elegant leafage more or less that of Astilbe japonica on a large scale; the flowering stems 6 or more ft. high, the upper portion some 2 ft. in length, densely covered with deep rose-violet, mauve-coloured flowers; these first opened in July 1901.

ASTILBE GRANDIS, *Stapf.*

Gard. Chron. 1905, vol. xxxviii. p. 426, suppl. illus.

This beautiful herbaceous perennial resembles in habit the well-known Astilbe Davidii, having similar foliage and the same erect tall flower-spikes, but differs in pure white flowers, and in some other technical details.

A native of Central China, quite hardy, the plant will prove of very great value.

ASTILBE THUNBERGII, *Miq.*

Fl. Mag. t. 457.

This Astilbe from Japan in 1878, first exhibited in flower in 1881, has elegant tufted foliage above which the scapes bear on the upper third densely clustered greenish-white flower.

BLUMENBACHIA CHUQUITENSIS, *Hook.*

Syns. *B. coronata*, Haage & Schmidt.

Fl. Mag. t. 139; Gard. Chron. 1873, p. 1211, fig. of stinging hairs; Bot. Mag. t. 6143; The Garden, 1875, vol. vii. p. 173, fig.; Fl. des Serres, 1877, p. 189.

A beautiful biennial from Peru through Pearce, possessing unfortunately one drawback, every part of the floral organs being covered with numerous stinging hairs.

When first sent to be named held to be Blumenbachia (Caiophora coronata, it was as this unfortunately distributed.

BLUMENBACHIA CONTORTA, *Hook. f.*

Syns. *Caiophora contorta*, Presl.

Bot. Mag. t. 6134.

A native of Peru and Ecuador raised from seed collected in the first-named country, and flowered in July 1874.

The leaves elegantly lobed, the flowers bright scarlet: the plant now probably lost to cultivation.

CALANDRINIA UMBELLATA, *DC.*

Paxt. Mag. Bot. vol. xii. p. 271; Nich. Dict. Gard. p. 236, fig. 316; Fl. des Serres, 1846, pl. v.

Discovered in rocky places in the regions around Concepcion by Ruiz and Pavon, and introduced to cultivation through William Lobb; a charming half-hardy biennial with dazzling magenta-crimson flowers about the size of a sixpence.

CALCEOLARIA ERICOIDES, *Juss.*

Gard. Chron. 1863, p. 659 (advt.).

A hardy herbaceous plant found by Richard Pearce on mountains of considerable elevation near Cuença, Ecuador, in habit and foliage not unlike a free-growing Erica.

The flowers are bright yellow in great abundance.

CALCEOLARIA PLANTAGINEA, *Sm.*

Gard. Chron. 1863, p. 695; Bot. Mag. t. 2805; The Garden, 1879, vol. xv. p. 261.

Re-discovered by Richard Pearce near the line of perpetual snow on the Andes of Chili, and sent to this country in 1860.

Previously in cultivation, sent home by Mr. Cruickshanks in 1826, it was subsequently lost.

CARDIANDRA SINENSIS, *Hemsl.*

Gard. Chron. 1903, vol. xxxiii. p. 82.

A perennial herb from the Province of Hupeh, with a creeping rootstock, alternate, oblong, or ovate-lanceolate leaves, and an inflorescence resembling that of the Hydrangea.

CELOSIA CRISTATA, *Linn.*, var. COCCINEA, *Hort.*

Fl. Mag. 1861, t. 49.

Raised from seed sent from China as a species of Amaranth.

The figure (*l.c. supra*) represents what would now be considered a poor specimen of the well-known annual, since improved by selection to an unusual degree. The possibilities of the plant were recognized by Thomas Moore Esq., the Editor of the Floral Magazine, who wrote:—"It is not improbable that the more branched of the spicate forms, if carefully selected, might in time yield a plumy crimson variety, analogous to the golden one we already possess; and this is the result at which growers should aim, rather than to obtain large expanded combs which would take away from the elegant aspect of the plant."

HERBACEOUS PLANTS

CHELIDONIUM LASIOCARPUM, *Oliver*.

Hooker's Ic. Pl. t. 1739.

A yellow-flowered herbaceous perennial introduced through Wilson from Central China, and previously collected by Dr. Henry.

The root known by a name signifying "Man's-Blood Herb," from the red juice in the root and stem, is used as a drug by the Chinese.

CHIONOGRAPHIS JAPONICA, *Maxim*.

Bot. Mag. t. 6510; Gard. Chron. 1881, vol. xv. p. 720; figs. 128-130.

This interesting plant, described by Thunberg in his Flora Japonica, p. 152, had been known to science for a century prior to introduction by Messrs. Veitch in 1880.

A glabrous perennial herb, the only species in the genus, closely related to the East North American genera Helonia and Chamælirium.

From seed sent by Maries from Japan, plants were raised and flowered in April 1880 for the first time.

CODONOPSIS TANGSHEN, *Oliver*.

Hooker's Ic. Pl. t. 1966.

A member of the Campanula family with twining stems, raised from seed from Central China, where, considered an important drug, and used amongst the very poor as a substitute for the costly ginseng, it is known as the t'ang-shên.

COLLINSIA BARTSIÆFOLIA, *Benth*.

Gard. Chron. 1852, p. 689 (advt.).

A pretty annual herb sent from California by William Lobb, distributed in 1852.

The plants from 6 to 9 in. in height, with pale lilac-coloured flowers in profusion, are useful for the borders or bed edging during the summer months.

COLLINSIA MULTICOLOR, *Lindl*.

Gard. Chron. 1852, p. 690 (advt.); Paxt. Fl. Gdn. 1851-1852, p. 89, t. 55.

A pretty Californian annual about 1½ ft. high, through William Lobb, first distributed in 1852.

The flowers, in whorled inflorescences in the axils of the uppermost leaves, are of variable colour, mostly white with lilac, rose, or violet stripes and markings.

For beds or borders in the summer a charming subject, growing on a poor soil equally as in a rich.

CONANDRON RAMONDIOIDES, *Sieb. & Zucc.*

Masters in Gard. Chron. 1879, vol. xii. p. 232; Bot. Mag. t. 6484; The Garden, 1897, vol. li. pl. 1099, p. 6.

An interesting plant, at home in moist rocks in the mountains of Nippon and Kiusiu, Japan.

It is an aberrant member of a group of Gesnerads which span the middle mountain regions of the old world from Spain to Japan, closely related to the genera Ramondia, Haberlea, Wulfenia, and Shortia, botanically connecting the regular-flowered five-membered corollifloral orders with superior two-celled ovaries, with the Gesneraceæ and Scrophulariaceæ.

The regular corolla is a most remarkable botanical character, exceptional not only in the order to which it belongs, but in the whole group of Personales.

CORYDALIS CHEILANTHIFOLIA, *Hemsl.*

Jour. Linn. Soc. 1892, vol. xxix. p. 392; Gard. Chron. 1902, vol. xxxii. p. 288.

A Chinese species from Hupeh, first distributed in 1904; a hardy herbaceous perennial with elegant fern-like foliage and spikes of pale yellow flower.

Dr. A. Henry and Wilson both discovered the herb among stones on the banks of streams in the higher mountains; it is the least interesting of all the Chinese Corydalis.

CORYDALIS THALICTRIFOLIA, *Franch.*

Bot. Mag. t. 7830; Gard. Chron. 1902, vol. xxxii. pp. 288, 289, 320, fig. 94 and suppl.

The Rev. E. Faber was probably the first to find this plant in the mountains of Ningpo. Dr. Henry sent specimens from Ichang, and Père Ducloux and W. Hancock Esq. collected some in Yunnan, though only first introduced to gardens by Wilson, who sent seed in 1900, from which plants raised flowered in June 1901.

CORYDALIS TOMENTOSA, *N. E. Brown.*

Gard. Chron. 1903, vol. xxxiv. p. 123; Gardening World, 1903, p. 757, fig.

This handsome half-hardy species from China has leaves finely divided covered with white silky hair. The flowers yellow, tubular, in erect racemes, hardy in favoured localities are, during the summer, adapted for cool greenhouse decoration.

CORYDALIS WILSONI, *N. E. Brown.*

N. E. Brown in Gard. Chron. 1903, vol. xxxiv. p. 123; Bot. Mag. t. 7939; Gard. Chron. 1904, vol. xxxv. p. 306, fig. 130.

A Chinese species with handsome glaucous-green foliage and spikes of deep golden-yellow, raised from seed and flowered for the first time at Coombe in 1903.

HERBACEOUS PLANTS

DEINANTHE BIFIDA, *Max.*

Jour. R.H.S. 1903, vol. xxviii. p. 62.

A rather tall-growing herbaceous perennial with unisexual, hermaphrodite, and sterile flowers on one and the same inflorescence, in general appearance resembling those of the Hydrangea, of which it is a near ally. The leaves large, usually in whorls of four, broadly ovate, serrate margins have often a bifid apex.

DELPHINIUM CARDINALE, *Hook.*

Bot. Mag. t. 4887.

A handsome species of Larkspur from California, due to William Lobb.

The rich scarlet colour of this much-cultivated, well-known plant was, prior to introduction, unknown in the genus, and is still unique.

In a description in the Botanical Magazine (*l.c. supra*) Sir Joseph Hooker writes:—"Blue or purple or white Larkspurs are familiar to us in our gardens. We have now the pleasure of making known a species of Delphinium equalling if not surpassing any other in the size and symmetry of the plant, and excelling in the brilliancy of colour of the flower, and that as rich a scarlet as can well be looked upon. It is one of the many novelties detected by William Lobb in California, and introduced to our gardens by Messrs. Veitch & Sons of the Exeter and Chelsea nurseries."

DELPHINIUM DAVIDII, *Franch.*

This species, from seed collected on Mount Omei, in the extreme West of China, grows some 2 ft. high, developing flowers of a very rich rose-purple. The foliage is typical of the genus, but the flowers do not compare in size with those of the finer garden forms.

DICENTRA CHRYSANTHA, *Walp.*

Syns. *Dielytra chrysantha.*

Fl. des Serres, tom viii. p. 193; Bot. Mag. t. 7954.

First detected by David Douglas, but not introduced, and flowered at Exeter 1852, from seed sent from California by William Lobb.

Long branching panicles of yellow blooms appear in August and September, and, though the plant is hardy and very beautiful, it is short-lived.

The leaves large, glaucous-green, pinnately divided; the ultimate lobes small linear or cuneate, somewhat acute.

DICENTRA MACRANTHA, *Oliv.*

Hooker's Ic. Pl. pl. 1937.

This is an important addition to a favourite garden genus, which includes the familiar Dicentra spectabilis, known as the "Bleeding Heart," valuable for forcing in early spring and for the hardy border.

D. macrantha, a Chinese species rare in its native habitat, was first met with by Dr. Henry, who detected the herb in a wood in Hupeh, and only obtained one plant.

Rare on Mount Omi, in the extreme West, Wilson obtained a few seed and succeeded in introducing it. The flowers droop, are narrower and longer than those of D. spectabilis, pale yellow in colour.

DRACOCEPHALUM RUYSCHIANA, *Linn.*, var. JAPONICUM, *Asa Gray.*

Masters in Gard. Chron. 1879, vol. xii. p. 167, fig. 29.

This old inhabitant of gardens, Dracocephalum Ruyschiana, was introduced as long since as 1699, and is now rarely seen.

In Central Europe and Southern Asia the species widely distributed, was only noticed in 1859 by Professor Asa Gray to be a variety known to be in Japan. A pretty herbaceous perennial introduced through Charles Maries.

ECHINOCYSTIS LOBATA, *Torr. & Gray.*

Gard. Chron. 1897, vol. xxii. pp. 270, 271, fig.; Rev. Hort. 1896, p. 9, fig.

A trailing annual of the Cucurbitaceæ, native of the North-Eastern States of America, with inconspicuous flowers, of elegant habit and green prickly fruit.

Flowered and fruited at Coombe Wood during the summer of 1897, and from this material the figure in the Gardeners' Chronicle was prepared.

ESCHSCHOLTZIA CÆSPITOSA, *Benth.*

Syns. *E. tenuifolia*, Benth.

Bot. Mag. t. 4812; Gard. Chron. 1854, p. 451 (Notice of Exhibit); Bentham in Trans. of the R.H.S. 2nd ser. i. p. 408.

In figuring this plant, sent from California by William Lobb, in the Botanical Magazine as Eschscholtzia tenuifolia, a doubt is expressed as to the advisability of keeping one so variable apart from E. cæspitosa as a distinct species, and modern authorities, with a fuller knowledge, hold it synonymous, or merely a form of that species.

GENTIANA DETONSA, *Fries.*, var. BARBATA, *Griseb.*

An erect-growing biennial 1½ to 2 ft. in height with dark violet-blue flowers 1 in. in diameter.

The bearded margin of the corolla lobe suggested the varietal name.

The type species is Himalayan, and the variety common in the grasslands of the Chino-Tibetan borderland at elevations of 8,500-11,500 ft.

Plants raised from seed flowered at Coombe Wood for the first time in 1905.

GERANIUM PLATYANTHUM, *Duthie.*

Gard. Chron. 1906, vol. xxxix. p. 51.

A handsome hardy herbaceous perennial with three to five lobed leaves on erect stems 2 to $2\frac{1}{2}$ ft. in height, bearing in profusion large rosy purple flowers; raised from seed collected in Central China 1901.

GERARDIA LANCEOLATA, *Benth.*

Syns. *Virgularia lanceolata*, Ruiz & Pav.

Gard. Chron. 1849, p. 564.

A herbaceous plant with rose-coloured Pentstemon-like flowers from North Chili, distributed in 1849.

Of a genus difficult to cultivate, not perfectly hardy, it was soon lost to cultivation.

GILIA CALIFORNICA, *Benth.*

Syns. *Leptodactylon californicum*, Hook. & Arn.

Bot. Mag. t. 4872; Gard. Chron. 1855, p. 423 (Notice of Exhibit); The Florist, 1855, pl. 105; Fl. des Serres, 1856, tom. i. 2nd ser. p. 79.

A half-hardy annual through William Lobb, by seed from San Bernardino, South California, though David Douglas first made known the plant by dried specimens from the same locality; Coulter had also previously met with it.

Exhibited in flower for the first time in June 1855.

GILIA DIANTHOIDES, *End.*

Syns. *Fenzlia dianthiflora*, Benth.

Bot. Mag. t. 4876; Fl. des Serres, 1856, tom. i. 2nd ser. p. 89.

Discovered in California by Douglas, later introduced to our gardens from the same locality through William Lobb, and flowered during the summer of 1855.

Excellent as a bedding-out plant, closely ramified and spreading, furnished with filiform branches, completely covered and concealed by numerous blossoms of a delicate lilac.

GILIA LUTEA, *Steud. Nomencl.*

Syns. *Leptosiphon luteus*, Benth.

Bot. Mag. t. 4735; Fl. des Serres, tom. ix. p. 97.

Seed of this species, first detected by Douglas about the year 1833, was sent by William Lobb in 1852 from California.

It remained unknown in cultivation till raised from Lobb's seed; first flowered in 1853, and exhibited at a meeting of the Horticultural Society held at Chiswick.

GOMPHRENA PULCHELLA, *Mart.*

Bot. Mag. t. 4064.

Raised from seed imported from Monte Video, and flowered for the first time in July 1844.

The flowers, in dense heads, rose-purple in colour, are not unlike the common Globe Everlasting, the bracts highly coloured.

HELONIOPSIS JAPONICA, *Maxim.*

Bot. Mag. t. 6986.

This interesting liliaceous plant, first flowered in the spring of 1881, was long known to botanists prior to introduction from Japan by Maries; it is widely spread in the mountains, and has been detected in Corea. In general habit a Scilla, the leaves persistent, but not fleshy in texture.

HEMIBŒA HENRYI, *C. B. Clarke.*

Hooker's Ic. Pl. t. 1798.

A hardy herbaceous perennial Gesnerad, a family poorly represented amongst hardy garden plants, first discovered by Dr. Henry at Ichang, and subsequently introduced.

It forms a plant about 1 ft. high with dark green shiny leaves and white waxy-looking tubular flowers, yellow in the throat. A decoction of the root in alcohol is said to be efficacious in cases of snake-bite.

INCARVILLEA VARIABILIS, *Batalin*, var. LATIFOLIA, *Batalin.*

A form of the type with much-divided leaves and lax panicles of small tubular rose-pink flowers; a herbaceous perennial not perfectly hardy, from Western China, common in gorges at elevations of 3,000-4,500 ft.

LINUM MACRÆI, *Benth.*

Syns. *L. Chamissonis*, Hort.

Bot. Mag. t. 5474; Fl. Mag. t. 214.

This species, a half-hardy perennial, has in abundance heads of orange-yellow flowers, detected and introduced through Pearce from Lota in 1860: first flowered in 1864, and exhibited at a meeting of the Royal Horticultural Society under the name of Linum Chamissonis.

HERBACEOUS PLANTS

LOASA PICTA, *Hook.*

Bot. Mag. t. 4428.

A very pretty annual, of which seeds were first sent from Chacapoyas, in the Andes, by William Lobb.

As with other members of the genus, the species is covered with stinging hairs, a drawback from a horticultural standpoint.

LOBELIA TENUIOR, *R. Br.*

Syns. *L. heterophylla*, Labill.

Paxt. Mag. Bot. vol. vi. p. 197; Gard. Chron. 1901, vol. xxix. 3rd ser.; Bot. Mag. t. 3784.

Introduced from Western Australia in 1835, lost to cultivation till recently, and a very fine plant, in the way of Lobelia ramosa.

It is a large-flowered species of the Erinus section, of an attractive blue, worthy of greenhouse culture.

LYSIMACHIA CRISPIDENS, *Hemsl.*

Hemsley in Jour. Linn. Soc. 1889, vol. xxvi. p. 50, pl. 1; Bot. Mag. t. 7919; The Garden, 1903, vol. lxiii. p. 389, fig.

A pretty little pink-flowered herb, raised from seed collected in Hupeh, the plants opening pink flowers for the first time in the spring of 1900 at Coombe Wood.

One of the numerous species of Lysimachia from the mountainous regions of Central China, differing from all species hitherto in cultivation.

LYSIMACHIA HENRYI, *Hemsl.*

The Garden, 1903, vol. lxiv. p. 269, with fig.; Gard. Chron. 1903, vol. xxxiv. p. 187; Bot. Mag. t. 7961.

A pretty species first discovered by Dr. Henry in the Province of Hupeh, Central China, in 1885, and later sent to Messrs. Veitch from the same locality.

The plant of tufted trailing habit has large yellow flowers in the angles of the crowded leaves, is of vigorous constitution, and an excellent subject for carpet beds or for the rock garden.

LYSIMACHIA STENOSEPALA, *Hemsl.*

The Garden, 1903, vol. lxiv. p. 269.

A free-growing Chinese species, with numerous white flowers in long racemes, introduced from the Province of Hupeh.

MECONOPSIS HENRICI, *Franch.*

A dwarf-growing species with scapes seldom exceeding 1 ft. in height.

The flowers 3 to 4 in. in diameter, of an intense violet-purple colour, usually borne singly, but occasionally two or more are produced on one flower-stem. The leaves narrow, oblong-obtuse, glabrous, more or less covered with brownish spinescent hair.

A native of the grasslands of the Chino-Tibetan frontier, at elevations of 11,000-14,000 ft., first met with by Prince Henri d'Orléans and subsequently introduced to cultivation from the same locality.

MECONOPSIS INTEGRIFOLIA, *Franch.*

Gard. Chron. 1904, vol. xxxvi. p. 240, fig. 97, and suppl. illus.; *id.* vol. xxxvii. p. 291, fig. 121; Bot. Mag. t. 8027; Die Gartenwelt, 1905, Jahrgang, ix. p. 534, col. pl.

Meconopsis integrifolia was first discovered by the Russian, Przewalski, who sent herbarium specimens to Maximowicz, and he named them Cathcartia integrifolia; Franchet subsequently, however, noticed that the dehiscence of the capsule excluded the plant from that genus, and he named it Meconopsis integrifolia; the honour of first flowering in Europe this plant is to M. Maurice de Vilmorin, a principal in the great seed house on the Quai de la Mégisserie.

A Meconopsis of which there is a coloured plate in Flora & Sylva, 1905, vol. iii. p. 80, named M. integrifolia, is not the plant of which Franchet writes in *Plantæ Delavayanæ*:—

"La plante du Yunnan ressemble absolument à celle du Kansu, dont M. Maximowicz a communiqué un exemplaire à l'herbier du Muséum. Le mode de déhiscence de la capsule ne permet pas de la considérer comme une espèce du genre Cathcartia, ainsi que l'avait pensé M. Maximowicz auquel du reste les fruits du M. integrifolia étaient demeurés incornus.

"La plante ressemble singuilièrement au M. simplicifolia, Hook. et Thomps.; mais les tiges sont constamment pluriflores et la portion rétrécie constituant le style est très raccourcie."

The plant referred to as La plante du Yunnan was received with 301 other packets by M. de Vilmorin on March 28th 1895 from Abbé Farges Tchein-keon—Eastern Schezuan: packet 1,102 was distributed in three directions, and the seed differently treated, and at Les Barres, the country mansion of M. de Vilmorin par Nogent-sur-Vernisson, Loiret, a plant flowered in April 1896—a photograph of that plant is before the writer. Others kept in pots remained small and did not bloom till 1897; in every case, hand pollination proved unsuccessful—and the plant was lost until distributed in 1905 by the Veitchian firm.

MECONOPSIS PUNICEA, *Maxim.*

Gard. Chron. 1904, vol. xxxvi. p. 282, fig. 130; Flora and Sylva, 1905, vol. iii. p. 84.

A hardy Alpine from the Tibetan frontier through Wilson, first flowered at Langley in September 1904.

MECONOPSIS INTEGRIFOLIA

The plant, of tufted habit with lanceolate or ovate-lanceolate leaves, tapering at both ends, is covered with straggling yellow hair. The flowers, deep reddish-purple in colour, nod on scapes 7 to 8 in. long, and are 6 to 7 in. in diameter.

Found in company with Meconopsis integrifolia on a pass 12,500 ft. high, between the Provinces of Szechuan and Kansu, the watershed of the Yellow and Yangtsze rivers.

Plants remain in flower for several weeks—usually three blooms open at one time; the most congenial treatment is as for M. integrifolia.

MIMULUS LUTEUS, *L.*, var. ALPINUS.

Syns. *M. cupreus*, Regel; *M. variegatus*, Loddiges.

Gard. Chron. 1861, p. 530 (Note on Exhibit); Veitch in Gard. Chron. 1864, p. 2; Bot. Mag. t. 5478; Fl. Mag. 1862, pl. 70.

This species, found at elevations of 6,000-7,000 ft., sent from the Chilian Andes by Richard Pearce, is the origin of a beautiful race of hybrids known as Mimulus maculosus.

MIMULUS RADICANS, *Hook.*

Gard. Chron. 1883, vol. xx. p. 21, fig.

An attractive perennial bog plant from New Zealand: the stems creeping with short leafy branched densely packed leaves; the flowers white with a violet blotch.

MONARDELLA MACRANTHA, *A. Gray.*

Bot. Mag. t. 6270.

A highly aromatic plant from seed received from California in 1875, of a spreading tufted habit; the bright scarlet flowers in close terminal heads.

NEPETA WILSONI, *Duthie.*

A hardy herbaceous perennial 2 to 2½ ft. high, produces flower in whorls of a rich deep blue colour, introduced from Sungpan in the extreme west of China by Wilson, and flowered for the first time at Coombe in August 1905.

NIEREMBERGIA RIVULARIS, *Miers.*

Bot. Mag. t. 5608; Veitchs' Catlg. of Pl. 1867, with fig.

A trailing perennial, commonly known as the "Trailing Cup Plant," introduced from La Plata in 1866.

Common on the banks of the Plate River, within high-tide mark, the flowers rise above the dwarf grass and may be discerned from a great distance.

NIEREMBERGIA VEITCHII, *Berkeley.*

Bot. Mag. t. 5599; Fl. and Pom. 1872, p. 141, fig.

A dwarf hardy perennial from Tucuman in South America, with campanulate flowers on a tube ½ to ¾ in. long, white or pale lilac, a charming dwarf border subject.

NOLANA LANCEOLATA, *Miers.*

Syns. *Sorema lanceolata*, Miers.

Bot. Mag. t. 5327; Hooker's Lond. Jour. of Bot. 1845, vol. iv. p. 493.

This handsome compact-growing annual, with large flowers bright blue in colour with a white centre, is a form of the Chilian Bellflower discovered by Mr. Cuming at Coquimbo, in Chili, and later introduced from the same locality.

ŒNOTHERA BISTORTA, *Nutt. MS.*, var. VEITCHIANA.

Bot. Mag. t. 5078.

A native of South California introduced through William Lobb in 1858.

Of dwarf habit, the flowers large, in copious succession on the racemes, bright yellow in colour with a deep bronze-orange or blood-red spot at the base.

OMPHALODES KRAMERI, *Franch. & Sav.*

The Garden, 1881, vol. xix. p. 411 (Report of R.H.S. Meeting).

A beautiful little herbaceous plant with flowers in clusters, Forget-me-not in shape, deep blue in colour, about ½ in. in diameter; a native of Japan, first flowered in 1881.

OSTROWSKYA MAGNIFICA, *Regel.*

Bot. Mag. t. 7472; The Garden, 1888, vol. xxxiv. p. 604, pl. 681; Gard. Chron. 1888, vol. iv. pp. 16, 65, fig. 6.

Discovered in Eastern Bokhara at an elevation of 7,000 ft. by Albert, son of Dr. Regel, long the able Director of the Imperial Botanic Gardens, St. Petersburg. The honour of first flowering this magnificent plant in Europe is to Herr Max Leitchlin, in whose garden at Baden Baden it bloomed in 1887, and also for the first time in England at Coombe in the following year.

OURISIA COCCINEA, *Pers.*

Bot. Mag. t. 5335; Gard. Chron. 1849, p. 564; *id.* 1862, p. 398 (advt).; Nich. Dict. Gard. fig.

A hardy perennial with bright scarlet, drooping, tubular flowers, native of the Andes of Chili introduced through William Lobb in 1849.

This truly lovely plant, now seldom seen in gardens, other than in choice collections, requires a not too full sun and a moist, well-drained soil: it is best in masses.

OURISIA PEARCEI, *Phillip.*

Fl. Mag. t. 154; Proc. R.H.S. vol. iii. p. 227; Gard Chron. 1863, p. 439.

Introduced from the Chilian Andes through Richard Pearce, and flowered in May 1863.

In general as the better-known Ourisia coccinea, the leaves are flatter, more regularly ovate and crenate; the flowers larger, with a longer tube, more spreading lip, in colour crimson, the limb streaked with the deepest blood red.

OXALIS ELEGANS, *H. B. K.*

Bot. Mag. t. 4490; Paxt. Mag. Bot. vol. xvi. p. 258.

A native of the Andes of Loxa in Columbia, bordering on Peru, usually found at elevations of nearly 7,000 ft. First detected by Humboldt in this locality, and subsequently introduced to cultivation by William Lobb.

OXALIS VALDIVIENSIS, *Veitch.*

Gard. Chron. 1862, p. 550 (Notice of Exhibit); Regel's Gartenflora, 626.

A charming annual from Chili, with yellow flowers streaked with red on the outside, hardy during the summer months and suitable for greenhouse culture in the winter.

PÆONIA OBOVATA, *Maxim.*

Baker in Gard. Chron. 1884, vol. xxi. p. 779.

This fine Pæony, first discovered by the Russian explorer, Dr. Alexander Tartarinov, near Pekin, and afterwards collected by several later travellers, was introduced to cultivation through Wilson, who sent seeds collected in the Province of Hupeh in 1900.

A strong-growing herbaceous Pæony, a perennial, about 2 ft. high, with red-purple flowers as large as those of the common garden type (Pæonia officinalis).

PALAVA DISSECTA, *Benth.*

Syns. *P. flexuosa*, Mast.

Gard. Chron. 1866, p. 435; Bot. Mag. t. 5768.

A half-hardy annual discovered in the valley of San Lorenzo, in Peru, by Mr. McLean, of Lima, about the year 1810. Nothing further was known until Pearce sent home seed from which plants raised flowered at Chelsea in 1868.

PENTSTEMON JAFFRAYANUS, *Hook.*

Bot. Mag. t. 5045; Fl. des Serres, 1858, tom. iii. 2nd ser. p. 1.

This lovely species first found by a Mr. Jaffray, after whom it is named, at Clear Creek, North California, in 1853, was not introduced until William Lobb sent home seeds from which plants raised flowered for the first time in 1857.

PEREZIA VISCOSA, *Less.*

Syns. *Homoianthus viscosus*, DC.

Bot. Mag. t. 5401.

A native of the Southern Provinces of Chili, common in Valdivia, introduced through Richard Pearce.

It was first exhibited at the meeting of the Royal Horticultural Society in June 1863, the blue flowers (in form resembling the Tagetes) proving an admirable subject for the open summer border.

PHACELIA WHITLAVIA, *A. Gray.*

Syns. *Whitlavia grandiflora*, Harv., *Cosmanthus grandiflorus*.

Bot. Mag. t. 4813; Fl. des Serres, 1856, tom. i. 2nd ser. p. 19; *id.* 1861, 2nd ser. p. 153.

This beautiful annual, with bell-shaped flowers of a bright blue colour, discovered by Douglas, was later introduced from California through William Lobb.

Exhibited at the Chiswick Exhibition held in July 1854, it attracted great attention, and as a plant for the open border is still unequalled for the very unusual porcelain blue of the blossom.

PHLOMIS UMBROSA, *Turcz.*, var. AUSTRALIS, *Hemsl.*

Hemsley in Jour. Linn. Soc. vol. xxvi. p. 306.

A variety of the type collected in many districts of Central China by Dr. A. Henry, and introduced to cultivation from the Province of Hupeh in 1901.

The plant is a herbaceous perennial with opposite cordate or often nearly orbicular leaves coarsely toothed along the margin, sometimes bifid at the apex.

The flowers are two-lipped cream-white in colour with a reddish base, in false whorls in the axils of the uppermost leaves.

PHYGELIUS CAPENSIS, *E. Meyer.*

Bot. Mag. t. 4881; Fl. des Serres, 1856, tom. i. 2nd ser. p. 73; Gard. Chron. 1855, p. 471 (Notice of New Plants).

A beautiful herbaceous perennial with scarlet tubular drooping flowers, yellow in the throat, discovered in Caffreland at Witbergen, on the sides of streams, by Drège, and first flowered at Exeter during the summer of 1855; still in general cultivation.

PHYSALIS FRANCHETI, *Mast.*

Masters in Gard. Chron. 1894, vol. xvi. p. 434, fig. 57, p. 441; The Garden, 1896, vol. xlix. p. 232, pl. 1059.

A gigantic species of the "Bladder Cherry," raised from seed brought from Japan by James H. Veitch. It differs from Physalis Alkekengi, to which it is closely allied, in much larger fruit of a vivid brilliant scarlet, and as a winter decorative subject has supplanted that species.

PLATYCODON GRANDIFLORUM, *A. DC.*, var. MARIESII.

Gard. Chron. 1893, vol. xiv. p. 163, fig. 33; The Garden, 1885, vol. xxvii. p. 216, pl. 483.

This variety differs from the typical species introduced by Fortune, in being dwarfer and in having larger flowers of a more intense shade of blue; there is no other blue quite like it.

Introduced through Maries from Japan, it is still more common on the low hills in the districts north and east of Sœul in Corea, and found throughout the whole of the Far East; a charming plant with a profusion of flowers during the early summer months.

PRATIA ANGULATA, *Hook. f.*

Syns. *Lobelia littoralis*, R. Cunn.

Masters in Gard. Chron. 1879, vol. xii. p. 136.

A little creeping plant with white flowers in great profusion and angular sub-orbicular leaves, a native of New Zealand, introduced through Peter C. M. Veitch.

It is a favourite for the rock garden, where its neat foliage, forming a dense carpet, on which the small white flowers, followed by small violet-coloured berries about the size of peas, show to advantage.

PRIMULA COCKBURNIANA, *Hemsl.*

Gard. Chron. 1905, vol. xxxvii. pp. 331, 345, fig. 137.

A remarkable species of Primula of an entirely unique colour introduced from Western China, first flowered at Coombe Wood in May 1905.

Previously collected by the English naturalist A. E. Pratt during the year 1889-1890 when exploring the mountains of Western China, it had not been sent to England. The orange-scarlet colour is very striking, quite unlike any other Primrose in cultivation.

The leaves obovate-oblong, irregularly toothed, narrowed at the base, are slightly farinose, the flowers in successive whorls as in P. japonica, but the plant is of more slender growth.

PRIMULA CORTUSOIDES, *L.*, var. AMŒNA.

Gard. Chron. 1862, p. 1218; Bot. Mag. t. 5528; Fl. Mag. 1865, t. 249.

A variety with larger flowers of a deeper, brighter shade of colour than the type, introduced from Japan through the late John Gould Veitch, and a very pretty form.

PRIMULA CORTUSOIDES, *L.*, var. GRANDIFLORA, *Hort. Veitch.*

The Florist, 1870, p. 193.

Introduced from Japan by the late John Gould Veitch, this Primula in style and habit of growth resembles P. cortusoides amœna. The flowers, however, larger than those of that variety, are concave or cup-shaped, slightly drooping. The exterior is of a pleasing deep rose colour, the interior nearly white.

PRIMULA CORTUSOIDES, *L.*, var. STRICTA.

Gard. Chron. 1862, p. 1218.

A variety with smaller flowers than those of Primula cortusoides amœna, delicate pink in colour striped with crimson.

PRIMULA DEFLEXA, *Duthie.*

Gard. Chron. 1905, vol. xxxvii. p. 332 (Report of R.H.S. Floral Committee).

A charming alpine species raised from seed collected in the high mountainous regions of Talien-lu, Western China, in 1903.

It forms a small tufted plant with leaves about 4 in. in length, hairy on both surfaces, irregularly dentate along the margin. The slender flower-scapes about 1 ft. high, bear at the apex a dense head or capitulum of dark blue or purple flowers, about ½ in. in length, pointed downwards. The interior of the corolla limb, some ¼ in. in diameter, is dusted with a white farina.

PRIMULA DENTICULATA, *Sm.*

Bot. Mag. t. 3959.

A native of Northern India, whence seeds were sent to the Veitchian people by the Directors of the Honourable the East India Company, through the medium of Dr. Royle, and plants first flowered in a cool greenhouse, were exhibited in March 1842 before the Horticultural Society of London.

HERBACEOUS PLANTS

PRIMULA NIVALIS, *Pall.*, var. FARINOSA, *Schrenck.*

Gard. Chron. 1905, vol. xxxvii. p. 332 (Report of R.H.S. Floral Committee).

A form of the type differing in having the stems and the under surfaces of the leaves mealy, and flowers of a beautiful blue colour, dotted with white farina on the limb. A native of Western China, discovered and introduced through Wilson, the flowers first opening at Coombe Wood in May 1905.

PRIMULA OVALIFOLIA, *Franch.*

Jour. R.H.S. 1903, vol. xxviii. pt. i. p. 64; Gard. Chron. 1905, vol. xxxviii. p. 62, suppl. illus.

A beautiful species, native of the mountainous regions of Central and Western China, introduced through Wilson.

Seed, sent in quantity, failed to germinate, but plants brought home in 1901 flowered in March 1905.

In a wild state in masses in moist, shady woods, carpeting the ground with flower as soon as the snow melts, as does the common Primrose in England.

The species remarkable for the deep violet-purple flowers, approaching blue, is a very fine introduction.

PRIMULA PULVERULENTA, *Duthie.*

Syns. *P. japonica*, A. Gray, var. *pulverulenta*, Duthie.

Gard. Chron. 1905, vol. xxxvii. p. 301 (Report of R.H.S. Floral Committee).

A very magnificent Primula with flower stems 2 to 3 ft. high, occupying a high place in the ranks of plants of recent introduction. The scape, pedicels, and calyx thickly coated with white farina, greatly enhance the appearance of the rich rose-purple flower.

From the mountainous regions of Western China on the borders of Tibet, flowered for the first time at Coombe Wood in May 1905.

PRIMULA PYCNOLOBA, *Bur. & Franch.*

This remarkable Primrose found by Prince Henri d'Orléans at Tatien-lu, in the extreme west of China, on the frontier of Tibet, was later introduced to cultivation from the same locality.

Of botanical rather than of horticultural interest, the flowers small, are almost entirely hidden by the large leafy green calyx. These were produced for the first time in this country at Coombe Wood in the spring of 1905.

PRIMULA TANGUTICA, *Duthie.*

Gard. Chron. 1905, vol. xxxvii. p. 301 (Report of R.H.S. Floral Committee); *id.* vol. xxxviii. p. 42, fig. 17; Bot. Mag. t. 8043.

A remarkable Primrose introduced to cultivation from the same district and first flowered at Coombe Wood in May 1905.

The flowers in whorls on an erect scape about 1 ft. in height, are remarkable for a deep maroon colour, unlike anything in the genus previously known. The petals long, strap-shaped, are, when the flowers have fully expanded, reflexed, as in the Cyclamen. Although of small size, they are strongly perfumed.

The plant was discovered by Wilson growing abundantly in open grassy places, at elevations of 11,000-13,000 ft. on the mountains of Szechuan, Western China.

PRIMULA VEITCHII, *Duthie.*

Gard. Chron. 1905, vol. xxxvii. p. 301 (Report of R.H.S. Floral Committee); *id.* p. 344, suppl. illus.; Bot. Mag. t. 8051.

A species from Western China introduced through Wilson, flowered at Coombe Wood for the first time in May 1905.

In general appearance this Primrose somewhat resembles Primula cortusoides, but is botanically quite distinct. The leaves are petiolate, elegantly crenate along the margin, densely hairy on the under surface.

The flower-scape is 1 ft. high, terminated by a cluster of richly coloured rose-purple blooms, similar in appearance to the improved varieties of P. obconica, about twenty in each cluster.

PRIMULA VIOLODORA, *Dunn.*

Dunn in Gard. Chron. 1902, vol. xxxii. p. 129.

A species closely related to the Himalayan Primula mollis, from which species however readily distinguished by the colour and shape of the calyx. The flowers mauve, have, especially in the evening, a delicate scent, resembling the violet.

It was introduced from Central China.

PRIMULA VITTATA, *Bur. & Fr.*

Gard. Chron. 1905, vol. xxxvii. p. 332 (Report of R.H.S. Floral Committee); *id.* p. 390, fig. 165.

A species from Western China of the same habit of growth as Primula sikkimensis, from which it differs in having rich rose-purple drooping flowers.

The appearance of the calyx, a striking feature, suggested the specific name. The whole of the upper part of the scape, the pedicels, and calyx are covered with a thick white farina, except for five longitudinal stripes on the latter corresponding to the sepals, of a brownish hue and very conspicuous.

Introduced to cultivation and first flowered at Coombe Wood in May 1905.

HERBACEOUS PLANTS

PRIMULA WILSONI, *Dunn.*

Gard. Chron. 1902, vol. xxxi. p. 413.

Discovered in Yunnan, South China, by Père Delavay and Dr. Henry, and introduced to cultivation from the same locality.

Of the section of the genus of which Primula japonica is an example, it is intermediate in habit between that species and the Javanese P. imperialis.

The flowers are about 1 in. in diameter, reddish-purple in colour, sweet-scented, in whorls of five or six each.

RANUNCULUS LYALLII, *Hook. f.*

Bot. Mag. t. 6888; Masters in Gard. Chron. 1881, vol. xv. p. 724, fig. 131; The Garden, 1879, vol. xv. p. 391, with fig.

This beautiful buttercup, justly called the monarch of the genus, commonly known as the "Rockwood Lily," or "Mountain Lily," was first discovered by Dr. Lyall on the west coast of the Southern Island of New Zealand, during an exploring expedition, between the years 1847-1849. A difficult plant to cultivate, many thousands of seeds sent have failed to germinate, nor will they grow on the plains of New Zealand.

The late Mr. Anderson Henry, said to have flowered the plant before 1864, remarked that the seeds remained dormant for four or five years.

Messrs. Veitch imported it in quantity from New Zealand through Peter C. M. Veitch, succeeded in flowering it in May 1879, and exhibited it before the Royal Horticultural Society.

RICHARDIA HASTATA, *Hook.*

Syns. *Calla ? oculata*, Lindl.

Gard. Chron. 1859, p. 788; The Garden, 1880, vol. xviii. pl. cclxi. p. 596; Bot. Mag. t. 5176; Watson in Gard. Chron. 1892, vol. xii. p. 123.

A hardy aroid received from Natal in 1857, smaller but resembling in foliage the well-known Richardia africana. The spathe, which rises above the leaves, is about 3 in. long and $1\frac{1}{2}$ in. across the mouth: in form bell-shaped with an oblique limb; in colour a clear rich yellow with a deep purple eye.

RODGERSIA PINNATA, *Franch.*, var. ALBA.

Gard. Chron. 1905, vol. xxxvii. p. 398 (Report of R.H.S. Floral Committee).

A white flowered form of the type introduced from Mount Wa in Western China.

It forms a strong-growing herbaceous plant with handsome foliage showing the pinnate arrangement of the leaflets very prominently, and with a general resemblance to those of the Horse-chestnut.

The flower stem rises to a height of from 2 to 2½ ft. bearing numerous small pinkish white flowers very showy in a mass and of Hawthorn-like fragrance. An excellent subject for bog gardens.

RODGERSIA PODOPHYLLA, *A. Gray.*

Gard. Chron. 1902, vol. xxxii. p. 131; Bot. Mag. t. 6691.

A favourite inhabiting the mossy woods of Japan, introduced through Maries, who sent seed from which plants raised flowered for the first time in June 1882.

At the time of introduction the only species of the genus known, but the botanical exploration of China during recent years has revealed several new to science, more than one promising to be unusually valuable garden plants.

Rodgersia podophylla, like the newer species, is essentially a plant for the bog garden where moisture is abundant and space can be given for the large handsome foliage, which, when mature, assumes a deep bronzy-green, forming an effective background.

SALVIA SOULIEI, *Duthie.*

This new species of Salvia was discovered by Wilson in the high mountain ranges and gorges of Tatien-lu, Western China, and introduced to cultivation.

It is a handsome herbaceous perennial with shining dark green rugose leaves of triangular outline, cordate at the base, with irregularly serrate margin.

The flower stems attain a height of about 2 ft. and produce in quantity large tubular two-lipped flowers of a bluish-white colour.

It flowered for the first time at Coombe Wood during the summer of 1905, and is a very promising herb.

SAUSSAUREA LAMPROCARPA, *Hemsl.*

Hemsley in Jour. Linn. Soc. vol. xxiii. p. 465.

Discovered by Dr. A. Henry in the Patung district, Central China, and introduced to cultivation by Messrs. Veitch.

Mr. Hemsley (*l.c. supra*) states, "This is very distinct from all the other Chinese species that we have seen, and closely resembles the Indian Saussaurea hypoleuca which has smaller heads, quite entire bracts and a rusty pappus."

SAXIFRAGA CORTUSIFOLIA, *Sieb. & Zucc.*

Bot. Mag. t. 6680.

A herbaceous perennial, native of South China and Japan, sent from the last-named country by Charles Maries.

Closely allied to Saxifraga sarmentosa, the old Strawberry Saxifrage of gardens, and still closer to S. Fortunei, of which, in the view of some authorities, it is but a form.

SCABIOSA BRETSCHNEIDERI, *Natal.*

A species first discovered by the Russian traveller Kaskarov in Tibet in 1893, and named in compliment to Dr. Bretschneider, formerly Physician to the Russian Legation at Pekin, and author of "A History of European Botanical Discoveries in China."

It was introduced to cultivation through Wilson, who sent seed collected in Western China in 1903, at elevations of 4,500 ft.

The plant is a hardy herbaceous perennial with deeply pinnatisect radical leaves and scapigerous flower-stems about 1 ft. in height bearing deep rose-coloured flower-heads during September.

SCOPOLIA SINENSIS, *Hemsl.*

Hemsley in Jour. Linn. Soc. vol. xxvi. p. 176.

A Solanaceous herbaceous plant 4 to 5 ft. high with obovate leaves and solitary axillary greenish-purple flowers, having a large green foliaceous involucre.

A native of Central China, much valued for the sake of a drug extracted, used as a cardiac.

Plants raised from seed flowered at Coombe in 1902.

SENECIO CLIVORUM, *Max.*

Gard. Chron. 1902, vol. xxxii. p. 217, suppl. illus.; Bot. Mag. t. 7902.

This, the largest flowered of all the hardy herbaceous species of the genus at present in cultivation, was introduced from the Province of Hupeh, through Wilson, and first flowered at Coombe Wood in July 1901.

The bold striking foliage and handsome orange-yellow flowers at once won a permanent and quite unassailable position in the wild garden and the herbaceous border.

SENECIO MOSOYNENSIS, *Franch.*

One of the numerous species of Senecio from Central China, with elongated cordate obtuse leaves and erect flower-stems, on the apices racemose inflorescences of yellow flowers; it is of little horticultural merit.

SENECIO TANGUTICUS, *Maxim.*

Syns. *S. Henryi*, Hemsl.

Bot. Mag. t. 7912.

Collected in the Province of Hupeh by Dr. Henry, by the Rev. E. Faber in the Province of Szechuan, and introduced to cultivation from the former locality by the collector, Wilson.

The stems, furnished with deeply cut leaves of a light, graceful appearance, reach a height of 6 or 7 ft.; the flowers, bright yellow in colour and very showy, in dense panicles at the apices of the branches, opened for the first time at Coombe Wood in the autumn of 1902.

SENECIO VEITCHIANUS, *Hemsl.*

Syns. *S. Ligularia*, Hook. f., var. *speciosa*, Hemsl.

Gard. Chron. 1905, vol. xxxviii. p. 212; *id.* p. 455, suppl. illus.

This strong-growing herbaceous perennial introduced from Western China flowered at Coombe during the summer of 1905.

At first considered only a form of the widely distributed Senecio Ligularia, further study revealed sufficient characters to justify specific rank.

For the wild or bog garden it is an effective subject, the bold foliage and strong erect spikes of yellow flowers some 6 to 8 ft. in height particularly striking.

SENECIO WILSONIANUS, *Hemsl.*

Syns. *S. Ligularia*, Hook. f., var. *polycephalus*, Hemsl.

Gard. Chron. 1905, vol. xxxviii. p. 212.

A species named in compliment to Wilson by whom it was introduced to cultivation from Western China.

The bold reniform leaves form a large tuft from which rise erect spikes thickly covered with small capitula of bright yellow flowers, resembling in the manner in which they are borne the tall spikes of the Eremurus.

SERRATULA ATRIPLICIFOLIA.

Veitchs' List of New Chinese Plants, 1905, p. 7, fig.

This striking herbaceous plant, a member of a genus poorly represented in gardens, was introduced from Central China from seed collected in the Province of Hupeh in 1901.

The plant forms a large tuft of triangular or cordate leaves dark green above, white beneath, on long petioles, and produces erect branched flower-stems some 5 to 6 ft. high, with numerous globular heads of purple flowers enclosed in overlapping involucral persistent bracts.

HERBACEOUS PLANTS

SPRAGUEA UMBELLATA, *Torr*

Bot. Mag. t. 5143.

A curious half-hardy perennial from California, first discovered by Colonel Fremont, from whose specimens a new genus constituted by Dr. Torrey was dedicated to Isaac Sprague Esq., of Cambridge, Massachusetts, the well-known botanical draughtsman.

Introduced to this country through William Lobb, the plant first flowered at Exeter in July 1859.

SWERTIA BIMACULATA, *Clarke.*

A herb of the Gentian family 2 to 2½ ft. high with greenish-yellow flowers covered with small black dots.

A native of grassy mountain-sides at elevations of 4,000-6,000 ft. in Central and Western China, and also found on the Himalayas; plants from seed collected by Wilson flowered at Coombe Wood in the summer of 1905.

TEUCRIUM ALBO-RUBRUM, *Hemsl.*

Hemsley in Jour. Linn. Soc. vol. xxvi. p. 311.

A hardy herbaceous perennial, made known from specimens collected in the neighbourhood of Ichang by Dr. Henry, the two-lipped flowers white and red, in axillary racemes.

Plants raised from seed collected in Central China flowered at Coombe Wood during the summer of 1905.

TEUCRIUM ORNATUM, *Hemsl.*

Hemsley in Jour. Linn. Soc. vol. xxvi. p. 313.

This species, also discovered by Dr. Henry, was introduced to cultivation through Wilson, from seed in 1900.

It is a hardy herbaceous perennial with lanceolate dark green leaves serrate along the margin, purplish beneath.

The small rose-coloured flowers, in erect racemes in the axils of the uppermost leaves, opened for the first time at Coombe Wood in 1904.

THALICTRUM DIPTEROCARPUM, *Franch.*

An attractive addition to a genus of hardy herbaceous plants, characterized by neat glaucous-green foliage and purple flowers, in light graceful panicles.

Sent to cultivation from Western China through Wilson, who detected it at elevations of 4,000-5,000 ft.

The specific name is in allusion to the flattened wing-like appearance of the ripe carpel.

TRICYRTIS LATIFOLIA, *Maxim.*

A broad-leaved species of the "Toad Lily," a herbaceous perennial, attaining a height of from 2 to 2½ ft., with white flowers spotted with purple and broadly ovate leaves distinctly marked with seven longitudinal veins.

A native of Central China, introduced to cultivation in 1900, but known to science previously from specimens collected by Maximowicz in Northern Japan.

TROPÆOLUM LOBBIANUM, *Hort. Veitch.*

Syns. *T. peltophorum*, Benth.

Hemsley in The Garden, 1878, vol. xiii. p. 442; Fl. des Serres, 1846, January, pl. iv.; Bot. Mag. t. 4097.

Found about the same time both by William Lobb and Hartweg, but to the former is the honour of introduction to gardens. Native of the lofty mountains of Columbia, Lobb detected it in 1843, and sent seed to Exeter.

TROPÆOLUM SMITHII, *DC.*

Bot. Mag. t. 4385; Hemsley in The Garden, 1878, vol. xiii. p. 444.

This hardy annual or greenhouse perennial, with dull, brick-red sepals and spur, bright orange-yellow petals striped red and curiously fringed, is a native of the high mountain ranges of Columbia, sent by William Lobb in 1848.

TROPÆOLUM SPECIOSUM, *P. & E.*

Paxt. Mag. Bot. vol. xiv. p. 173; Bot. Mag. t. 4323; The Garden, 1878, vol. xiii. pl. 127, p. 442; Rev. Hort. 1904, n.s. tom. iv. p. 88, col. pl.

A magnificent species, native of the southern provinces of Chili, from Concepcion to Chiloe, probably the handsomest of the genus, well meriting the common name of "Flame Flower," introduced to this country through William Lobb, and flowered for the first time in June 1847.

Though perfectly hardy, the plant will not succeed in all positions, and like others from the same locality dislikes strong sunlight, a dry atmosphere and draught, and failures in connection with its cultivation are often doubtless due to the want of these simple precautions.

A brilliant garden picture.

VIOLA PEDUNCULATA, *Torrey & Gray.*

Bot. Mag. t. 5004.

One of the many discoveries of David Douglas in California during his last journey, and but little before the accident in the Sandwich Islands which resulted in his death.

Subsequently introduced to cultivation through William Lobb, who sent seed from the same locality, from which plants were raised.

WAHLENBERGIA TUBEROSA, *Hook. f.*

Bot. Mag. t. 6155.

A campanulaceous plant, native of Juan Fernandez, introduced through Downton in 1873.

Remarkable for a tuberous root-stock which grows above the surface of the soil as a cluster of potato tubers, contrasting strongly with the graceful thread-like stems and pearl-white rose-streaked blossoms so very freely produced.

BULBOUS PLANTS

BULBOUS PLANTS

ALLIUM ANCEPS, *Kellogg.*

Bot. Mag. t. 6227.

An inhabitant of the Sierra Nevada portion of the Rocky Mountains, at elevations of 4,000-5,000 ft. Bulbs received proved quite hardy in England, and flowered in the open in May 1875.

From these the plate in the Botanical Magazine (*l.c. supra*) was prepared.

ALSTRŒMERIA INODORA, *Herb.*

Syns. *A. nemorosa*, Gardn.

Bot. Mag. t. 3958; Herbert's Amaryllidaceæ, p. 90, t. 2, fig. 1.

Introduced from the Organ Mountains of Brazil through William Lobb, and first flowered under glass during the winter of 1841-1842, at Exeter. The flowers, in loose clusters at the end of long stems, yellow and margined with red, have dark markings on the three inner floral segments.

BLANDFORDIA AUREA, *Hook. f.*

Bot. Mag. t. 5809; Fl. Mag. 1867, t. 403; Veitchs' Catlg. of Pl. 1870, p. 5, fig.; Fl. and Pom. 1872, p. 113.

This, the sixth species of the Australian genus Blandfordia introduced, was imported from New South Wales and flowered for the first time in this country at Chelsea in July 1869.

A cool greenhouse evergreen plant with narrow linear bluish-green leaves from the base of which the flower-scape rises from 1 to 2 ft. in height, bearing at the apex umbellate clusters of from three to five pure golden-yellow bell-shaped drooping flowers.

BOMAREA ANDINAMARCANA, *Baker.*

Syns. *Collania andinamarcana*, Herb.

Bot. Mag. t. 4247; Gard. Chron. 1846, p. 54 (Notice of Exhibition of New Plants).

Originally gathered by Mr. Mathews on the lofty mountains of Andinamarca in Peru, and from his dried specimens described by Dean Herbert for the first time.

William Lobb collected seed in the locality where the bulb was first detected, and from these, plants raised flowered for the first time in April 1846.

It is a singular-looking object with a climbing stem terminating in a much-branched leafy corymb, pendant from the weight of the flowers, remarkable for their coloration,—the three outer perianth pieces orange-red tipped with black, the three inner yellow flaked green, and the centre a tuft of golden-yellow anthers.

BOMAREA CALDASIANA, *Herb.*

Syns. *Alstrœmeria Caldasii*, Humb. & Rth.

Bot. Mag. t. 5442.

A lovely species first discovered on the Quitinian Andes by Humboldt and Bonpland, and later introduced from the same locality through Richard Pearce.

The climbing stems, terminated by drooping umbels of bright orange-yellow flowers, are thickly spotted with reddish-brown.

BOMAREA MULTIFLORA, *Mirb.*

Fl. des Serres, t. 2316; The Florist, 1864, p. 97, col. pl.; Gard. Chron. 1863, p. 627 (Report of Exhibition of New Plants).

A greenhouse climber with lance-shaped leaves and terminal umbels of handsome, drooping, orange-red flowers dotted inside with crimson.

A native of Peru, introduced through Richard Pearce.

BOMAREA TOMENTOSA, *Herb.*

Syns. *B. densiflora*, Herb.; *Alstrœmeria densiflora*, Herb.

Bot. Mag. t. 5531; Herbert's Amaryllidaceæ, p. 399, t. 46, fig. 4.

A richly-coloured Peruvian species, originally described by Dean Herbert from specimens collected by Mathews in the vicinity of Chachapoyas in Peru, and introduced to cultivation from the same locality by Richard Pearce.

The bright red flowers, dotted black in the throat of the perianth-tube, are in dense umbels terminating twining stems.

CRINUM PURPURASCENS, *Herb.*

Syns. *C. bracteatum*, var. *purpurescens*.

Bot. Mag. t. 6525; Fl. and Pom. 1879, p. 108, fig.

A very distinct Crinum of the star-flowered set, native of West Tropical Africa, from Old Calabar by Kalbreyer. It had previously been sent to this country by the Rev. H. Goldie.

The flowers white, tinted with rose, on purple foot-stalks, have conspicuous purplish-crimson stamens, terminated by yellow anthers.

EUCOMIS BICOLOR, *Baker.*

Baker in Gard. Chron. 1878, vol. x. p. 492; Bot. Mag. t. 6816; W. W. in The Garden, 1893, vol. xliii. p. 185.

Sent to Messrs. Veitch by Christopher Mudd, the son of a former

Curator of the Cambridge Botanic Gardens, from Natal, and flowered in this country in the autumn of 1878. The plant suited·to greenhouse culture, is, when in bloom, very ornamental. The flowers on the upper portion of a spike 1 ft. or more high, are greenish-white margined purple.

EURYCLES SYLVESTRIS, *Salisb.*

Syns. *E. australasica*, Loud.

Veitchs' Catlg. of Pl. 1879, p. 25, fig. p. 13.

A bulbous plant allied to Pancratium, re-introduced from the South Sea Islands by P. C. M. Veitch in 1877.

The pure white flowers in umbels open during the months of February and March.

FRITILLARIA VERTICILLATA, var. THUNBERGII, *Baker.*

Syns. *F. Thunbergii*, Miquel.

Gard. Chron. 1880, vol. xiii. p. 532, with fig.; The Garden, 1880, vol. xvii. p. 313, with fig.

A Japanese species introduced through Charles Maries, remarkable for a long, narrow, linear leaf, terminating in a tendril, the flowers small, bell-shaped, greenish and mottled with pale purple.

HÆMANTHUS CINNABARINUS, *Desne.*

Veitchs' Catlg. of Pl. 1878, p. 23; Fl. and Pom. 1878, p. 155, fig.; The Florist, 1877, col. pl.

A gorgeous flowering bulbous plant from the West Coast of Tropical Africa, with very dense globular heads of flower of an unusually brilliant cinnabar-scarlet, with the lustre of the familiar Guernsey lily, remaining in perfection for a long period.

HÆMANTHUS (DIACLES) HIRSUTUS, *Baker.*

Baker in Gard. Chron. 1878, vol. ix. p. 756.

Sent from the Transvaal by Mudd and flowered in April 1878. It resembles the hairy variety of Hæmanthus virescens, but has leaves of a different shape, longer pedicels, and a differently constructed bract.

HÆMANTHUS KALBREYERI, *Baker.*

Baker in Gard. Chron. 1878, vol. x. p. 202; Veitchs' Catlg. of Pl. 1880, p. 30, fig.; The Garden, 1879, vol. xvi. p. 438, col. pl.

Discovered on the West Coast of Africa by Kalbreyer, after whom Mr. Baker, late of the Herbarium, Kew, named the bulb.

The splendid flower-head fully 8 in. in diameter, has a somewhat dense and remarkably striking umbel of upwards of one hundred flowers of the brightest vermilion-red.

HEMEROCALLIS FULVA, *L.*, var. FLORE PLENO.

Syns. *H. disticha*, Dom.

Gard. Chron. 1860, p. 482 (Report of Exhibition of New Plants); Fl. Mag. 1861, pl. 13.

A semi-double form of the common "Day Lily," with spikes of orange flower similar in colour to the type, possessing the important quality of remaining longer on the plant.

Sent to Messrs. Veitch by the Rev. W. Ellis, by whom it was brought to this country from Mauritius; the type a native of China.

HIPPEASTRUM LEOPOLDII, *Dombrain.*

Syns. *Amaryllis Leopoldii*, Hort.

Gard. Chron. 1870, p. 733, fig. 140; Fl. Mag. 1870, tt. 475-476; Veitchs' Catlg. of Pl. fig. p. 3.

This beautiful species, introduced from Peru through Richard Pearce, first flowered during May 1869, and was named, by permission, in honour of the King of the Belgians, on the occasion of His Majesty's visit to an exhibition held in the Royal honour in the gardens of the Royal Horticultural Society at Kensington.

It was described by Pearce as the finest of any Amaryllid he had met with, and the plant has fully borne out this description.

The perianth of the flower widely spreading with a short tube and broad segments is tipped deep crimson and cream with a green star in the centre.

Largely used as a parent in the florists' varieties of the present day, its influence has been very great.

HIPPEASTRUM PARDINUM, *Dombrain.*

Syns. *Amaryllis pardina*, Hook.

Bot. Mag. t. 5645; Veitchs' Catlg. of Pl. 1867, fig.; Gard. Chron. 1867, p. 458 (advt.); Fl. Mag. t. 344.

Introduced from Peru through Richard Pearce to Chelsea, where it flowered for the first time in March 1867.

This striking, distinct species is remarkable for spreading flowers, with scarcely any tube, the whole inner surface displayed; the colour suggests a spotted variety of Calceolaria or Tydæa, so closely is it covered with small dots of crimson-red on a cream ground more or less confluent.

It has been used by the hybridist but sparingly, in the production of the florists' forms of the Hippeastrum.

HIPPEASTRUM PRATENSE, *Baker*, var. QUADRIFLORA.

Syns. *Habranthus pratensis*, Herb., var. *quadriflora.*

Bot. Mag. t. 3961.

Sent by William Lobb from Valdivia and about the same time to Mr. Bevan by Bridges, and first flowered in May 1842 in the open.

LILIUM AURATUM

THE ROYAL GARDENS, KEW

BULBOUS PLANTS

LILIUM AURATUM, *Lindl.*

Gard. Chron. 1862, p. 644, b.; Bot. Mag. t. 5338; Fl. des Serres, 1862-1865, tom. xv. p. 57; Fl. Mag. 1862, pl. 121.

The most splendid creation of the temperate zone—unrivalled in pure distinction—aristocrat in every line—and when the day comes that the House of Veitch must pass, it may well and safely leave its laurels with the "Golden-rayed Lily of Japan."

This lily was introduced to Europe from the "Land of the Rising Sun" by the late John Gould Veitch in 1862, and about the same time Robert Fortune sent bulbs to Mr. Standish, a nurseryman of that day.

Flowered for the first time in July of that year, shown at the great exhibition of the Horticultural Society in the Gardens at South Kensington, the interest and surprise were quite unusual.

At home in any ordinary garden in any ordinary peaty soil, it need but be left alone to improve and multiply.

Idle to attempt description of one of Nature's happiest efforts, it is yet permissible to quote Dr. Lindley in the *Gardeners' Chronicle*—the issue of July 12th, 1862:—

"If ever a flower merited the name of glorious, it is this, which stands far above all other Lilies, whether we regard its size, its sweetness, or its exquisite arrangement of colour. Imagine, upon the end of a purple stem no thicker than a ramrod, and not above 2 ft. high, a saucer-shaped flower at least 10 in. in diameter, composed of six spreading, somewhat crisp parts, rolled back at their points, and having an ivory-white skin thinly strewn with purple points or studs, and oval or roundish prominent, purple stains. To this add in the middle of each of the six parts a broad stripe of light satiny-yellow losing itself gradually in the ivory skin.

"Place the flower in a situation where side-light is cut off, and no direct light can reach it except from above, when the stripes acquire the appearance of gentle streamlets of Australian gold, and the reader who has not seen it may form some feeble notion of what it is.

"Fortunately some ten thousand eyes beheld it at South Kensington on the 2nd instant (July 2nd, 1862) and they can fill up the details of the picture."

LILIUM AURATUM, *Lindl.*, var. PLATYPHYLLUM, *Baker.*

Baker in Gard. Chron. 1880, vol. xiv. p. 198.

Introduced through Maries from a small island on the south-east side of Japan, and quite the finest of the type.

In form close to the typical species, it differs in a habit more dwarf and broader leaves; the perianth segments are more massive, with spots as in Lilium auratum, less copious and more concentrated in the centre of the segments.

LILIUM AURATUM, *Lindl.*, var. TRICOLOR, *Baker.*

Baker in Gard. Chron. 1880, vol. xiv. p. 198.

The distinctive marks are a robust habit, broad sub-erect leaves, and large flowers, without any brown dots, but with copious spots and papillæ of the same colour as the lamina.

LILIUM DUCHARTREI, *Franch.*

A handsome lily introduced to cultivation by Wilson, from Western China, found growing in bogs and marshes at elevations of 4,000-5000 ft.

The flowers at Coombe Wood in the summer of 1905, of a creamy-white ground colour with purple spots, had the perianth pieces reflexed.

The plant attains a height of from 2 to 4 ft., and carries from six to eight flowers on a stem.

LILIUM GIGANTEUM, *Wall.*

Syns. *L. cordifolium*, Don.

Gard. Chron. 1851, p. 513 (advt.); *id.* 1853, pp. 324, 327 (Notice of Exhibit); Bot. Mag. t. 4673.

Bulbs of this striking lily were collected by Thomas Lobb in Nepal, the native habitat, where it had originally been discovered by Dr. Wallich in moist shady places.

These arrived in England in 1851, flowered, and were exhibited in May 1853.

According to the Botanical Magazine (*l.c. supra*) it bloomed for the first time in this country in July 1852 in the nursery of Messrs. Cunningham, Comely Bank, Edinburgh, from seed collected by Major Madden.

A majestic object, attaining 6 to 10 ft. in height with large glossy heart-shaped leaves clothing the stem, and white trumpet-shaped fragrant flowers on terminal spikes; suitable for the bog garden and bamboo-planted lake-sides.

BULBOUS PLANTS

LILIUM LEICHTLINII, *Hook.*

Bot. Mag. t. 5673; The Garden, 1882, vol. xxi. p. 236, pl. cccxxxi.

A charming lily, somewhat difficult to cultivate, imported by Messrs. Veitch from Japan with Lilium auratum.

It resembles in some respects L. tigrinum, but differs in colour and in a more graceful habit, scattered leaves, and crested inner segments of the perianth. The flowers have a bright yellow groundwork, spotted with clear red-purple or maroon-brown.

In naming this lily, Sir Joseph D. Hooker writes (*l.c. supra*):—"I have named this plant after the zealous cultivator of the genus to whom the Royal Gardens are indebted for many rare species, M. Max Leichtlin, of Carlsruhe, a gentleman who is especially devoting himself to the elucidation, by culture, of the numerous species and races of this noble genus of bulbous plants."

LILIUM LONGIFLORUM, var. FORMOSANUM, *Baker.*

Baker in Gard. Chron. 1880, vol. xiv. p. 524; The Garden, 1880, vol. xviii. p. 458, fig. p. 459; Veitchs' Catlg. of Pl. 1881, p. 10, with fig.

A variety with more numerous, longer, and narrower leaves than the type, a red keel at the back of the perianth segments.

It is mentioned in Elwes' Monograph, but at the time of writing had not been seen alive in this country.

Bulbs were sent home by Maries from Formosa.

LILIUM MYRIOPHYLLUM, *Franch.*

Gard. Chron. 1905, vol. xxxvii. p. 328, suppl. illus.; Flora and Sylva, 1905, vol. iii. p. 328, col. pl. fig. 1, also p. 331, fig.

A charming lily with numerous narrow slightly twisted lanceolate leaves clothing the stem from base to apex. The flowers large, trumpet-shaped, of substance, have a pure white ground colour stained with rose-purple on the exterior of the perianth and inside, as far as the throat a rich golden-yellow; the anthers large, the pollen orange-yellow. The flowers open during the latter part of August to the beginning of September.

Originally discovered by Père Delavay in Central China, introduction by Messrs. Veitch did not take place till bulbs were collected in 1904.

LILIUM NEILGHERRENSE, *Wight.*

Syns. *L. Wallichianum*, Wight; *L. neilgerricum*, Hort. Veitch.

Baker in Jour. Linn. Soc. vol. xiv. p. 230; Fl. Mag. n.s. t. 237; Bot. Mag. t. 6332; l'Illus. Hort. vol. x. t. 253; Fl. des Serres, 2266-2267.

This, the only lily of the mountains of Southern India, inhabiting the Neilgherries and Pulnies, at an elevation of some 8,000 ft., was introduced by Thomas Lobb in 1862, but failed to become established, and later was again imported in considerable quantity.

The flowers from 6 to 10 in. long, narrowly funnel-shaped, cream-white with a yellow throat, change to almost pure white when mature.

LILIUM PHILIPPINENSE, *Baker.*

Gard. Chron. 1873, p. 1141, fig.; Veitchs' Catlg. of Pl. 1878, frontispiece; Bot. Mag. t. 6250; Jour. Linn. Soc. vol. xiv. p. 228.

A very distinct species, sent to Chelsea by Gustav Wallis from the Island of Luzon, found at an elevation of 7,000 ft.

The flowers pure white with the exception of a green tinge at the base of the tube, very long and funnel-shaped, are remarkably sweet-scented.

Botanically interesting as showing the extreme development of the peculiarities distinguishing the Eulirion group.

LILIUM SPECIOSUM, *Thunb.*, var. GLORIOSOIDES, *Baker.*

Baker in Gard. Chron. 1880, vol. xiv. p. 198.

Discovered by Maries in the mountains of Central China and introduced by him to this country.

It differs from previously known varieties of Lilium speciosum in more narrow leaves, much reflexed perianth-segments, resembling those of Gloriosa superba, and by scarlet spots and papillæ mainly confined to the third quarter of the segment.

LILIUM SUTCHUENENSE, *Franch.*

Bot. Mag. t. 7715; Gard. Chron. 1905, vol. xxxviii. p. 91, suppl. illus.; Flora and Sylva, 1905, vol. iii. p. 328, col. pl. fig. 2, also p. 329, fig.

The figure of this species in the Botanical Magazine (*l.c. supra*) was prepared from plants which flowered in the Royal Botanic Gardens, Kew, in July 1899, from bulbs procured from M.M. Vilmorin of Paris.

Seed was first sent to M. Maurice de Vilmorin by the Abbé Farges, collected in the Province of Szechuan, Western China, in which locality it was also found by Prince Henry d'Orleans.

Bulbs in quantity introduced to this country through Wilson to Coombe Wood flowered in July 1905.

The lily belongs to the Martagon section, orange-scarlet flowers thickly spotted with purple and narrow linear leaves resembling those of Lilium tenuifolium, to which species it is closely allied: it is possible that its constitution will prove suitable to the English climate.

BULBOUS PLANTS

NARCISSUS TAZETTA, *L.*, var. ORIENTALIS, FLORE PLENO.

Gard. Chron. 1874, p. 385; Bot. Mag. t. 1001.

Exhibited before the Floral Committee of the Royal Horticultural Society on March 18th 1874 and awarded a First-class Certificate.

An old garden plant, figured in the Botanical Magazine as early as t. 1001, but probably lost to cultivation for many years.

TRISTIGMA NARCISSOIDES, *Benth. & Hook.*

Syns. *Stephanolirion narcissoides*, Baker.

Baker in Gard. Chron. 1875, vol. iv. p. 234.

An interesting bulbous plant from Chili, first flowered at Chelsea in September 1874.

At a first glance the flowers have the general appearance of Narcissus Tazetta, with a white limb and orange crown, but further examination shows the flower not an Amaryllid, but Liliaceous, more closely allied to the Millas than to the Narcissus.

Mr. Baker founded the new genus Stephanolirion on this plant, but it has now been referred to Tristigma, a genus confined to only three or four species.

URCEOLINA AUREA, *Lindl.*

Syns. *U. pendula*, Herb.

Gard. Chron. 1864, p. 627; *id.* p. 890, fig.; Bot. Mag. t. 5464; Fl. Mag. 1866, p. 307; The Garden, 1888, vol. xxxiii. p. 436, pl. 648.

This interesting bulbous plant, a native of the Andes of Peru, was first detected by the travellers Ruiz and Pavon and described by them in Flora Peruviana, iii. p. 58, t. 287.

Introduced to this country through Richard Pearce in 1863, from Muña, and first flowered during the spring of the following year, it has proved unusually amenable in cultivation.

As an Amaryllid the flowers are remarkable both in colour and shape, in an umbel at the apex of a scape about 1 ft. high, each resembling an inverted pitcher suspended by a slender stalk; the floral segments bright yellow, tipped with green and margined white.

WATSONIA DENSIFLORA, *Baker.*

Bot. Mag. t. 6400.

Introduced by us from Natal through Christopher Mudd, who sent home living plants collected when in the Veitchian service in that country; first flowered at Chelsea in August 1878.

ZEPHYRANTHES CITRINA, *Baker.*

Bot. Mag. t. 6605.

This, the first yellow Zephyranthes to be introduced to cultivation, was imported by the Veitchian people, it is believed from Demerara, but Mr. Baker, in describing it in the Botanical Magazine (*l.c. supra*), suggests the probability that the bulb is a native of Mexico.

ZEPHYRANTHES MACROSIPHON, *Baker.*

Baker in Gard. Chron. 1881, vol. xvi. p. 70.

A species imported from Mexico, somewhat close to Zephyranthes carinata, with which it agrees in the bright red colour of the flower, readily distinguished by short stigmatic lobes and an expanded tube as long as the segments.

ZEPHYRANTHES TREATIÆ, *S. Wats.*

W. W. in The Garden, 1890, vol. xxxvii. p. 155.

Introduced from the swamps of Florida about the year 1880, and very closely approaching Zephyranthes Atamasco.

BEGONIAS

BEGONIAS

ANDEAN OR TUBEROUS-ROOTED SPECIES.

The numerous garden varieties of this section of the genus are among the most popular of the summer-flowering plants of the present day, and probably no race has so quickly gained favour or become so widely distributed.

This pre-eminence is due to the rich and varied colours of the flowers, the many forms they take, and the ease with which they can be grown.

To the student of the evolution of our garden plants, and to the intelligent cultivator, there is not a more interesting group to study, as it is within a comparatively short period, not more than five-and-thirty years, that the work of hybridizing Begonias was commenced, and the foundation laid of that magnificent race common in our gardens to-day.

The modern varieties with their large flowers have entirely ousted the early hybrids from cultivation, whilst the original species from which they were derived are now unfortunately seldom seen outside Botanic Gardens.

Seven wild species were employed in the production of the modern summer-flowering Begonias, all natives of the Andes of Peru, and all, with but two exceptions, introduced to this country and distributed by Messrs. Veitch.

The exceptions mentioned are Begonia Clarkei and B. cinnabarina, both introduced by E. G. Henderson of Pine Apple Place.

Of the five other species—Begonia boliviensis, B. Pearcei, B. Veitchii, B. rosæflora, and B. Davisii—four were introduced through Richard Pearce and the last-named through Walter Davis.

The first species, *Begonia boliviensis, was found by Pearce in the Bolivian Andes in 1865, though it had previously been discovered by Weddell in the same region but not introduced.

It was exhibited for the first time at the International Horticultural Show in Paris, in May 1867, on which occasion it is stated to have "attracted more of the attention, both of botanists and horticulturists, than any other plant then brought to that magnificent exhibition."

The plant is occasionally met with in cultivation at the present day in summer bedding, and its bright scarlet flowers and graceful drooping habit merit such attention, though more brilliant effects are produced by its descendants, the modern garden varieties.

The stems of Begonia boliviensis spring from a tuberous root-stock, and attain a height of from 12 to 18 in.; the flowers produced in pairs or threes on short stems in the angles of the obliquely lanceolate leaves, are bright in colour, composed of four pointed segments.

* Begonia boliviensis, Gard. Chron. 1867, p. 544, fig.; Bot. Mag. t. 5657; Fl. Mag. 1867, t. 354.

The stamens differ from those of the typical Begonias in that they are in an elongated column, instead of a globular cluster, and the stigma also differs in some minor technical points. These characters considered by Klotzch sufficient to merit a new genus, he proposed the name of Barya, subsequently merged into Begonia by De Candolle.

Begonia boliviensis will always be of special interest to gardeners, from the fact that it was one of the species used by Seden in the production of the first hybrid tuberous Begonia raised in England, B. × Sedenii.

The next great find was *Begonia Pearcei, a beautiful species with yellow flowers, native of La Paz, where it was discovered by Pearce, after whom it was named.

The plant is of tufted habit, has the unusual quality of possessing ornamental foliage and showy flowers. The leaves are dark velvety green above, traversed by straw-coloured veins: the under surface dark red with the exception of a prominent venation. Well above the handsome leaves are borne the bright yellow flowers, usually three in number, on slender scapes.

This species entered largely into the production of garden varieties, developing colours entirely absent prior to its introduction. Traces of its ornamental foliage can still be detected in the yellow-flowered varieties of the present day.

It was distributed in 1866, and figured amongst other new Plants, in a coloured plate in the Veitchian Plant Catalogue of that year, associated with Calathea (Maranta) Veitchiana, Lapageria alba, and Urceolina pendula, all of which we owe to Pearce, and all of which remain favourite garden plants.

† Begonia Veitchii, another introduction due to Richard Pearce, was discovered near Cuzco in Peru at an elevation of 12,000-12,500 ft. in 1866, and first flowered in this country in 1867, soon after B. boliviensis had produced its scarlet blooms.

In the Botanical Magazine (sub t. 5663), Sir J. D. Hooker writes:—"Of all species of Begonia known, this is, I think, the finest. With the habit of Saxifraga ciliata, immense flowers of a vivid vermilion-cinnabar red, that no colourist can reproduce, it adds the novel feature of being hardy, in certain parts of England at any rate, if not in all.

"Unwilling as I am to pronounce on the probable or possible adaptation of exotic plants to an English climate, I cannot but believe that in south-western counties and in the south of Ireland, the Begonia Veitchii will certainly prove one of the most ornamental of garden plants."

This expectation of the probable hardiness of Begonia Veitchii was not realized, as it was found of too delicate a constitution to withstand the combined effects of cold and damp of English winters, but tubers stored in a dry place flowered freely out-of-doors during the summer months; this is the case with all the Andean species of Begonia.

Walter Davis, who traversed Pearce's ground some years later, found Begonia Veitchii inhabiting rocky positions by waterfalls, in company with Masdevallia Veitchiana.

In association with Begonia boliviensis and B. Pearcei, this species has been

* Begonia Pearcei, Hook. f. Bot. Mag. t. 5545; Veitchs' Catlg. of Pl. 1866, col. pl.

† Begonia Veitchii, Hook. f. Gard. Chron. 1867, p. 734, fig.; Bot. Mag. t. 5663; Fl. Mag. 1867, vol. vi. pl. 365.

BEGONIA PEARCEI

much used by the hybridist, and in the varieties of to-day may still be seen traces of the stout circular leaves and short thick stem.

A closely allied species, named Begonia Clarkei, in honour of Colonel Trevor Clarke who first flowered it, had been introduced from Peru a few years prior to the introduction of B. Veitchii, but the plants did not flower for several years, and when they did, were found to differ from B. Veitchii in habit as well as in colour, size, and shape of flower.

A higher temperature is required by Begonia Clarkei than by the majority of the genus, and the presumption is that it is a native of a lower altitude than B. Veitchii. A figure of the latter is given in the Veitchian Plant Catalogue for 1868, in which year it was distributed.

Another of Pearce's Andean introductions, *Begonia rosæflora, a species in many respects resembling B. Veitchii, from which, however, it differs in several important points.

The petioles of the leaves coloured red, as is the villous flower-scape; the bracts and stipules of a different form to those of Begonia Veitchii—bright rose-coloured; the colour of the flowers, the most marked difference, is that of the Wild Rose, not the cinnabar-red of B. Veitchii.

Seden early made use of this species in the production of hybrids, and it may be looked upon as one of the most active progenitors of the garden tuberous Begonias of to-day.

Another interesting fact regarding this species is that from seedlings obtained by crossing light-coloured varieties, the first white-flowered tuberous Begonia was obtained, and sent out under the varietal name Queen of the Whites.

Davis found that the colours of the flowers in this species varied greatly, and a white-flowered form sent home by him some years later, proved identical with the seedling known as Queen of the Whites.

The last of the Andean species to be introduced to cultivation was †Begonia Davisii, named after the discoverer, who detected it near Chupe, in Peru, when collecting in that region.

Like its congeners, Begonia Veitchii and B. rosæflora, it is a stemless species with broadly ovate-cordate leaves, glossy green above and purplish beneath, with a slightly lobed serrated margin; the flowers rich orange-scarlet in colour, are in threes on erect red-coloured scapes 6 to 8 in. high.

The dwarf habit and erect flowers characteristic of this species were taken advantage of by Seden, who rapidly evolved several garden forms possessing most desirable qualities and of a very high order.

TUBEROUS-ROOTED HYBRIDS.

The first hybrid Begonia of the tuberous-rooted section was raised at Chelsea by John Seden as the result of a cross between B. boliviensis and an unnamed Andean species.

The Royal Horticultural Society awarded it their Silver Floral Medal as the "best new plant shown for the first time in bloom" at their New Plant Show held on June 2nd, 1869.

* Begonia rosæflora, Bot. Mag. t. 5680; Fl. and Pom. 1869, col. pl. p. 1.

† Begonia Davisii, Hort. Veitch; Bot. Mag. t. 6252; Fl. Mag. t. 6252; Veitchs' Catlg. of Pl. 1879, fig.; The Garden, 1878, vol. xiii. p. 208, pl. 118.

*Begonia × Sedenii was the name given in compliment to the raiser.

Plants distributed in 1870, and figured in the Plant Catalogue of that year, are described as "of the same upright habit as Begonia boliviensis but with larger leaves, the veins of which are slightly rose-coloured; the flowers are of the richest magenta and of large size."

It is interesting in that the first double-flowered Begonia was obtained from seeds produced by self-fertilized flowers from this hybrid.

This was followed by † Begonia × Chelsoni, another hybrid raised by Seden by fertilizing the hybrid B. × Sedenii with B. boliviensis, thus crossing the progeny with one of its parents. The appearance of B. × Chelsoni greatly resembled B. boliviensis, but the flowers larger, were a glossy red in colour.

‡ Begonia × intermedia was the next, a product of the two Andean species B. Veitchii and B. boliviensis, distributed in 1872 and described in the Veitchian Plant Catalogue for that year:—"In habit it partakes strongly of B. boliviensis, being a strong upright-growing plant, branching freely, and attaining an average height of 15 to 18 in. The leaves have much the form and substance of B. Veitchii, but are toothed like B. boliviensis. The flowers are of the size and form of B. Veitchii and resemble it also in colour, but are of a rather darker shade."

In 1874 two more were offered, § Stella and § Vesuvius: the former the result of crossing Begonia × Sedenii with B. Veitchii, and the latter the progeny of B. Clarkei crossed with B. × Sedenii. Vesuvius had bright orange-scarlet flowers, combined with a robust habit, a useful and popular subject for summer-bedding. || Excelsior and || Model next raised, were distributed in 1875. The variety Excelsior was obtained from Begonia × Chelsoni, crossed with B. cinnabarina, and is described as a very free-blooming variety, with large flowers of the form of B. Veitchii and the colour of B. cinnabarina. It was one of the best bedding Begonias of its time. Model was the product of a cross between B. × Sedenii and B. Pearcei, the flowers of a delicate rosy blush colour, and of the finest shape.

The hybrids Begonia × Sedenii and B. × intermedia were next used as parents, and these produced the variety named ¶ Acme, sent out in 1876. A figure of the flower is given in the Plant Catalogue of that year, and in an accompanying description it states:—"The flowers are of a delicate orange-pink, tinged with a deeper shade of orange-rose. In well-grown plants the flowers are large, the male or staminate ones from 3 to 4 in. between the extremities of the alternate narrower petals; the petals of both staminate and pistillate flowers are beautifully veined symmetrically with the edges."

** Kallista, derived from Begonia × Sedenii crossed with the variety Stella, was also sent out in 1876. The flowers, of a rich vermilion-scarlet, were the darkest shade of that colour known at the time Later Kallista gave rise to many fine varieties from self-fertilized seed.

* Veitchs' Catlg. of Pl. 1870, col. pl. and fig. p. 4; Fl. and Pom. 1869, p. 169, col. pl.
† Veitchs' Catlg. of Pl. 1871, col. pl. and fig. p. 2.
‡ Veitchs' Catlg. of Pl. 1872, p. 2, fig.; Fl. Mag. Feb. 1872.
§ Veitchs' Catlg. of Pl. 1874, pp. 4, 5, figs.
|| Veitchs' Catlg. of Pl. 1875, pp. 4, 5, figs.
¶ Veitchs' Catlg. of Pl. 1876, p. 6, fig.; The Garden, 1878, vol. xiii. pl. 118.
** Veitchs' Catlg. of Pl. 1876, p. 7, fig.

In 1877 appeared *Emperor, from Begonia Clarkei crossed with the hybrid B. × Chelsoni, which proved a marked advance on the forms then existing, having brilliant orange-scarlet flowers of large size.

Following Emperor came †Monarch, from crossing the two hybrids Begonia × Sedenii and B. × intermedia. This, with bright vermilion-scarlet flowers, was, when distributed, undoubtedly one of the finest of all the earlier successes.

Up till 1877 the colours of the flowers of the Hybrid tuberous-rooted Begonias were not of a wide range, mostly shades of red or varying from magenta to orange-scarlet, but a break occurred when ‡Queen of the Whites developed among a batch of seedlings of B. rosæflora.

In habit Queen of the Whites resembled Begonia rosæflora, dwarf and furnished with orbicular leaves and deeply-sunk veins; the flowers, 2 to 2½ in. in diameter, milk-white in colour, were freely produced on erect scapes, and the variety largely used as a parent, became the forerunner of all the white and light-coloured forms now so numerous.

In 1880 two varieties, named §Mrs. Charles Scorer and §roseo-superba, were distributed. Mrs. Charles Scorer, a splendid crimson-scarlet flowered variety, was produced by crossing Viscountess Doneraile (a hybrid not then sent out) with another seedling.

It is thus described:—"A splendid variety with large well-formed flowers of a brilliant glowing crimson-scarlet, unequalled in this particular shade of colour by any Begonia of its class. The plant is of robust habit, free-flowering and vigorous, furnished with a neat dark green foliage, which, together with brilliant flowers, render it one of the best Begonias yet obtained."

The variety roseo-superba, the progeny of Begonia rosæflora crossed with an unnamed seedling, produced flowers of a clear bright rose-colour suffused with white, at that time a unique tint amongst Begonias.

‖ Admiration and Viscountess Doneraile were the next two, catalogued for the first time in 1881. The first-named was raised from the Excelsior crossed with Begonia Davisii, and showed the influence of the latter parent in its dwarf, compact habit and vivid orange-scarlet flowers.

Viscountess Doneraile, one of the most brilliant of Seden's achievements with the Begonia, resulted from crossing the two hybrids Monarch and B. × Sedenii; the flowers, on stout erect scapes, well above the light green leaves, were freely produced, and rich vermilion-red in colour.

An important *rôle* played by this variety was its use in connection with Begonia socotrana in the production of that entirely new and remarkable race of Begonias which has become such a popular winter-flowering section, and of which the variety John Heal was the first to be distributed.

In 1882 two varieties, the results of experiments made with the dwarf Andean species, Begonia Davisii, with a view to the production of dwarf compact plants for bedding or for small pot culture, were distributed. They were named Miss Constance Veitch and Mrs. Arthur Potts, and both had the desired compact habit, and bore flowers of varying shades.

* The Garden, 1878, vol. xviii. p. 208, pl. 118; Fl. Mag. 1876, t. 194.
† Veitchs' Catlg. of Pl. 1878, p. 8, fig.
‡ The Garden, 1878, vol. xiii. p. 208, pl. cxviii.; Veitchs' Catlg. of Pl. 1878, p. 9, fig.; Gard. Chron. 1877, Dec. 15, col. pl. fig. 4.
§ Veitchs' Catlg. of Pl. 1880, p. 6, 7. ‖ Veitchs' Catlg. of Pl. 1881, p. 49, fig.

With the introduction of this dwarf race of Begonias, Seden ceased experimenting. The hybrids produced had become widely distributed, and many hybridists, both in England and on the Continent, had engaged in the work of improvement, and new varieties appeared each year, but the eighteen hybrids, with the five original species introduced by Messrs. Veitch, form the foundation of the Begonia of to-day.

The following table gives in a concise form the history of the eighteen hybrids raised by Seden, with the order in which they were produced and the dates of their distribution:—

Order in which the hybrid was raised.	Name.		Parents.	Date of Introduction to Commerce.
1.	B. Sedenii	=	(B. boliviensis × unnamed species)	1870
2.	B. intermedia	=	(B. boliviensis × B. Veitchii)	1872
3.	B. Chelsoni	=	(B. boliviensis × B. Sedenii)	1871
4.	B. Stella	=	(B. Sedenii × B. Veitchii)	1874
5.	B. Vesuvius	=	(B. Clarkei × B. Sedenii)	1874
6.	B. Excelsior	=	(B. Chelsoni × B. cinnabarina)	1875
7.	B. Model	=	(B. Sedenii × B. Pearcei)	1875
8.	B. Acme	=	(B. intermedia × B. Sedenii)	1876
9.	B. Monarch	=	(B. Sedenii × B. intermedia)	1878
10.	B. Viscountess Doneraile	=	(B. Monarch × B. Sedenii)	1881
11.	B. Mrs. Chas. Scorer	=	(B. V'tess Doneraile × B. seedling)	1880
12.	B. Emperor	=	(B. Clarkei × B. Chelsoni)	1877
13.	B. Kallista	=	(B. Sedenii × B. Stella)	1876
14.	B. Queen of the Whites		a sport from B. rosæflora	1878
15.	B. Admiration	=	(B. Excelsior × B. Davisii)	1881
16.	B. roseo-superba	=	(B. rosæflora × B. seedling)	1880
17.	B. Miss Constance Veitch	=	(B. Davisii × B. seedling)	1882
18.	B. Mrs. Arthur Potts	=	(do. do.)	1882

WINTER-FLOWERING VARIETIES.

Since the introduction of Begonia Davisii from Peru probably no species has played a more important part in hybrids and garden varieties than has B. socotrana.

The credit of introducing this species is due to Professor Balfour, of the Edinburgh Botanic Gardens, who discovered it during botanical explorations in Socotra, a small island off the Arabian coast, in the Indian Ocean, and one of the most improbable places in the world in which to find a Begonia.

Tubers, or more correctly scaly rhizomes, of Begonia socotrana were sent, among other plants, to Kew in April 1880, and produced plants which flowered in December of the same year, and the great horticultural merit of the species at once became patent.

From Kew, by arrangement with the late Director, the stock passed into Messrs. Veitchs' hands and was distributed in 1882.

Begonia socotrana is a dwarf-growing plant with stout orbicular leaves nearly 1 ft. in diameter, and rose-pink flowers of a very pleasing shade, nearly 2 in. across, freely produced in mid-winter.

It was at once perceived that, could this latter property of flowering in mid-winter be combined with large flowers having the brilliant and varied colours of the summer-flowering tuberous kinds, a valuable race of garden plants would ensue, and extend the flowering period of this beautiful genus through practically the whole year.

Several hybridists at once commenced work, but the merit of being the first to produce a hybrid, of which Begonia socotrana was one parent, is to John Heal, who succeeded in crossing B. insignis (incarnata), a Mexican species with rose-coloured flowers, with the pollen of B. socotrana. The hybrid plant from this cross first flowered in 1882, and was named Autumn Rose, from the colour of the flowers and in allusion to the flowering season; it was not distributed, and probably is not now in existence.

The variety * John Heal, now well known, was the next raised, and the first sent out. Flowered in 1883, the produce of Begonia socotrana crossed with the pollen of the summer-flowering Viscountess Doneraile, the latter hybrid raised by Seden and previously referred to.

Only one seedling obtained, all plants of this variety now in cultivation are the produce of this one plant.

Distributed in 1885, it is still largely grown as a winter-flowering decorative subject, the neat, compact habit of growth, rich rosy-carmine flowers gracefully borne in terminal panicles, and bright emerald-green leaves, combine to make a useful plant for the decoration of the table or conservatories.

The next Adonis was raised from an orange-flowered tuberous variety crossed with the pollen of the hybrid John Heal, a marked advance in the size of the flower which, upwards of 3 in. in diameter, resemble those of the summer-flowering varieties. In colour a bright scarlet to red with carmine, distributed in 1887.

Following Adonis came † Winter Gem, the result of crossing Begonia socotrana and a very dark crimson tuberous variety, the former the female parent.

In habit resembling Begonia socotrana, dwarfer and more compact, the leaves smaller and neater. The flowers, the darkest of all the group, are rich deep crimson, 2 to 2½ in. in diameter, held erect on stout peduncles well above the foliage.

A semi-double rose-coloured tuberous variety was next used as a seed parent crossed with the pollen of Begonia socotrana, and from a single seed-pod, the varieties Ensign, Winter Perfection, ‡ Ideala, and Success, were selected.

These all flowered in 1891, showed a marked deviation from each other both in leaf and habit and in the size and colour of the flowers.

Ensign, exhibited for the first time in November 1896, was the first of this group to be distributed. The flowers are semi-double, of a pleasing shade, of light rose-carmine, with the petaloid stamens yellow or yellow-green, and the foliage intermediate between that of the two parents.

Winter Perfection, a taller-growing form, produces semi-double rose-pink flowers, the outer petals spreading, the metamorphosed stamens remaining in various stages of development, the outer ones rose-pink and the inner more or less streaked with yellow.

* The Garden, 1889, vol. xxxv. p. 218, col. pl. 691.
† The Garden, 1891, vol. xxxix. p. 504, col. pl. 807.
‡ Gard. Chron. 1901, vol. xxx. p. 411, fig. 124.

The variety Success also rather tall-growing, bears numerous semi-double flowers fully 2 in. in diameter, and in colour bright carmine toned with scarlet; the petaloids are yellow tipped with green.

The most distinct is the variety Ideala,—unusually neat and compact, some 9 in. in height; the flowers large for so small a subject, are semi-double, of a brilliant rose-colour, freely produced.

Another cross of the year, between a single scarlet-flowered variety crossed with the pollen of Begonia socotrana, proved unusually prolific. From the plants raised of one capsule were selected * Mrs. Heal, Myra, and Winter Cheer.

Mrs. Heal is by far the finest and most distinct of the set, the flowers, 2 to 3 in. in diameter, in colour brilliant rose-carmine toned with scarlet, are freely borne, and gracefully disposed.

The variety Winter Cheer, from the same seed-pod, is clearly distinct, having leaves which resemble those of Begonia socotrana, and semi-double flowers of an effective shade of rose-carmine.

The third of the set, Myra, produces single flowers of a bright rose-carmine colour, the pistillate blooms smaller and of a lighter shade.

A decided break in colour was obtained by using a semi-double white variety as seed-bearer and Begonia socotrana the pollen parent. From the one seed-capsule which this cross produced two distinct varieties were selected, Julius and Sylvia.

Julius is the most distinct of any in point of colour, a rose-pink suffused with white, and flowers more truly double than those of any other of the section. Sylvia has semi-double rosy carmine flowers 3 in. in diameter, which open in a peculiar flat manner.

Venus was the result of crossing Begonia socotrana with a single-flowered variety of the tuberous section—single flowers, in colour resembling those of Winter Cheer.

In 1903 followed two other hybrids, Agatha and Agatha compacta, the first-named from Begonia socotrana with a hybrid long known in gardens under the name of Moonlight, the product of B. Pearcei with B. Dregii, a small-flowered species from South Africa. Agatha is not only a charming garden plant, but interesting, as the flowers closely resemble those of Gloire de Lorraine, a hybrid raised by M. Lemoine of Nancy from the supposed cross-fertilization of B. socotrana and B. Dregii.

The correctness of the supposition is confirmed by the close resemblance between the two plants—Agatha showing a slight difference in the shape and colour of the leaves, as well as in a more compact growth, features probably due to traces of Begonia Pearcei in its composition.

Agatha compacta, a dwarf form of the last-named, was obtained from the two species Begonia socotrana and B. natalensis, the latter a small white-flowered South African species, resembling B. Dregii.

The flowers of this hybrid bear a close resemblance to those produced by Agatha, but are of a deeper shade of rose, and slightly larger. The great distinction, however, lies in the compact habit, unusually dwarf and neat, requiring no tying to make a shapely plant, as in the case of Gloire de Lorraine.

* Gard. Chron. 1895, vol. xviii. p. 585, fig. 101.

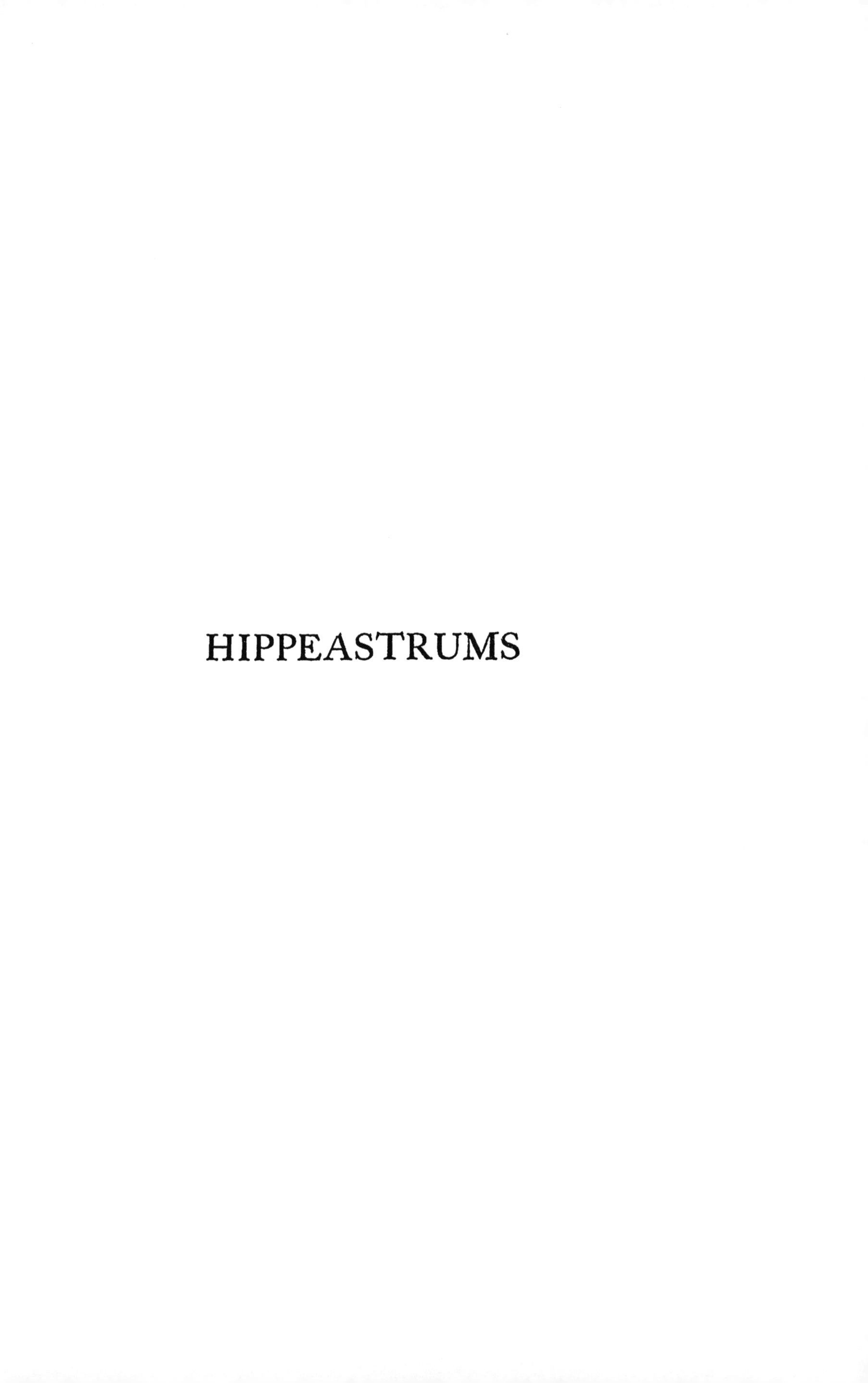

HIPPEASTRUMS

THE HIPPEASTRUM (Amaryllis)

The plants long known in gardens as Amaryllis have been in cultivation for a number of years. The generic name in use before the time of Linnæus was applied to many species of plants now referred to as Hippeastrum, and to a number of others, since separated under various generic designations, as Vallota, Griffinia, Sprekelia, and Lycoris.

Parkinson figured the old Amaryllis formosissima in his Paradisus at p. 71 as Narcissus indicus, or the Indian Daffodil with a red flower. This plant subsequently became distributed in gardens as the Jacobæa Lily, and is now botanically known as Sprekelia formosissima, this name having been given by the German botanist Heister, in honour of Dr. Sprekel.

The native habitat of Sprekelia formosissima long remained uncertain, but it is now known to be Mexico and Guatemala. It appears to have reached England by way of Spain, and to have become generally grown in this country about the year 1593.

Of the early species in cultivation in England, Hippeastrum reginæ seems to have been the first. It is stated in the Botanical Magazine, t. 453, that this species first flowered by Mr. Fairchild in his garden at Hoxton, in 1728, and was in full beauty on March 1st, the birthday of Queen Caroline, wife of George II., in whose honour it was named.

Hippeastrum equestre, another of the early species, is said to have been introduced from the West Indies by Mr. William Pitcairn in 1778. Figured in the Botanical Magazine, t. 305, the Editor states that "The spatha is composed of two leaves, which standing up at a certain period of the plant's flowering, like ears, give to the whole flower a fancied appearance to a horse's head; whether Linnæus derived his name of equestris from this circumstance or not, he does not condescend to inform us."

Other early species cultivated were Hippeastrum vittatum (Bot. Mag. t. 128) introduced in 1788, and H. reticulatum (Bot. Mag. t. 657) in 1777 through Dr. E. W. W. Gray.

With the generic name Amaryllis that of Dean Herbert is indissolubly connected. This enthusiastic horticulturist and Churchman laboured long at the genus, and assiduously cultivated and experimented on every species of the family he could procure, publishing the results of his investigations from time to time in the Botanical Magazine, the Botanical Register, and the Transactions of the Royal Horticultural Society. His famous collection of Amaryllids formed at Spofforth, in Yorkshire, attained a world-wide reputation, and the views he put forth regarding them (the first of which appeared in the Botanical Magazine in 1820, t. 2117) assisted considerably towards their class definition, and formed the basis of the classification adopted by botanists of the present day.

Another collection exceptionally rich and well-cared for was that belonging to Mr. Griffin of South Lambeth, from which the Botanical Register largely drew material for those figures and descriptions of such importance to modern cultivators and historians of the genus.

The labours of this gentleman have been commemorated by the genus Griffinia, proposed by Ker in compliment to him.

Dean Herbert was the first to cross the different species and to raise seedlings, and in so doing incurred no small amount of reproach from his contemporaries for tampering with Nature.

The good Dean was likewise the first to notice the difference between the seeds of the Cape Amaryllis Belladonna and those of the South American and Central American species, and to demonstrate by repeated experiments that those from America would not cross with those from South Africa. On these grounds, but technically on the appearance of the seeds, the Dean separated the species Belladonna from the rest of the Amaryllis, leaving it as the sole representative of the genus, and proposed for the more numerous South American species the generic term Hippeastrum, or the Knight's Star Lily, from the resemblance the radiating segments of the flower bear to the stars of some of the orders of knighthood. The new characters which separated the genus Amaryllis from Hippeastrum were published in 1822, and first adopted in the Botanical Magazine in the same year when H. puverulentum, the Bloom-leaved Knight's Star Lily, was described under t. 2273. In explanation of this change of nomenclature it states, "The name Amaryllis having been given by Linnæus originally to Belladonna, with a reason assigned, it has been thought expedient to leave the name Amaryllis to that plant and its congeners, and to detach the occidental group (to which, as more numerous, it has been proposed to preserve the known appellation) under the name Hippeastrum or Knight's Star Lily, following the idea which suggested the name equestre for one of the species."

The name Hippeastrum thus given was kept until the introduction, in 1867, through our collector Pearce, from Peru, of the beautiful H. pardinum, figured in the Botanical Magazine, t. 5645, and described by Sir Joseph Hooker under the name of Amaryllis pardina.

In justification Sir Joseph states, "The genus Hippeastrum, of Herbert, which includes many American species of Amaryllis, differs from this latter by such slight and variable characters that it cannot be regarded as of any practical value, and I therefore follow Endlicher in regarding it, together with its allies Zephyranthes Nerine, Vallota, as sections of the great and widely diffused and very natural genus Amaryllis." This arrangement, however, did not last long, for with the publication of the "Genera Plantarum," the standard work of botanical reference, the Dean's genus Hippeastrum was again revived and has remained in use to the present day, though Amaryllis still lingers in gardens as a name for the beautiful flowers botanically Hippeastrum. Besides H. reginæ and H. equestre, before alluded to, the ancestry of the present race includes H. vittatum, a native of Central America, the influence of which may be still traced in the long bands of colour on the segments of some of even the latest hybrids (it was named vittata by M. L'Heritier, "from the gaiety of its flowers, which, from their stripes, appear like an object decorated with ribbands"); H. reticulatum from Brazil, which imparted the crimson veinings and reticulations to the floral segments, and the variety striatifolium, which gave the white stripes to the leaf.

Many years later came Hippeastrum psittacinum, also from Brazil, a species

HIPPEASTRUM "EGLAMOR"

with a prominent green centre and deep crimson veining on the apical portion of the floral segments. About the same time H. aulicum, from the Organ Mountains of Brazil, introduced by the celebrated horticulturist Mr. William Harrison, was described by Ker in the second volume of the Journal of Science and the Arts, p. 353. It has an irregular flower, red in colour, a prominent green centre, and segments thin and pointed.

Another remarkable species, named Hippeastrum solandræflorum, has flowers with a long greenish-white tube and almost regular segments, which, when seen individually, resemble some of the Japanese species of Lilium. The Dean crossed this species with the male H. reginæ-vittatum, and obtained a similar result to that which is figured in the Botanical Magazine, t. 3542, as Hippeastrum ambiguum longiflora, afterwards used as a parent in the production of the forms with long-tubed flowers.

As early as 1824 Dean Herbert had thirty-five recorded hybrids, all, with the exception of five, raised by himself.

In his "Amaryllidaceæ," Herbert states that the first cross made was between the two species Hippeastrum vittatum and H. reginæ or equestre, by a Mr. Johnson, a shoemaker, or, according to others, a watchmaker of Lancashire, in the year 1811, and later by Herbert himself at Mitcham in 1811.

This hybrid received the name of Johnsoni from the first-named raiser, and from Herbert, a compound of the two specific names—H. regio-vittatum. The correctness of the recorded parentage of Johnsoni was at one time doubted, but Mr. Gowen, at that time gardener to the Earl of Carnarvon, by raising seedlings from the same cross, proved it identical with Johnsoni.

A remarkable fact in connection with this hybrid is the length of time it retained its character under cultivation, and its potentiality with other species and varieties when used as a breeder, influenced the offspring to a great extent.

Amongst many hybrids that existed but for a short time, and exerted little influence on succeeding generations, one may be singled out as of more importance than the rest.

This received the name of Acramanni (and also the erroneous one of Ackermanni), and was the result of a cross by Messrs. Garraway & Sons of Bristol in 1835 between Hippeastrum aulicum, H. platypetalum, and H. psittacinum. It was named Acramanni in compliment to G. Acraman Esq., of the city of Bristol.

This was unquestionably the finest hybrid yet raised, but a few years later was eclipsed by a seedling of the same firm, flowered in 1850, from Hippeastrum aulicum crossed with the hybrid Johnsoni, named Acramanni pulcherrima, from a resemblance it bore to the original hybrid. This last is of great interest, as one of the parents used by Messrs. Veitch some years later, in the production of the first Hippeastrum raised at Chelsea.

Van Houtte and other horticulturists in Belgium and France took up the culture of these plants soon after the appearance of Acramanni pulcherrima, and produced many fine seedlings remarkable for brilliance of colouring, though usually deficient in form, with narrow pointed petals of unequal size.

The elder de Graaff of Leyden afterwards surpassed Van Houtte's productions, especially with one named Graveana. This fine form afterwards used by his sons in connection with a dark form of Hippeastrum psittacinum produced the fine Empress of India, still in cultivation.

The introduction of the beautiful species Hippeastrum pardinum, through the collector Pearce, in 1866 gave a new impetus to the cultivation of Hippeastrums and supplied hybridizers with new material.

The magnificent flowers are spotted all over with vermilion on a yellowish ground, as in the skin of a leopard. The flower-tube short or nearly absent, the floral segments broad, recurved, and spreading, form a flower nearly 7 in. in diameter.

The Veitchian *employés* commenced the work of hybridizing in 1867, using Hippeastrum pardinum as one parent and the hybrid Acramanni pulcherrima as the other. The cross was effected by John Seden, and from the one capsule obtained three distinct plants originated, named Chelsoni, Brilliant, and maculata respectively, the former the best. With all its fine qualities H. pardinum did not realize the expectations which were held of it as a breeder, and it was with the introduction of H. Leopoldii discarded.

Hippeastrum aulicum was also used in some of our early experiments, but the results were always disappointing, and did not justify any further use of that species as a parent, most of the flowers produced by its progeny being thin and ill-shapen, with narrow pointed segments and a large amount of green in the centre.

Hippeastrum Leopoldii was the species most successfully used as a parent, and one which exerted the most influence in producing the present race. The species first flowered in 1870, was shown for the first time at the Royal Horticultural Society's Exhibition held at South Kensington in honour of Leopold King of the Belgians, at the time paying a visit to this country. His Majesty was much struck by the beauty of the flower of the new species, and courteously granted permission to name the plant in commemoration of his visit. The flowers are large and widely expanded, the tube is short and the segments broad and of good substance, the habit and general contour of the plant possesses points which it became desirable to infuse into those at that time in cultivation.

Many were raised of the Leopoldii type, the highest degree of perfection in form and size being attained in John Heal.

Beyond this the plants of this type refused to break into any form that could be considered an improvement, and, as the constitution became weakened by continued in breeding, new blood became necessary. Moreover the varieties produced by Hippeastrum Leopoldii seldom bore more than two flowers on a spike.

The next step was to increase the number of flowers on a spike to five or six, and to retain the symmetry and refinement characteristic of the Leopoldii type. With this in view de Graaff's hybrid Empress of India was crossed with the best Leopoldii forms, and resulted not only in an increase in the number of flowers on a scape, but in decided breaks of colour, and shades and tints previously unknown.

A further aim to obliterate or reduce as far as possible the green centre from the blooms met with a like success; the modern flowers seldom show sufficient green to be objectionable, and in certain self-coloured forms it is usually replaced by a darker shade of the ground colour. In other varieties the green centre assumes a yellowish or orange tinge, a happy contrast to the prevailing colour of the segments.

Among the most striking colours of the past few years, the yellowish-green tinge of the Veldt is probably the most novel. It is by no means so pleasing a shade as that possessed by many of the older forms, but to the breeder it has considerable attractions, and indicates an approach to an entirely new break which may eventually culminate in a yellow-flowered Hippeastrum.

Other points always kept in view were the shortening and expansion of the flower-tube, the widening of the floral segments, especially the lowermost one, the intensifying of the natural colour, and the production of new tints.

The work of improving the Hippeastrum, carried on for thirty-five years, has resulted in the production of numerous beautiful forms, nearly 200 Certificates having been awarded the Veitchian house since 1867 by the Royal Horticultural and Royal Botanic Societies.

ORCHID HYBRIDIZATION

ORCHID HYBRIDS

THE exceptional attention devoted to orchid hybridization for some thirty years, whether with species or varieties, has given rise to a numerous progeny as varied in form and colour as many Florists' flowers, and still scarcely a year passes without adding a new genus amenable to artificial hybridization.

Probably the first attempt to raise hybrid orchids is due to Dean Herbert, known for his work among the Amaryllideæ, but little apparently resulted. Robert Gallier, a gardener, sent to the Gardeners' Chronicle in 1849 an account of his attempt to raise hybrid orchids by crossing Dendrobium nobile and D. chrysanthum, and of the short life—only a few weeks—of the young plants reared from the cross-fertilized seed. The first successful hybridizer of Orchids, John Dominy, acting on the suggestion of Mr. John Harris, a surgeon of Exeter, worked there about the year 1853. The first result was Calanthe × Dominii, obtained from C. masuca and C. furcata, which flowered for the first time in October 1856, the plants being two years old. About three years later a hybrid Cattleya followed, C. × hybrida, the result of crossing C. guttata and C. intermedia, and was exhibited for the first time in August 1859.

In November of the same year Cattleya × Dominiana flowered, a very grand hybrid named after the raiser. Still another appeared in 1859, and this, perhaps the most brilliant result ever obtained by the hybridist, named Calanthe × Veitchii, is probably the most widely grown of any hybrid orchid, the beautiful rose-coloured flowers especially valuable during the winter season. It resulted from a cross between C. rosea (then called Limatodes rosea) and C. vestita.

In June 1861, what was held to be the first bigeneric hybrid appeared, and was exhibited under the name of Goodyera × Dominii, a product of a cross between G. discolor and Anœctochilus Lowii, now known as Hæmaria discolor and Dossinia marmorata respectively. It appears to be now lost, but a similar hybrid, Goodyera × Veitchii, flowered the following year, the result of crossing Hæmaria discolor and Macodes petola, or Anœctochilus Veitchianus, as it was then called.

Cattleya × Brabantiæ (at first named Cattleya × Aclandi-Loddigesii) flowered in July 1863, and received a silver Banksian Medal from the Royal Horticultural Society.

The next, a remarkable advance, the forerunner of the numerous Lælio-cattleyas since raised, was exhibited under the name of Cattleya × exoniensis, the parents being either C. Mossiæ or C. labiata and Lælia crispa.

Following these came Lælia × Pilcheriana, from L. crispa and L. Perrinii, one of the earliest of all crosses, the seed sown in 1853; Anœctochilus × Dominii, derived from Goodyera (Hæmaria) discolor and Anœctochilus xanthophyllus, in May 1865; Cattleya × quincolor in June 1865; C. × Manglesii in August 1866;

Phaius (Phaiocalanthe) × irroratus, the result of crossing P. grandifolius and Calanthe vestita, early in 1867, followed by Phaius × inquilinus in May of the same year.

It was not till 1869 that the first hybrid appeared in the genus Cypripedium, a genus which proved unusually prolific.

This hybrid was named Cypripedium × Harrisianum, a name perpetuating the memory of Dr. Harris who first suggested to Dominy the possibility of obtaining hybrid orchids. The parents were C. villosum and C. barbatum. In 1870 C. (Selenipedium) × Dominianum from Selenipedium caricinum and S. caudatum flowered, and in the same year Cypripedium × vexillarium made an appearance. This was derived from C. barbatum and the at that time rare C. Fairieanum, the influence of the latter predominating.

Among other good seedling hybrids due to Dominy may be mentioned Lælia × Veitchiana, derived from Cattleya labiata and Lælia crispa in 1874; Cattleya × Felix, from Lælia crispa and Cattleya Schilleriana in 1876; Lælia × caloglossa, from Cattleya labiata and Lælia Boothiana or L. crispa in 1877; Dendrobium × Dominianum, the first hybrid of the genus, from D. Linawianum and D. nobile in 1878. Dominy's last and very remarkable hybrid Læliocattleya Dominiana, a gorgeous seedling from Cattleya Dowiana and Lælia purpurata, was described by Reichenbach under the name of L. × Dominiana.

The twenty-four hybrids raised by Dominy during his fifteen years' work, trifling in view of later results, cover a wide field, and were the foundation of all future work—Dominy was the pioneer. They comprise six Cattleyas, one Lælia, six Læliocattleyas, two Calanthes, one Phaiocalanthe, three Cypripediums, one Dendrobium, one Aërides (probably lost), and several Goodyeras and Anœctochili.

John Seden succeeded Dominy, and continued the work so successfully commenced, and from the appearance of his first hybrid in 1873 till retirement in 1905 he added hundreds of the most distinct to collections, and for a series of years experimented over a wide field.

Cypripedium (Selenipedium) × Sedenii, the name given to the first hybrid, was the result of a cross between Selenipedium longifolium and S. Schlimii. This flowered in 1873, and is still frequently met with; the flowers valued for a rose-pink colour, are produced with freedom.

The same year that Seden flowered his first cross he also established a record by flowering a secondary hybrid, i.e. a cross between a pure species and a hybrid. This, Læliocattleya Fausta, derived from Cattleya Loddigesii as seed-bearer and Dominy's hybrid Læliocattleya exoniensis as pollen-parent, developed several forms from the same seed-pod, a variability that has proved usual in secondary hybrids.

The year 1874 saw two genera, the species of which would consent to cross, added to the list of hybrids:—Chysis and Zygopetalum. The hybrid in the first-named genus, Chysis × Chelsoni, was derived from C. bractescens and C. lævis (at first recorded as C. Limminghei), and flowered early in the year.

The hybrid Zygopetalum, named in compliment to the raiser, had as parents Z. Mackayii and Z. maxillare.

Lælia × flammea, a secondary hybrid, from L. cinnabarina and Dominy's hybrid L. × Pilcheri, also flowered in 1874, as well as a fine hybrid Cypripedium, named C. × Arthurianum in compliment to the late Mr. Arthur Veitch, from C. insigne and the at that time rare C. Fairieanum.

ORCHID HYBRIDS

During 1875 three new hybrids were flowered, the result of Seden's work; Cypripedium × tesselatum, obtained from C. concolor and C. barbatum; C. × Euryandrum from C. barbatum and C. Stonei, the first of the group with racemose flowers; and C. × Marshallianum from C. venustum pardinum and C. concolor, a charming little plant.

In 1876 the first secondary hybrid in the genus Cypripedium was also due to Seden, derived from Dominy's C. × Harrisianum and C. insigne Maulei, the former the seed-bearer, establishing a further record, the result having a hybrid as seed-bearer.

A charming Dendrobe, from Dendrobium moniliforme (D. japonicum) and D. aureum, flowered about the same time, and during that same year two new hybrids were added to the list—Cypripedium × pycnopterum, obtained from C. Lowii and C. venustum, and C. × superciliare from C. barbatum and C. superbiens.

In 1877 Cypripedium (Selenipedium) × atropurpureum flowered, the first secondary hybrid amongst the Selenipediums, the parents S. Schlimii and Dominy's cross S. × Dominianum.

Other hybrids of the year were Læliocattleya Sedenii, the only plant raised from a cross between Cattleya superba and Læliocattleya Devoniensis; Cypripedium × patens from C. barbatum and C. Hookeræ, and C. × lucidum from C. Lowii and C. villosum.

Early in 1878 Calanthe × Sedenii made its appearance, a hybrid from C. × Veitchii and C. vestita rubro oculata. The flowering established another record, as the first instance among hybrid orchids of one of the original parents being used with its hybrid offspring. Others of the year were Cypripedium × selligerum from C. barbatum and C. philippinense; C. × nitens from C. villosum and C. insigne Maulei, and Cattleya × Marstersoniæ from C. Loddigesii and C. labiata.

Six appeared in 1879, comprising one Cattleya, two Dendrobiums, two Cypripediums, and one Lælia.

In 1880 the handsome Masdevallia × Chelsoni, the first seedling in the genus to be raised from M. amabilis and M. Veitchiana, flowered for the first time, and added a new genus to the list of hybrids.

Chysis × Sedenii, derived from C. Limminghei and C. bractescens, the second hybrid in the genus, also flowered in 1880, and the handsome Cypripedium × Morganiæ from C. superbiens and the remarkable C. Stonei, and C. × calanthum, were others of that year.

In 1881 four hybrids flowered, including the remarkable Cypripedium (Selenipedium) × grande, the product of a cross between Selenipedium Roezlii and S. caudatum.

The year 1881 completed a quarter of a century's work in this absorbing branch of Horticulture—Dominy's first hybrid having flowered in 1856, and during the following years to the present time results have been numerous, and very varied.

In 1882 the beautiful Læliocattleya callistoglossa flowered, remarkable for an unusually large and richly-coloured lip, from a cross between Lælia purpurata and Cattleya Warscewiczii.

Cypripedium × microchilum, from C. niveum and C. Druryi; C. × macropterum from C. Lowii and C. superbiens, C. (Selenipedium) × cardinale from Selenipedium × Sedenii and S. Schlimii were other successes.

In 1883 three interesting seedlings made their first appearance; namely,

Calanthe × lentiginosa from C. labrosa and C. × Veitchii; Cypripedium (Selenipedium) × Schröderæ from Selenipedium caudatum and the hybrid S. × Sedenii, one of the finest of this group of Slipper Orchids, and Cattleya × triophthalma, from C. superba and Læliocattleya exoniensis, a rare hybrid, two plants only being raised.

In 1884 five flowered at Chelsea, comprising two Cypripediums, two Lælias and a Masdevallia. This last-named was M. × Gairiana obtained by crossing the yellow-flowered M. Davisii with the scarlet-flowered M. Veitchiana. It received the name Gairiana in compliment to John Gair Esq., of Falkirk.

A hybrid of great interest in a new genus flowered for the first time in 1885, Thunia × Veitchiana, a charming seedling from T. Marshalliana and T. Bensoniæ. Six others followed: Dendrobium × euosmum from D. × endocharis and D. nobile; Cypripedium (Selenipedium) × leucorrhodum from Selenipedium Roezlii and S. Schlimii; Zygopetalum × pentachromum from Z. Mackayi and Z. maxillare; Læliocattleya Canhamiana from Lælia purpurata and Cattleya Mossiæ; Cypripedium × radiosum from C. Lawrenceanum and C. Spicerianum, and Cattleya × porphyrophlebia from C. × intermedia and C. superba.

The most remarkable plant of 1886 was Lælia × Batemanniana, so called when first exhibited, but now known under the more appropriate title of Sophrocattleya Batemanniana, a bigeneric hybrid raised by Seden from Sophronitis grandiflora and Cattleya intermedia, described by Professor Reichenbach as "a lovely gem; a miniature Lælia," and dedicated to James Bateman Esq., of Worthing, an enthusiastic orchidist.

The five following hybrids also appeared in 1886, viz.:—Cypripedium × Winnianum derived from C. villosum and C. Druryi; C. × orphanum, of unrecorded parentage; Zygopetalum × leopardinum, also unrecorded; and Masdevallia × glaphyrantha from M. infracta and M. Barlæana.

The year 1887 was remarkable for several striking crosses which made their appearance at Chelsea—first Zygocolax Veitchii, a bigeneric seedling from Zygopetalum crinitum with the pollen of Colax jugosus, the plants but a little over five years old when they flowered.

* "The appearance of this plant marked a new era in the history of Orchid hybridization, as it led to an extension of the principle of compounding a new name derived from the joint names of the parent genera for all hybrids between species of different genera raised in gardens," as proposed by Dr. Masters in 1872 when naming Philageria Veitchii, a bigeneric hybrid between Philesia buxifolia and Lapageria rosea.

Phaiocalanthe Sedenianus, another bigeneric cross, from Phaius grandifolius and Calanthe × Veitchii, flowered the same year and named after the raiser. Other hybrids of note, Phalænopsis × Rothschildiana, the second hybrid in the genus, was obtained from P. Schilleriana and P. grandiflora (amabilis), and dedicated to Lord Rothschild; P. × Hariettiæ from P. amabilis and P. violacea, named after Hariett, daughter of the late Hon. Erastus Corning of Albany, U.S.A.; Dendrobium × Cybele from D. Findlayanum and D. nobile; Læliocattleya × Zenobia from Cattleya Loddigesii and Lælia elegans Turnerii; and Cypripedium × obscurum, of unrecorded parentage.

* Rolfe in Jour. Linn. Soc. vol. xxiv. pp. 156-170.

ORCHID HYBRIDS

In 1888 the genus Epidendrum was added to the list of those genera, the species of which would intercross.

The first hybrid, Epidendrum × O'Brienianum, was from E. evectum with the pollen of E. radicans, but the most beautiful of the year was undoubtedly Phalænopsis × John Seden, a very rare plant obtained from P. grandiflora (amabilis) and P. Lueddemanniana.

Phalænopsis × Leda and P. × F. L. Ames, the latter dedicated to the late Hon. F. L. Ames, of North Easton, Mass., U.S.A., were flowered in 1888.

In 1889 the first of a series in the parentage of which the fringed Lælia Digbyana participated, flowered, as also did Læliocattleya Digbyano-Mossiæ, the parentage indicated by the name. The original plant is in the collection of Baron Sir Henry Schröder, The Dell, Egham.

In addition to these seedlings Seden undertook experiments to verify the supposed parentage of natural hybrids, i.e. supposed hybrids occurring in a wild state.

Masdevallia splendida and M. Parlatoreana are two such, and in his description Professor Reichenbach spoke of them as being probably "mules" between M. Veitchiana and M. amabilis, and M. Veitchiana and M. Barlæana respectively.

By making crosses and reverse crosses with the two last-named species artificial hybrids were obtained identical in appearance with the type, and these flowered during 1889.

There flowered in 1890 a noteworthy hybrid, a totally new and unexpected departure, the bigeneric Epiphronitis Veitchii, from a cross of the dwarf Sophronitis grandiflora with the tall-stemmed Epidendrum radicans, two plants quite unlike in habit. The seedling has intermediate characters, and the flowers a combination of the brilliant colours of both parents.

Sophrocattleya Calypso, three Læliocattleyas, and five Cypripediums were the other new hybrids of the year.

During 1891 the first between two species of the genus Disa flowered, Disa × Veitchii, raised from D. racemosa crossed with the pollen of D. grandiflora, and but one year and a half old, the youngest of any hybrid orchid to flower. Another interesting seedling of 1891 was Odontoglossum × excellens, hitherto supposed a natural hybrid between O. Pescatorei and O. triumphans, and Seden's plant, raised artificially from the two species, proved to be identical with the wild form.

The most interesting result of 1892 was probably Sophrolæliocattleya Veitchiana, a complicated hybrid derived from Sophronitis grandiflora and Læliocattleya Schilleriana, involving three distinct genera in its parentage. Epidendrum × Endresio-Wallisii, Zygopetalum × leucochilum, Phalænopsis × Artemis, three Cattleyas, Lælia × Latona, four Dendrobiums, two Selenipediums and eight Cypripediums were among the other hybrids of the year.

Twenty-two new hybrids flowered during 1893, but no new genus shared in their production.

In 1894 probably the most interesting of many was Sobralia × Veitchii, the first in the genus, obtained by Seden from S. macrantha and S. xantholeuca.

Others of the year were Sophrolælia Læta, a bigeneric hybrid from Sophronitis grandiflora and Lælia pumila Dayana; Cypripedium × James H. Veitch, from C. Curtisii and C. Stonei platytænium, the most remarkable of the whole group, in which the peculiar and handsome broad petals of the variety of C. Stonei are transmitted—the only instance; and in addition there were sixteen other hybrids of various genera.

The novelties of 1895 included two in the difficult genus Phalænopsis—P. × Ariadne, obtained from P. Aphrodite and P. Stuartiana and P. × Luedde-violacea, the parentage is expressed by the name.

Amongst several Læliocattleyas probably the best was Lady Rothschild, from Cattleya Warscewiczii and Lælia Perrinii.

The most interesting of the year was Dendrobium × illustre from D. chrysotoxum and the remarkable D. Dalhousieanum, a very rare hybrid, of which the only specimen is in the collection at The Dell, Egham.

Another success in the genus Chysis added in 1896, Chysis × Langleyensis, was derived from C. bractescens and the hybrid C. × Sedenii, a secondary hybrid and the first in the genus.

The most striking novelty of the year was the beautiful Læliocattleya Decia alba, an exquisite form with clear white sepals and petals, with a delicate rose-pink lip. To the Cypripediums was added C. × Baron Schröder, in the parentage of which the at that time rare C. Fairieanum combines with C. × œnanthum superbum.

During the year 1897 numerous fine winter-flowering Cattleyas and Læliocattleyas which it had been the especial endeavour to produce were flowered, many crosses between new species, and others from choice forms or varieties of species previously employed.

A hybrid in a new genus also followed in 1897, Spathoglottis × aureo-Veillardii, derived from the two species indicated by the name. The flowers of a rich yellow are effectively marked with crimson.

As finer forms of the favourite species became available they were at once made use of, and many of the old crosses reproduced with more brilliant results:—Læliocattleya Wellsiana Langleyensis, L.-c. Eudora splendens, L.-c. Cahamiana superba, L.-c. callistoglossa ignescens, L.-c. Dominiana Langleyensis are improvements on the original seedlings.

These and many others of the same class flowered in 1898, as did also the following noteworthy hybrids:—Disa × Diores, L.-c. Clio, the curious Epicattleya radiato-Bowringiana, Phalænopsis × Stuartiano-Mannii and Masdevallia × Imogene.

Several more appeared during 1899, one a great advance in a new genus—the remarkable Angræcum × Veitchii, derived from A. sesquipedale and A. superbum. The flowers are waxy white, and show intermediate characters between those of the two parents. Epicattleya Mrs. James O'Brien, a bigeneric hybrid from Cattleya Bowringiana and Epidendrum × O'Brienianum, is also interesting. Phaiocalanthe Niobe, Phalænopsis × Cassandra, and Phalænopsis × Mrs. James H. Veitch are among the other notable productions of the year.

Hybrids amongst species of the fore-named genera continued to flower, with greater variety in colour and improved form each year.

In 1902 a new bigeneric cross appeared, from Leptotes bicolor and Lælia cinnabarina, received the name of Leptolælia Veitchii, and although the flowers of a pleasing colour are small, they are very interesting. The last bigeneric hybrid was Dialælia Veitchii, a cross between Diacrium bicornutum and Lælia cinnabarina, followed early in 1905.

DENDROBIUM × EUOSMUM EXCELLENS

THE DELL, EGHAM

ORCHID HYBRIDS

From the numbers mentioned, and by reference to a more complete list given in another portion of this work, it will be readily seen that Seden extended the practice of hybridizing among many species and genera, and raised several between genera apparently quite distinct.

True hybrids, or crosses between two species belonging to one genus, were raised in the following:—Angræcum, Anguloa, Cattleya, Calanthe, Chysis, Cymbidium, Cypripedium, Dendrobium, Disa, Epidendrum, Lælia, Masdevallia, Miltonia, Odontoglossum, Phaius, Phalænopsis, Sobralia, Spathoglottis, Thunia, and Zygopetalum.

The bigeneric hybrids or crosses between species belonging to different genera are:—

Epilælia,	from species of	Epidendrum and	Lælia
Epicattleya	"	"	Cattleya
Epiphronitis	"	"	Sophronitis
Dialælia	"	Diacrium	Lælia
Læliocattleya	"	Lælia	Cattleya
Leptolælia	"	Leptotes	Lælia
Phaiocalanthe	"	Phaius	Calanthe
Sophrocattleya	"	Sophronitis	Cattleya
Sophrolælia	"	"	Lælia
Sophrolæliocattleya	"	"	Læliocattleya
Zygocolax	"	Zygopetalum	Colax

NEPENTHES

NEPENTHES *

SPECIES.

THE first introduced to British Gardens was Nepenthes distillatoria, the type species on which Linnæus founded the genus. A native of Ceylon, this Pitcher had formed the subject of some remarkable writings by early travellers in the East before its introduction to this country, and which, according to Aiton, was in 1789. N. distillatoria was not the first species known to science, that distinction probably belonging to N. madagascariensis, which nevertheless was one of the latest to reach England. N. distillatoria was followed in 1826 by N. gracilis (under the name phyllamphora) and in 1828 by N. Khasiana (as distillatoria) three species before the year 1830. These soon lost to cultivation, their cultural requirements being little understood.

Later followed Nepenthes Rafflesiana and N. ampullaria, discovered by Dr. William Jack in Singapore, the former named in honour of Sir Stamford Raffles, but an attempt to introduce it by the discoverer was unsuccessful. Better fortune attended the labours of Captain Bethune, who brought living plants to Kew in a Wardian case, which he so well cared for that they practically all lived.

In 1847, Nepenthes Hookeriana was introduced to the Clapton Nursery by Mr. (later Sir Hugh) Low, and shortly afterwards Thomas Lobb sent to Exeter N. Rafflesiana, N. ampullaria, N. albo-marginata, N. Veitchii, N. phyllamphora, N. sanguinea, and other unnamed species. These formed the nucleus of the large collection since cultivated at Chelsea.

Nepenthes lævis, sent from Singapore, is figured and described by Lindley in the Gardeners' Chronicle, 1848, p. 655, from specimens growing in the houses at Chelsea at that time.

Sir Hugh Low made known four new species in 1851, having discovered them in making the ascent of Kina Balu, a well-known mountain in Borneo. These respectively Nepenthes Rajah, N. Edwardsiana, N. Lowii, and N. villosa, were not introduced to cultivation at that time, but sufficient material was sent home to enable Sir Joseph Hooker to write a lucid account of the genus, in a paper read before the Linnæan Society in 1859, subsequently printed in the Transactions.

Ten species and four hybrids are enumerated and described by Dr. Masters in the Gardeners' Chronicle for 1872, as growing at Chelsea in that year. The species mentioned were Nepenthes ampullaria, N. Veitchii, N. Rafflesiana, N. phyllamphora, N. Khasiana, N. albo-marginata, N. gracilis, N. lævis and N. sanguinea.

* Harry J. Veitch, F.L.S., F.R.G.S., &c., in Jour. R.H.S. vol. xxi. pt. ii.

During the following ten years some of the finest were introduced by the Veitchian collectors, including, *Nepenthes Rajah and N. bicalcarata from Borneo through Burbidge; N. hirsuta from the same region through another agency; N. madagascariensis from Madagascar through Curtis; N. Kennedyana from North Australia, and N. Viellardii, a native of New Caledonia, through the Botanic Gardens at Sydney; and N. Northiana from Borneo, also through Curtis. With a view of obtaining some of those remarkable plants made known through the discoveries of Mr. Low in Borneo, Thomas Lobb, acting under the direction of Mr. James Veitch junior, reached the foot of Kina Balu in 1856, but was prevented from ascending the mountain by the hostility and extortion of the natives. He was followed in 1877 by Burbidge and P. C. M. Veitch, who met with a like failure, and again by the first-named, eight months later, on which occasion some seed of N. Rajah was obtained, despatched to Chelsea, and plants raised, but few lived. N. Rajah has the largest pitchers of any known species, and these are described in Mr. Spencer St. John's book "Life in the Forests of the Far East":—"This morning, while the men were cooking their rice, as we sat before the tent enjoying our chocolate, observing one of our followers carrying water in a splendid specimen of Nepenthes Rajah, we desired him to bring it to us, and found that it held exactly four pint bottles. It was 19 in. in circumference. We afterwards saw others which were much larger, and Mr. Low, while wandering about in search of flowers, came upon one in which was a drowned rat." † N. albo-marginata one of the earliest introduced species, previously unknown to science, was one of Thomas Lobb's Bornean discoveries, sent home to Exeter in 1848. The species is difficult to cultivate, but repays trouble by the great beauty of the pitcher, light green at the base, rosy red at the apex, with a pale band edging the top below the peristome.

Nepenthes zeylanica, or, as it is sometimes called, N. hirsuta glabrescens, and N. zeylanica rubra, a red form of the type, were the next introductions, followed shortly afterwards by the handsome, and still rare, ‡ N. Veitchii. Much confusion has arisen regarding the nomenclature of this species, sometimes called N. lanata and also N. villosa, both in themselves good species.

Nepenthes Veitchii was first met with by Hugh Low Esq. junior, on Mount Kina Balu, Borneo, but not introduced. Later found by the collector Thomas Lobb in Sarawak, living plants sent to Exeter proved one of the most remarkable of all Nepenthes: large pitchers covered with hair, a remarkable peristome or frill round the mouth, resembling both in structure and appearance the gills of a fish; the frill cream-coloured, slightly reddish; the habit as in other of the family.

In 1879 Nepenthes Veitchii was followed by § N. Kennedyana, a species from Cape York, North Australia, sent from the Botanical Gardens, Sydney; the pitchers are over 5 in. in length by 1½ in. in width, reddish in colour, elongate-cylindrical in shape, slightly dilated below the middle and tapering at the base with two sharply fringed wings.

* *N. Rajah*, Masters in Gard. Chron. 1881, vol. xvi. p. 493, fig.

† *N. albo-marginata*, Lobb ex Lindl. in Gard. Chron. 1849, p. 580, fig. The Garden, 1880, col. pl. p. 542.

‡ *N. Veitchii*, The Garden, 1880, vol. xvii. p. 542, col. pl.; Bot. Mag. t. 5080 as villosa. Gard. Chron. 1881, p. 781, fig. 152.

§ *N. Kennedyana*, Gard. Chron. 1882, vol. xvii. p. 257, fig. 36.

NEPENTHES

In 1880 Messrs. Veitch distributed and figured in the Catalogue the Bornean species *Nepenthes bicalcarata, found in Borneo, by Low, Beccari, and other early travellers, but not sent to England till Burbidge found it in the neighbourhood of Lazas River. N. bicalcarata is peculiar in having two prominent spurs projecting from below the base of the operculum over the mouth of the pitcher, as the head of a snake with projecting fangs and head uplifted about to strike. It is one of the most robust and vigorous of all Pitcher-plants. The N. Dyak of Mr. S. Le Moore, figured in the Journal of Botany, is an immature form of this species.

In 1881 Dr. Masters described in the Gardeners' Chronicle a new species raised from seed sent from Sarawak by Curtis as †Nepenthes angustifolia. Of no value as a decorative plant, it was not distributed. A more important species, one of the first to be known, was also introduced through Curtis, ‡N. madagascariensis, not introduced till 1880. The name denotes the island in which it was discovered, the extreme western limit of the Nepenthes range. It is at home in fully exposed swamps, and has characters clearly distinguishing it from all other species. The pitchers from 6 to 8 in. long, remarkable for the richness of their coloration, rival in this the N. sanguinea of Borneo.

The next important find was Nepenthes Rajah, a magnificent species already alluded to, named in honour of Rajah Brook, whose services to its country it commemorates.

In the following year, 1883, §Nepenthes Northiana, a species as wonderful as the last named, was offered to European growers, the specific name commemorating Miss North, the lady through whom it was first made known.

Plant drawings executed by Miss North in Borneo were shown to Mr. Harry Veitch, and one of a curious Pitcher-plant, at that time unknown to science, greatly attracted his attention. Further information was obtained regarding the habitat, and Curtis, about to start on a collecting expedition to Borneo, was commissioned to go in search. After long and unsuccessful effort, Curtis gave up hope, under the impression that Miss North had been wrongly informed, but fortunately before leaving the district it occurred to him to look over a steep escarpment in the hill-side, accomplished by lying prostrate on the ground, when to his great joy he discovered the long-looked-for plant some distance below.

He succeeded in gathering ripe capsules, and lost no time in transmitting them to Chelsea, where the seed soon germinated. The pitchers of Nepenthes Northiana are flask-shaped, striped and spotted with purple on a greenish ground, when mature they are 1 ft. and more long, and $3\frac{1}{2}$ in. in width, with two dentate fimbriate wings. The mouth oblique, surrounded by a broad finely ribbed margin or peristome. The shape variable; the upper pitchers swing in mid-air unsupported, trumpet-shaped, whilst those on the ground are larger and more distended.

* *N. bicalcarata*, Masters in Gard. Chron. 1880, vol. i. p. 200, fig. 36; The Garden, 1880, vol. xvii. p. 542.

† *N. angustifolia*, Masters in Gard. Chron. 1881, vol. xvi. p. 524.

‡ *N. madagascariensis*, Poiret, Gard. Chron. 1881, p. 685, fig. 139; Veitchs' Catlg. of Pl. 1882, fig. p. 12.

§ *N. Northiana*, Hook. f. in Gard. Chron. 1881, vol. ii. p. 717, fig. 144 and suppl.

Another handsome species, Nepenthes * Curtisii, was sent from Borneo, and offered in 1888. The pitchers dull green, are thickly spotted with purple. A fine variety of this species, N. C. superba, possesses larger and more highly coloured urns than the type.

During a second mission to Malaysia, 1882-1883, Curtis sent seed of a species since named † Nepenthes stenophylla, with green pitchers mottled with red, and first held to be a form of N. Curtisii, but since raised to specific rank by Dr. Masters.

In the following year Burke collected plants and seeds of two species in the Philippines, one, ‡ Nepenthes Burkei, commemorating his labours. Handsome pitchers narrowed in the middle, devoid of the winged appendages common to the majority. A variety N. B. excellens is richly coloured and unusually handsome. During this trip Burke also collected a further supply of seed of N. Northiana, and from this was obtained a distinct plant distributed under the name of § N. cincta. It is a supposed natural hybrid between N. Northiana and N. albo-marginata (as these two plants grow in company), and from the resemblance N. cincta bears to the two species, the inference is reasonable. It resembles N. Northiana in leaf and habit of growth, and the pitchers have the white band round the mouth characteristic of N. albo-marginata.

Nepenthes Pervillei, obtained from seeds sent by a correspondent in the Seychelles, has since been sent to Kew through Mr. Griffith, the Administrator of the Islands.

Another handsome species, obtained through Mr. Ford, late of the Hong Kong Botanic Gardens, and also from the Royal Gardens, Kew, 1891, is || Nepenthes ventricosa, a native of the Philippines, one of the most distinct of this remarkable genus. The pitchers, wholly without wings, are curiously contracted in the middle, with a transverse, not oblique mouth, surrounded by a bright red undulating peristome marked with numerous transverse ribs. The colour delicate pale green with a rosy suffusion at the base, without spots, the pitchers deepening in colour to a deep purplish-rose as they reach maturity.

Many of the species are from various causes now lost to cultivation, or supplanted by hybrids, in most cases easier to cultivate and more decorative.

HYBRIDS.

The artificial hybridization of Nepenthes was commenced by John Dominy at Exeter, and continued by Seden, Court, and Tivey at Chelsea.

The diœcious character of the Nepenthes renders the pollination of a female flower easy, as emasculation is unnecessary and self-fertilization an impossibility; but there are drawbacks to even these apparently advantageous conditions, the greatest the difficulty often experienced of procuring pollen when a female plant is in flower, and *vice versa*, as the sexes seldom flower at the same time.

* *N. Curtisii*, Masters in Gard. Chron. 1887, vol. ii. p. 681, fig. 133; *id.* 1889, vol. vi. p. 661, fig.; Bot. Mag. t. 7138; Veitchs' Catlg. of Pl. 1888.

† *N. stenophylla*, Masters in Gard. Chron. 1890, vol. viii. p. 240; *id.* 1892, vol. xi. p. 401, fig. 58.

‡ *N. Burkei*, Masters in Gard. Chron. 1889, vol. vi. p. 493, fig.

§ *N. cincta*, Gard. Chron. 1884, vol. xxi. p. 576, fig. 110.

|| *N. ventricosa*, Gard. Chron. 1898, vol. xxiii. p. 380, fig. 143.

The methods pursued in cultivating Nepenthes as decorative subjects are likewise unfavourable to the production of flowers, as pitchers being the desiderata, plants are subjected to severe pruning, with the object of their production.

The species used by Dominy in the first cross was an unnamed one with green pitchers from Borneo, and Nepenthes Rafflesiana, the result a plant producing pitchers fairly intermediate in character. Named N. × Dominiana after its raiser, and distributed in 1862, exhibited at the Royal Horticultural Society's Show, held in June of the same year at Kensington. The second of Dominy's hybrids, N. × hybrida, had as parents N. Khasiana (then known as N. distillatoria) and an unnamed Bornean species. The pitchers larger than those of N. distillatoria, were bright green. A variety N. × hybrida maculata has green pitchers thickly covered with red spots; both were distributed in 1866.

Seden followed Dominy in this interesting work, and obtained his first hybrid, which bears his name, from Nepenthes Khasiana (distillatoria) and an unnamed Bornean species, the same as Dominy employed in producing N. × hybrida. The pitchers of N. × Sedenii are vivid green, splashed with bright crimson spots.

This was followed by Nepenthes × Chelsoni,* also raised by Seden, from N. Hookeriana crossed with the pollen from N. × Dominii, a hybrid being used for the first time as a parent.

The work of hybridization has been carried on by succeeding growers whenever staminate and pistillate flowers have been available simultaneously, either of species or of hybrids.

Court, who succeeded Seden, produced several fine hybrids, the first Nepenthes × intermedia, the result of crossing an unnamed species with N. Rafflesiana, followed in 1877 by N. × Courtii, from the same parentage as N. × hybrida.

Nepenthes × Stewartii appeared in 1879, from N. phyllamphora crossed with N. Hookeriana, and N. × Ratcliffiana† in 1881 from a similar parentage. The latter dedicated to Alfred E. Ratcliff Esq., of Edgbaston, Birmingham, at that time a distinguished amateur of this interesting race.

A hybrid Nepenthes × rubro-maculata, the result of crossing N. × hybrida with an unknown species, was distributed in 1881. The varietal name was given in allusion to the marked claret-red spotting marking the pale green ground colour of the pitchers. The same year a hybrid, N. × Wrigleyana, named in compliment to Oswald Wrigley Esq., of Bridge Hall, Lancashire, was obtained, the product of N. Hookeriana and N. phyllamphora, and this was followed by a hybrid of American origin, N. × Morganiæ, in 1882, the result of either N. × Sedenii or N. phyllamphora with N. Hookeriana.

The year 1883 is noteworthy for one of the most ornamental and easily grown of all hybrid Nepenthes, N. × Mastersiana, "in compliment to Dr. Masters, of the Gardeners' Chronicle, as a slight recognition of his invaluable services to Horticulture." Raised by Court from N. sanguinea crossed with N. Khasiana (N. distillatoria, Glasnevin variety, of gardens), the seed remained dormant so long that it barely escaped destruction. Fortunately life was detected at the last moment, and a further trial resulted in a plant which

* Veitchs' Catlg. of Pl. 1874, fig.

† Veitchs' Catlg. of Pl. 1881, fig.

ranks amongst the finest of the genus. N. × Mastersiana produces pitchers remarkable for a fine coloration, rivalling that of N. sanguinea, with the characteristic blotches of N. distillatoria. The plant is of a robust constitution, dwarf in habit.

A hybrid with peculiar-shaped pitchers was next obtained by crossing Nepenthes hirsuta glabrescens with N. Veitchii. This, called N. × cylindrica, from the shape of its pitchers, 6 to 8 in. long, cylindric, has a slight dilation below the middle, is pale green in colour, with a few crimson spots and markings.

Nepenthes × Dicksoniana, offered in 1889, is the offspring of N. Rafflesiana flowering in the Botanic Gardens at Edinburgh, fertilized by pollen of N. Veitchii sent from Chelsea. Mr. Lindsay, late Curator of the Edinburgh Botanic Gardens, effected the cross, and in deference to his wish the seedling bears the name of Professor Dickson, formerly Professor of Botany at the University. The pitchers of the hybrid are fully 10 in. long, sub-cylindric, of a light fulvous green, densely spotted and speckled with bright crimson.

A very mixed variety named Nepenthes × *rufescens, sent out with N. × Dicksoniana, is remarkable in having the blood of three species and two hybrids in its composition. Raised at Chelsea from N. × Courtii crossed with N. zeylanica rubra, N. × Courtii, itself a hybrid from an unnamed Bornean species and N. × Dominiana, the latter also a hybrid between N. Rafflesiana and the same unnamed Bornean species.

George Tivey, to whose charge the Nepenthes were eventually entrusted, has produced some excellent crosses, the parentage indisputable, a statement which cannot be made without reserve of some of the earlier results, of which records are imperfect, and when the variability both in colour and shape of the pitchers, a marked characteristic of seedling Nepenthes, was not so well understood or appreciated.

Tivey's first hybrid, †Nepenthes × mixta, was from two beautiful species, N. Northiana and N. Curtisii, the latter the pollen parent. As might be expected N. × mixta is a fine cross, with pitchers 1 ft. or more in length, of a cream-yellow colour suffused with red and blotched as is N. Northiana. The wings shallow, are deeply laciniated; the ribs, which form the mouth of the pitcher, of a rich shining crimson. It was distributed in 1893. A handsome variety, N. × mixta sanguinea, has reddish-brown pitchers spotted with large blotches of chocolate-brown.

The next success was a superb cross from Nepenthes Veitchii and N. Curtisii, ‡N. × Tiveyi, named in compliment to the raiser. N. Veitchii is one of the grandest Pitcher-plants in cultivation, remarkable for hairy pitchers and a curious gill-like peristome, and many of the best characteristics have been imparted to the hybrid, the most conspicuous the broad rim round the mouth, richly coloured a deep mahogany-red, with occasional transverse bars of a deeper shade. The pitchers, larger than those of N. Veitchii, have much the same form, but are on finer lines.

Nepenthes × Balfouriana, an especially interesting hybrid, has for parents two hybrids, and four distinct species concerned in the production. It is the outcome of a cross between N. × Mastersiana and N. × mixta; N. × Mastersiana

* Masters in Gard. Chron. 1888, vol. iv. p. 699, fig. 95.
† Gard. Chron. 1893, vol. xiii. fig. 9.
‡ Gard. Chron. 1897, vol. xxii. pp. 200, 201, fig.; Gard. Mag. 1897, Sept. 18th.

NEPENTHES × MIXTA

resulted from N. sanguinea and N. Khasiana, and N. × mixta from N. Curtisii with N. Northiana. The pitchers are of a sub-cylindric shape, from 7 to 9 in. long, stained with carmine, marked with crimson spots and streaks on a yellowish-green ground.

Another beautiful Pitcher offered in 1903 is probably one of the finest hybrid Nepenthes in cultivation, a result effected by Tivey between N. × mixta and N. × Dicksoniana, both hybrids. It is named *Sir William T. Thiselton-Dyer in honour of the late Director of the Royal Gardens, Kew. The pitchers attain a length of 14 in. or more, are sub-cylindric in shape, with a handsome peristome or ribbed mouth, the colour bright crimson, and the form undulate as in N. × mixta. The ground colour of the body is green, the surface irregularly blotched with large spots of purplish or crimson brown. In addition to the ordinary slender spur at the back of the lid, the hump-like process characteristic of N. Curtisii is prominent.

Another offspring of the same parents, but from a separate cross, was exhibited in 1903 as †Nepenthes × picturata, so called from a highly-coloured, spotted pitcher and lid. It attains large dimensions, and is 10 to 12 in. in length, and the rim round the mouth is exceptionally fine and of a rich brownish-red. A further cross to the two preceding varieties has produced quite a distinct form, to which the name F. W. Moore has been applied, in compliment to the Curator of the Botanic Gardens, Glasnevin. The pitchers, somewhat globular in form and green in colour, have a moderately broad rim of deep reddish-brown.

The majority or all of the hybrids mentioned, raised under artificial conditions in this country, have proved more amenable to cultivation than many species from the equatorial regions, and, from a horticultural point of view, are very much superior.

* Gard. Chron. 1900, vol. xxviii. p. 256, fig. 76.
† Flora and Sylva, 1904, vol. ii. p. 68, fig.

GREENHOUSE RHODODENDRONS

GREENHOUSE RHODODENDRONS

EAST INDIAN SPECIES.

THE beautiful plants comprised in this section of the genus Rhododendron known in gardens by the above name are natives of the East Indies, inhabiting the islands of Java, Sumatra, and Borneo, Penang and Malacca in the Malay Peninsula.

The first species introduced to cultivation in this country was Rhododendron *jasminiflorum, sent to Exeter from Malacca by Thomas Lobb, flowered for the first time in September 1849, and exhibited at Chiswick on the occasion of the first show of the year, held by the Royal Horticultural Society in 1850. The flowers pure white with a deep pink eye, somewhat resemble the blooms of a Stephanotis or a Jessamine. The corolla salver-shaped, pure white, slightly tinted with pink below the limb, 2 in. in length, has five equal wavy lobes: the anthers are red, forming a deep rosy pink eye in the centre of the flower.

The next species named Rhododendron †javanicum, after the island in which it was discovered, Blume met with on Mount Salak, and Dr. Horsfield on the volcanic range extending through Java, in dense forests, at an elevation of 4000 ft. Received from Java through Thomas Lobb in 1845, with a much darker form previously introduced by Rollison; the flowers orange-coloured, ten or twelve in a head, here and there marked with red spots, have ten dark purple-coloured anthers.

Rhododendron javanicum was followed by a beautiful species R. ‡Brookeanum, also sent by Thomas Lobb, who met with it in Borneo, where it had previously been observed by Mr. Low. The name was given in honour of Sir James Brooke, the distinguished Rajah of Sarawak, in whose territory it was discovered. The first flowers produced by this species in 1855, when sent to the Royal Horticultural Society's Show, attracted great attention. In its native country epiphytal on trees overhanging water-courses, the plant, when lit by the sun's rays, is a gorgeous sight: closely allied to R. javanicum, it differs in having a crisped margin to the petals, more prominent stamens, and broader petioles to the leaves. There are two varieties, one having yellow flowers of slender growth received the varietal name *gracile*, whilst the other, as the typical form, has a more robust habit of growth and flowers of a full orange or buff-yellow colour.

* R. jasminiflorum, Bot. Mag. t. 4524.

† R. javanicum, Bot. Mag. t. 4336.

‡ Low in Jour. R.H.S. vol. iii. p. 82 with fig.; Gard. Chron. 1855, p. 454 with fig. Bot. Mag. t. 4935.

In 1855 Thomas Lobb sent a species with small cerise-crimson flowers from Sumatra, subsequently found to be the species Rhododendron *malayanum of Jack, figured in the Botanical Magazine with the following remarks:—"Dr. William Jack, of the late East India Company's Service, a very able botanist and author of 'Malayan Miscellanies,' was the first to make known this fine plant (in about 1823) which he discovered on the summit of Gunong Bunko, a remarkably insulated mountain, commonly called by the Europeans the Sugar-loaf, in the interior of Bencoolen, Sumatra. Dr. Jack observed of this mountain that, though estimated at only 3000 ft. in height, the character of its vegetation is decidedly alpine, which he attributed to the form and consequent exposure of the sharp conical peak. R. malayanum has since been gathered repeatedly on Mount Ophir, Malacca, at an altitude of 4000 ft., and is clearly the same plant as the Javanese R. tubiflorum.

"The Celebes Island R. celebicum differs only in the paler under surface of its leaves, and was originally introduced by Lobb in 1854, probably from Borneo where he was travelling: as that indefatigable collector had already visited Mount Ophir, whence he had sent excellent dried specimens, now in the Hookerian Herbarium, it is probable that the Bornean habitat is a mistake."

The plant is said to form a shrub or small tree with elliptic or elliptic-lanceolate leaves 3 to 4 in. long; the flowers, in terminal few-flowered umbels, are ¾ in. long by about ½ in. across at the mouth, dull-scarlet or cerise-crimson in colour.

Curtis followed Thomas Lobb as collector in the East, and sent home many plants, amongst them being two new species of Rhododendron, named respectively R. †Teysmanni and R. ‡multicolor.

Rhododendron Teysmanni, a lax shrub, is a native of Penang and Sumatra, with golden-yellow flowers 2½ in. in diameter. Exhibited for the first time in flower in March 1885, a certificate was awarded.

Though not in itself a first-class garden plant, the flowers reflexed and the trusses small, R. Teysmanni has for cross-breeding proved one of the best, and entered largely into the present race of hybrids.

Rhododendron multicolor is distinct from the Malayan species in the form of its funnel-shaped corolla, and approaching R. citrinum, differs from that species in the absence of calyx-lobes, and in the possession of twice as many stamens. The most marked characteristic of the species is the great variability in the colour of the flowers, in the type yellow, and in the variety § *Curtisii*, rich crimson.

This last-mentioned received a First Class Certificate at the meeting of the Royal Horticultural Society on November 13th, 1883, when it was greatly admired.

HYBRIDS.

From the fore-mentioned seven species of Rhododendron, all of which are natives of the various islands of the Malay archipelago, there have been produced, by hybridizing and cross-breeding in a variety of ways, several hundred new

* Bot. Mag. t. 6045.
† Bot. Mag. t. 6850, as R. javanicum, var. tubiflora.
‡ Bot. Mag. t. 6769.
§ Fl. and Pom. 1884, p. 113.

RHODODENDRON JAVANICO JASMINIFLORUM
"NE PLUS ULTRA"

forms, many, from a horticultural standpoint, exceed the original species in brilliant and varied colours, large size of truss and individual blooms, compact habit of growth, and the ease with which they can be cultivated.

A classic paper by Professor Henslow, on the Hybrid Rhododendrons raised in the Veitchian houses, read before the Royal Horticultural Society and afterwards printed in the Journal, vol. xiii. pt. ii. (1891), contains a full account of the various hybrids, with their genealogies, and the various phenomena exhibited by them as the result of cross-breeding and hybridizing. To this paper we are largely indebted for much of the following information.

With regard to the colours of the flowers of the original species, Professor Henslow remarks:—"They are all reducible to two, yellow and rose-red. The former is produced by the presence of yellow granules scattered within the cells of the epidermis or underlying tissue, while the reds are due to various degrees of concentration of the coloured fluid, both in individual cells, as well as by superposition of cells containing the rose-coloured fluid. The buffs or orange-colours are due to combinations of the pink fluid with yellow granules, either in the same cell, as occurs in some epidermides, or in adjacent cells, as occurs in orange-coloured anthers examined. If there be a pink throat with a yellow or orange border to the corolla, this is due to the epidermal cells containing a more concentrated solution of the pink fluid."

"The first hybrid raised was named Princess Royal, the product of a cross between R. jasminiflorum (white) and R. javanicum (yellow), and the result is remarkable. The flowers of Princess Royal show no trace of yellow, but are of a delicate pink or rose colour. Another hybrid produced later from the same cross, named jasminiflorum carminatum, resembles Princess Royal in all but colour, which approaches crimson. By combining Princess Royal (pink) with one of its parents (jasminiflorum, white) a white-flowered variety was produced, which received the name of Princess Alexandra."

"The dissociation of colours by crossing, in other plants, may give rise to a striped, flamed, or blotched appearance, as in Calceolaria, and in some varieties of the Snapdragon: but this has never occurred in these Rhododendrons. Sometimes, however, the flower has the interior of the tube or throat of a more strongly tinged hue than the lobes, and *vice versa*."

Our *employé*, the late George Taylor, accomplished much with the species of this genus, and raised several excellent varieties. His first success, named *Duchess of Edinburgh, a scarlet-flowered hybrid from Rhododendron Lobbii crossed with R. Brookeanum, was exhibited at the Royal Horticultural Society's meeting in March 1874, and distributed in 1877, in which year it was figured and described in Messrs. Veitchs' Plant Catalogue. A seedling, named in compliment to its raiser †Taylori, was sent out the same year, a pink-flowered variety of the third generation, the parents Princess Alexandra (white) and Rhododendron Brookeanum, a Bornean species with bright yellow flowers.

In 1879 two new varieties were distributed, named respectively Duchess of ‡Teck and §Prince Leopold. The first resulted from crossing Princess Royal with Rhododendron Brookeanum, and produced flowers of a light buff-yellow,

* Fl. and Pom. 1874, p. 145, col. pl.
† Veitchs' Catlg. of Pl. 1877, p. 19, fig.; Fl. Mag. 1877, pl. 242.
‡ Veitchs' Catlg. of Pl. 1879. § Fl. and Pom. 1876, p. 145.

shaded with orange-scarlet. R. × Prince Leopold, from R. Lobbii, a Bornean species with red flowers, and R. Brookeanum, had flowers of fawn suffused with rose, deepening in intensity towards the centre.

Duchess of Connaught, from the same parents as Prince Leopold and Duchess of Edinburgh, was sent out in 1882, the flowers, in shape those of the Duchess of Edinburgh, but the colour a rich glowing red.

From the commencement of hybridizing experiments with this genus up till the year 1882, the following varieties, as well as those already mentioned, had received certificates of merit from the Royal Horticultural and Royal Botanic Societies:—

Queen Victoria (R. Lobbii × R. Brookeanum), Princess Frederica (Princess Royal × R. Brookeanum), *Maiden's Blush (Princess Alexandra × R. Brookeanum), Excelsior (Princess Royal × R. javanicum), Monarch (Princess Alexandra × Duchess of Edinburgh), Favourite (Princess Alexandra × R. javanicum), Crown Prince of Germany (Princess Royal × R. Brookeanum), and Aurora (Crown Prince of Germany × R. javanicum). These hybrids were all raised by John Heal, who received charge of the Rhododendrons on the retirement of George Taylor. All possess large flowers, varying much in colour and form.

Heal's great success was achieved when varieties appeared with double flowers, which now constitute what is known as the Rhododendron † balsamæflorum hybrids. This section originated by impregnating the stigma of a flower which had one of the filaments slightly petaloid, the others being normal, with the pollen from its own anthers:—self-fertilization. From the seed capsule which resulted, about twenty plants were raised, and when these flowered they were found to produce double or semi-double blooms. The section received the name of balsamæflorum from the resemblance the flowers bore to those of the double Balsams.

The colours of the offspring show great variety, from white and pink to dark red or crimson, from pure yellow to various shades of orange; the foliage is also very variable, and frequently there is a want of symmetry between the two halves of the leaf-blade. The principal seedlings in this section are named album aureum, roseum, carneum, indicating the colour, and the handsome Rajah, with bright fawn-yellow segments tinted with rose towards the margin.

The balsamæflorum hybrids cross with other javanico-jasminiflorum hybrids or with species, but in all cases the progeny have single blooms, and the tendency to petaloidy is apparently overcome by the natural vigour of the true species or crosses.

In 1886 Favourite and ‡ Lord Wolseley were sent out. The former, a hybrid of the third generation, from Princess Alexandra (white) crossed with Rhododendron javanicum (orange-yellow), has flowers of a satiny rose colour, with a white tube and crimson filaments. Lord Wolseley raised from Duchess of Teck (buff-yellow) and R. javanicum (orange-yellow), has flowers of an orange-yellow, tinted with rose at the margin, in large trusses.

In the year 1887 Rhododendron album and R. aureum of the balsamæflorum section were distributed, and also a new single-flowered variety, a cross between

* L'Horticulture Belge, 1888, p. 141, col. pl.

† Fl. and Pom. 1883, p. 31.

‡ Veitchs' Catlg. of Pl. 1886.

Crown Prince of Germany and R. javanicum, called President. The flowers of this last-named resemble those of R. javanicum, but are an improvement, to be preferred for decorative purposes. These varieties were followed in 1888 by R. × balsamæflorum carneum, R. × jasminiflorum carminatum, and Princess Christian (Princess Frederica × R. javanicum), the last-named remarkable for large trusses of bright nankeen-yellow flowers.

Rhododendron × luteo-roseum distributed in 1889, as the name implies, had flowers satiny rose, toned with yellow,—a very charming tint.

Rhododendron × balsamæflorum Rajah, sent out in 1890, is well figured in the Plant Catalogue for that year. The figure typical of the whole of the balsamæflorum section, shows a high decorative value, but no mere picture can convey an idea of the durability of these flowers, perfect for several weeks, and, when cut, surpassing in lasting qualities many commonly grown plants. Souvenir de J. H. Mangles was distributed the same year, and figured in the same issue of the Catalogue of Plants. Derived from Crown Prince of Germany and R. javanicum, it was named in compliment to the late distinguished amateur of Rhododendrons, to whom experiments with his favourite genus were a source of great interest.

The varieties Ophelia (light-rose), Brilliant (brilliant scarlet), and Aphrodite (blush-white) sent out as novelties in 1890, were followed in 1891 by Minerva (nankeen-yellow) Militaire (orange-scarlet) and Princess Beatrice (yellow suffused with pink).

In 1893 appeared the superb variety * Exquisite, which originated from crossing the two species Rhododendron javanicum and R. Teysmanni. The flowers produced by this hybrid, nearly 3 in. in diameter, have a short broad funnel-like tube. The colour toned with a faint fawn-yellow with a slight tinge of rose-pink in the centre of each segment, bright crimson anthers, add greatly to the beauty of the cross.

With this novelty there were also described in the Catalogue upwards of fifty others then in cultivation, including such fine forms as † Ceres, ‡ Primrose, § Minerva, ‡ luteo-roseum, and ‡ R. jasminiflorum carminatum.

All the hybrids mentioned are the products of crosses between the five species Rhododendron jasminiflorum, R. Lobbii, R. javanicum, R. Teysmanni, R. Brookeanum gracile, and their offspring, but, with the introduction of R. multicolor from Penang through Curtis, fresh material came, which proved invaluable; promptly used in connection with the best forms of the javanico-jasminiflorum hybrids, many varieties, the flowers of which, though inferior in size to the last named group, are unparalleled in brilliance and depth of colouring.

Professor Henslow remarks with regard to the form of the small-flowered Rhododendron multicolor, that whenever this species is crossed with any other, or a descendant of them, it universally reduces the size, so that the offspring, though intermediate, are nearest to that of multicolor. The short funnel-shaped tube is more or less traceable in all offspring of the multicolor section.

One of the most beautiful of multicolor hybrids, that named Mrs. Heal is

* The Garden, 1889, vol. lvi. pl. 1232.
† The Garden, 1892, vol. xli. pl. 845. ‡ The Garden, 1892, vol. xli. pl. 852.
§ Cassell's Dictionary of Practical Gardening, vol. ii. p. 270, pl. 16.

one of the fourth generation, and a combination of five distinct species. Its genealogy is as follows:—

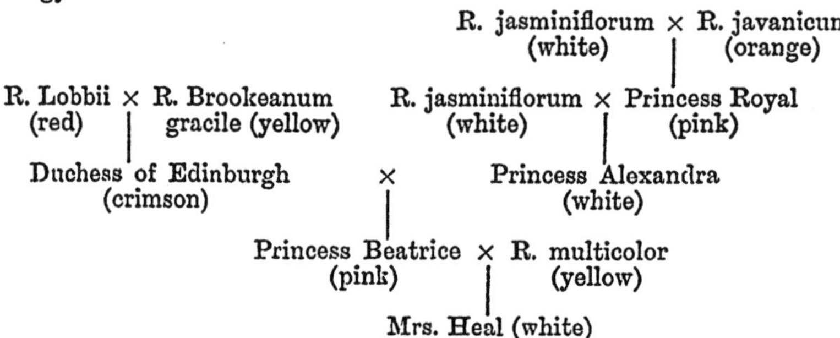

The flowers produced by Mrs. Heal are of a delicate pure white, funnel-shaped, and nearly 2 in. in diameter, and other distinct varieties belonging to the multicolor series are Ruby (R. jasminiflorum carminatum × R. multicolor Curtisii), dark coral red; *Rosy Morn, soft rose-pink, Latona, cream-yellow, Neptune, brilliant scarlet, *Hippolyta, crimson, and * Virgil, yellow.

The Malaysian species, with small cerise-crimson flowers, Rhododendron malayanum, has also been used as a parent, the result a charming variety † Little Beauty, which, apart from its interesting pedigree, has many excellent qualities, not the least the compact habit of growth; a fit subject for a small structure where the larger-growing varieties cannot be used.

Little Beauty is the product of a cross between the javanico-jasminiflorum hybrid known as Monarch and the Sumatran species Rhododendron malayanum, containing a combination of five species, and consequently a hybrid of the fourth generation. Its pedigree is as follows:—

R. jasminiflorum × R. javanicum

R. javanicum × Princess Royal R. Lobbii × R. Brookeanum gracile

Princess Alexandra × Duchess of Edinburgh

Monarch × R. malayanum

Little Beauty

The flowers are of a uniform-growing carmine-scarlet tint, the tube being somewhat less than 1 in. long, the segments of the limb spreading to little more than the length of the tube. It is one of the most brilliant of all Rhododendrons, and has the further advantage of a dwarf and compact habit.

Difficulty was found in crossing the East Indian section with the Sikkim and American species. Success, however, after many attempts was ultimately obtained by crossing R. Aucklandii, a Sikkim species with white flowers, with Princess Royal (R. jasminiflorum × R. javanicum), a pink-flowered hybrid, the former being the pollen parent. The result was a small white-flowered cross called Pearl, as the male parent had eliminated the pink colour from Princess Royal, scarcely affecting the shape or size of the flower, and leaving a hybrid closely resembling Princess Alexandra.

* The Garden, 1892, vol. x. lii. pl. 871. † The Garden, 1889, vol. lvi. pl. 1241.

Another interesting cross resulted from a variety of the Indian Azalea Stella with Lord Wolseley, a javanico-jasminiflorum hybrid, the latter being the female parent. This bigeneric hybrid did not flower until it was six years old.

Professor Henslow gives the following account of this hybrid in the paper above referred to:—"The female parent is of the third generation, the descent being as follows:—

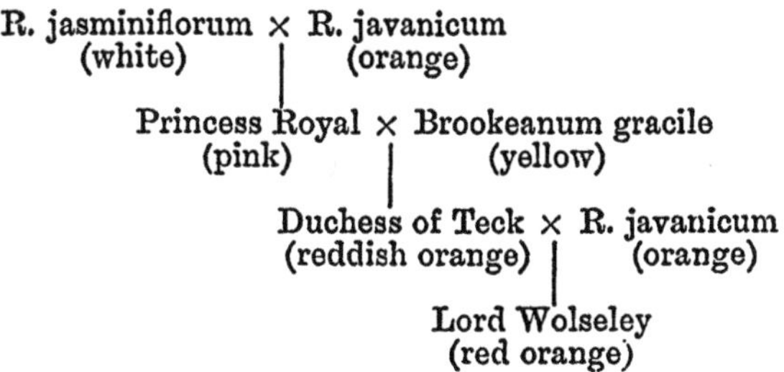

"The male-parent has a dark rose-coloured corolla, with crimson spots over the upper petals. The corolla of the cross is smaller than that of either parent, having a broadish, nearly straight tube, slightly bulging above; the lobes of the limb are much shorter than is the case with either parent. The colour is a rather redder orange than that of the female parent; the anthers are crimson, as well as in both parents. With regard to the foliage of the cross, though smaller in size, the leaf agrees, both in form and anatomical details, with that of the Rhododendron or female parent in every detail of importance. The leaf of Azalea is markedly different, being obovate instead of lanceolate; toothed, and not entire; covered with fibrous hairs instead of being glabrous above, with minute peltate scales below; the cell-walls of the epidermis being sinuate instead of straight; and the proportion of stomata being less than in the Rhododendron as well as the cross.

"The hairs of the Azalea are very peculiar in structure. They grow on the branches, petioles, midrib, and veins below, and are generally scattered over the upper surface of the leaf. They are composed of numerous fibres resembling short libre-fibres, graduated in length so that the longest form the point of the hair, like a fine camel's-hair brush.

"With regard to a dwarf Sister-plant (this plant is now twenty-one years old, but does not exceed five inches in height:—it has never flowered), the anatomical details exhibited a very considerable amount of arrest of structure, the number of cells, in consequence of their minute size, being nearly twice that of the Rhododendron in the same area, with fewer stomata. It also agreed in most other respects both with the sister-cross as well as with the Rhododendron, except that the shape was more elliptical, and possessed glandular hairs instead of scales. It is observable that this case followed the supposed rule in so far as that the female parent gave its likeness to the foliage. Of the numerous hybrids and crosses raised by Messrs. Veitch, amongst these seven East Indian species, the above rule was found to fail very generally, in that each parent would impart certain peculiarities either to the flowers or leaves, according to its own prepotency, but the cause of such power is unknown."

The great value of the javanico-jasminiflorum hybrids for the decoration of conservatories, corridors, and for cut bloom, is now generally recognized, and provided a sufficient number of plants be grown, it is not difficult to have flower throughout the year. This perpetual-flowering property was demonstrated in 1897, when at every fortnightly meeting held by the Royal Horticultural Society, throughout that year, a tray of cut blooms from the collection at Chelsea was exhibited.

Many of the early hybrids, which, on their first appearance, were valued as improvements, have now been discarded in favour of still more highly developed varieties, as King Edward VII., with magnificent bright yellow flowers borne in bold trusses, The Queen, a pure white-flowered form, and Cloth of Gold, pale yellow.

STREPTOCARPUS

STREPTOCARPUS ACHIMENIFLORUS

STREPTOCARPUS

SPECIES.

THE species composing the beautiful race of Gesneraceous plants, mostly natives of South Africa and Madagascar, are allied to the Gloxinia, Achimenes, Tydæa, and other favourite garden genera.

Some dozen are in cultivation which may be for horticultural purposes classed in three groups:—(1) The one-leaved or monophyllus, of which Streptocarpus Wendlandii is a familiar example; (2) the stemless or acaulescent, typified by S. Rexii, an old garden plant; and (3) the caulescent, of which S. caulescens is perhaps the best known; these have stems 6 to 8 in. high and opposite leaves, but are seldom seen in cultivation.

The acaulescent or stemless species which form the first two groups are the most frequently met with, the caulescent species of botanical interest only. The plants belonging to the latter have insignificant flowers, which refuse to cross with other species, and are hardly worth cultivating.

The first Streptocarpus to be introduced to British Gardens was * Streptocarpus Rexii of Lindley, figured in the Botanical Magazine as Didymocarpus Rexii, and sometime known in gardens as Streptocarpus biflorus or S. floribundus. This species, sent to Kew in 1824 by Mr. Bowie, His Majesty's collector in South Africa, was found on the estate of Mr. George Rex, after whom it was named. It belongs to the stemless group, has several prostrate leaves from 3 to 9 in. long, white flowers, with a tube 3 in. long of a delicate pale-blue inclining to purple; the lateral lobes have each two, and the lower three purple lines.

Following Streptocarpus Rexii came † S. polyantha, a native of Natal, accidentally introduced to Kew in 1853 in some material surrounding the trunks of tree-ferns sent from Natal by Captain Garden. With one, or occasionally two, pairs of leaves, one of which grows to a much larger size than the other, reaching 1 ft. or more in length, the plant produces from the base upright flower-scapes. The flowers, 1½ in. broad, are of a delicate pale-blue coloration.

‡ Streptocarpus Gardeni, also introduced in 1853 from the same country through the same channel, and named in compliment to Captain Garden, who sent seeds to Kew, bears a tuft of radical leaves from which the flower-scapes rise, each bearing at the apex a greenish-white flower with a lilac lip.

In 1860 a species was sent from Natal to Kew by Mr. Wilson Saunders, and received the name of § Streptocarpus Saundersii. It forms a solitary prostrate leaf 1 ft. long and 8 to 9 in. broad, flowers pale lilac in colour with two deep purple blotches separated by a yellow line at the throat, and closely allied to the above-mentioned S. polyantha.

* Bot. Mag. t. 3005. † Bot. Mag. t. 4850.
‡ Bot. Mag. t. 4862. § Bot. Mag. t. 5251.

In 1882 the caulescent species Streptocarpus Kirkii was sent to Kew from Zanzibar by Sir John Kirk, and, although an interesting plant, is of no value in the garden. S. caulescens followed in 1886, an allied species also of little horticultural merit.

*Streptocarpus parviflora, of the acaulescent group, the next introduced, was raised from seed brought from the vicinity of Grahamstown by Mr. Watson, of Kew, in 1887, and the plant apparently has a somewhat wide range in South Africa, from Cape Town to Natal. From the base of the tufted leaves, several in number, 4 to 6 in. long, arise the flower-scapes 6 to 10 in. high. The corolla-tube, purple within and without, with a white, slightly unequal, five-lobed limb, is $\frac{3}{8}$ in. long.

A somewhat closely allied species, figured in the Botanical Magazine as †Streptocarpus parviflora, raised from seed by Mr. Lynch, of the Cambridge Botanic Gardens, produces yellowish flowers with purple streaks: the plant is now found to be the S. lutea of Clarke.

The next in order of introduction, a handsome and interesting plant, ‡Streptocarpus Dunnii, played an important part in the production of the beautiful hybrids such ornaments to our conservatories at the present day. A native of Spitzkop in the mountains of the Transvaal Gold-fields, at elevations of 3,600-6,000 ft., whence seed was sent to Kew by Mr. E. G. Dunn, of Claremont, Cape Town, in 1884. It is a one-leaved species, producing a single prostrate leaf attaining a length of 18 to 36 in.: flower-scapes numerous, form a sheaf of much-branched panicles with many blooms of a rose or salmon red colour, varying in tint. S. Dunnii flowered first at Kew in 1886.

§Streptocarpus Galpini, another South African species, was introduced to Kew in 1890 by Mr. E. Galpin, who met with it on the "Bearded Man," a peak forming one of the boundaries of Swaziland. The flowers, rich blue in colour tinted with purple, are remarkable in that they are nearly regular at the mouth and not oblique as in other species.

Perhaps the most remarkable species of all is that named ‖Streptocarpus Wendlandii, sent from the Transvaal to Naples in 1887 to Messrs. Damman, and first made known in their Catalogue of Plants for the year 1890-1891. First seen in England in the Royal Gardens, Kew, where seed had been sent in some soil with tree-ferns from South Africa, it flowered in 1895. This notable plant produces only one huge prostrate leaf, from 30 in. in length to 24 in. in breadth; the numerous flowers blue and white in colour, with violet markings.

A hybrid named Streptocarpus × Dyeri raised at Kew by crossing S. Wendlandii with S. Dunnii.

An Australian species, Streptocarpus Faninii, closely allied to S. polyantha, and very noteworthy as having helped in the production of many of the more beautiful hybrids.

HYBRIDS AND VARIETIES.

The first hybrid Streptocarpus on record is ¶S. × Greenii, the progeny of two species, S. Saundersii and S. Rexii, the latter the pollen parent. It was

* Bot. Mag. t. 7036.
† Bot. Mag. t. 6636.
‡ Bot. Mag. t. 6903.
§ Bot. Mag. t. 7447.
‖ Bot. Mag. t. 7320; The Garden, 1892, vol. xli. pl. 849.
¶ Gard. Chron. 1882, vol. xvii. p. 303, fig.

raised by Mr. Charles Green, at that time gardener to Sir George Mackay, of Pendell Court, Bletchingly, and is described in the Gardeners' Chronicle as of tufted habit, bearing several leaves and pale lilac-blue flowers. This hybrid apparently never became widely grown, and had nothing whatever to do with the production of the present garden race.

The initial step which led to the formation of the beautiful forms now in cultivation was taken by the Curator of Kew, who raised the hybrid Streptocarpus × * Kewensis by crossing S. Rexii with the pollen of S. Dunnii. Before the introduction of S. Dunnii some had been raised at Kew by fertilizing S. parviflora with the pollen of S. Rexii, and to the most distinct of these the names White Pet, S. R. multiflorus, and S. R. albus were given. A coloured plate of these appeared in The Garden for May 22nd, 1886, and it is interesting to compare the present wide range of colour in the modern varieties with the limited area of that day.

The flowers produced by Streptocarpus × Kewensis were 2½ in. long with a spreading limb 1½ in. in diameter, and bright mauve-purple in colour striped with brownish-purple in the throat.

Another hybrid, flowered in 1887, raised at Kew by crossing Streptocarpus parviflora with the pollen of S. Dunnii, received the name of † S. × Watsoni, after Mr. Watson by whom it was raised. It has several tufted leaves and numerous flower-stems, each bearing ten to sixteen flowers 1¼ in. long by 1 in. in diameter, bright rose-purple with a white throat and brownish-purple stripes; like others, this hybrid fails to produce seed, but the pollen is potent when used on others.

In 1887, the two hybrids Streptocarpus × Kewensis and S. × Watsoni were crossed with each other and with their parents in all possible combinations, and a host of seedlings resulted. These showed a marked deviation amongst themselves in colour, size and form of flower, many decidedly attractive. A selection of these was obtained from Kew, and came under the care of Heal, who again crossed them amongst themselves and with the red-flowered S. Dunnii, obtaining many seedlings now known in gardens as Veitchs' Original Hybrids. These hybrids are remarkable for the abundance of bloom, the continued succession with which the flowers are produced, and for the long time the individual blooms remain in perfection; the flowers trumpet-shaped, widely open at the mouth, measure 1¾ in. in length.

A coloured plate, prepared from plants growing at Chelsea, appeared in The Garden for February 6th, 1892, p. 843, and may be compared with one published in 1886. The influence of Streptocarpus Dunnii in the production of the rich and varied colours of the modern varieties is very obvious.

By crossing the finest and most highly coloured varieties selected from Veitchs' Original Hybrids with the beautiful species from South Africa, a new race has been created, which for distinction and delicacy of shading are amongst the most valuable of modern plants for the decoration of conservatories or the cool greenhouse.

A variety of Veitchs' Original Hybrids, having magenta flowers with a deep blotch on the lip, was crossed with the red-flowered Streptocarpus Dunnii, and from this cross the strain known as gratus was derived. The flowers of many

* Gard. Chron. 1887, vol. ii. p. 247, fig. † Gard. Chron. 1887, vol ii. p. 215.

distinct shades of colour, magenta-rose, pink, light-red, mauve, pale violet, blue, all have crimson streaks on the lower segments.

In 1894, Virgil, a white variety of Veitchs' Original Hybrids with a deep blotch on the lower lip, was crossed with Streptocarpus Wendlandii, and from this cross originated the seedling Sylph; flowers of a soft light mauve, with two violet maroon bands in the throat, and a narrow line of the same colour between.

The charming variety Mrs. Heal had a similar origin, but in this case a magenta-rose shade was crossed with Streptocarpus Wendlandii. The flowers are 1½ in. in diameter, the tube broad, bluish-red above, whitish beneath; the limb violet-blue shaded with purple; the throat yellow with maroon spots on the basal half, and two to three maroon streaks and blotches on each side.

The pulchellus strain from a white-flowered variety of Veitchs' Original Hybrids crossed with the pollen of Streptocarpus Fanninii, an Australian species, has flowers of various shades of blue, from violet to almost white, all with a narrow blotch at the base of the lowermost segment of the limb.

A selection now much cultivated, and very popular, is one known as achimeniflorus, from the resemblance the flowers bear to those of an Achimenes, obtained by crossing a form of Veitchs' Original Hybrids, with white flowers and a purple blotch in the throat, with Streptocarpus polyantha, the latter the seed-bearer. The flowers of a light mauve tinted blue, in much-branched panicles, have a light canary-yellow and white throat.

By selecting the finest variety from seedlings of the achimeniflorus strain, and crossing with pollen from a pure white-flowered form of Veitchs' Original Hybrids, two distinct varieties were obtained, to which the names achimeniflorus albus and achimeniflorus giganteus were given, the former pure white and the latter lavender-blue of fine large form.

The strain known as pallidus was selected from a number of seedlings originated by crossing a white-flowered form of Veitchs' Original Hybrids with a white-flowered form of the achimeniflorus strain.

The latest variety selected from the achimeniflorus section is the beautiful rose-coloured achimeniflorus roseus, originally produced by crossing achimeniflorus albus with a magenta-flowered variety of Veitchs' Original Hybrids, and for a number of years rigorously selecting the rose-coloured seedlings.

Only a tithe of the crosses effected are here mentioned, many results worthless from a horticultural standpoint, others showing in no way the influence of the parents were discarded. An attempt to cross the caulescent with the acaulescent species has also been made, but without success, and the Gloxinia, closely related, has also proved no more amenable. Great improvement has been effected in size, form, and colour by selection, hybridization of late only being resorted to when a break in some desired direction was necessary.

FRUITS AND VEGETABLES

GOOSEBERRY "GOLDEN GEM"

FRUITS

APRICOT, KAISHA (*De Syrie*).

Jour. Hort. Soc. London, July, 1849; Gard. Chron. 1851, p. 451 (advt.); *id.* 1850, p. 487; Hogg's Fruit Manual, ed. v. p. 265.

A well-known, richly flavoured, early variety, introduced from Syria in 1842 by Mr. Barker, Consul at Aleppo, from whom the Chelsea house acquired the stock; now one of the standard sorts.

APPLE, H. BALLANTINE.

Gard. World, vol. xxi. n.s. 1904, p. 931, fig.

Raised from Peasgood's Nonsuch and St. Edmund's Pippin; the fruit large, oblate in form, with a yellowish-green skin, spotted with russet, and slightly splashed with red.

The flesh, greenish-white, firm, brisk, and juicy, is slightly acidulated when at its best; a culinary variety in use in October and November.

APPLE, LANGLEY PIPPIN.

Gard. Mag. Sept. 3rd, 1898, with fig.

Raised by Seden from Mr. Gladstone crossed with Cox's Orange Pippin, the latter the seed parent. The fruit of medium size, roundish, inclining to conical, the skin pale yellow, brighter on the side next the sun.

APPLE, MIDDLE GREEN.

Gard. Chron. 1903, vol. xxxiv. p. 279 (Report of R.H.S. Fruit Committee); *id.* p. 291, fig. 123; The Garden, 1903, vol. lxiv. p. 292, fig.

Raised by Seden at Middle Green Farm, Langley, from Frogmore Prolific crossed with Blenheim Orange.

The fruit, of medium size, is nearly round in shape, yellow, streaked with red on the exposed side, the flesh soft, of good quality.

APPLE, MR. LEOPOLD DE ROTHSCHILD.

Gard. Chron. 1899, vol. xxvi. p. 295, with fig.

The fruits are of globose-conic shape, clear yellow in colour on the shaded portion, brilliant orange-red on the basal half of the exposed side; the fruit crisp, sub-acid, of pleasant flavour. An ornamental as well as a useful dessert fruit, in season during the month of October.

APPLE, MRS. JOHN SEDEN.

Jour. R.H.S. vol. xxii. p. 189, figs.; Gard. Chron. 1898, vol. xxiv. p. 293 (Report of R.H.S. Fruit Committee).

Raised by Seden from a cross between Transcendent Crab and King of the Pippins Apple.

The tree is a marvellous bearer, the fruit ornamental, flavour slightly acid, but decidedly pleasant, and the flesh firm and crisp.

APPLE, REV. W. WILKS.

Gard. Chron. 1904, vol. xxxvi. p. 253, fig. 109.

A culinary sort raised at Langley from Peasgood's Nonsuch and Ribston Pippin, named in honour of the most energetic Secretary of his day.

The fruits resemble in shape those of Peasgood's Nonsuch, are of pale greenish-yellow colour, with faint streaks and markings of red, and the flesh firm, crisp, juicy, of excellent flavour.

In season during September.

An Award of Merit was granted by the Fruit Committee of the Royal Horticultural Society on September 29th, 1904, when first exhibited.

BULLACE, THE LANGLEY.

The Garden, 1902, vol. lxii. p. 399, fig.; Gard. Mag. 1902, p. 671, fig.

A hybrid raised at Langley, from Damson, The Farleigh, and Plum, Black Orleans, the latter the pollen parent; the tree is an enormous cropper, the fruits, about the size of an Orleans' Plum, have a distinct trace of the Damson flavour.

CRAB APPLE, MARSHAL OYAMA.

Gard. Mag. Oct. 15th, 1904, p. 669, fig. p. 673.

A handsome ornamental Crab raised by Seden from Apple American Mother and Crab John Downie.

The brilliantly coloured fruits, freely produced, are a beautiful ornament in the shrubbery in early autumn. The flavour that of American Mother, and the fruits make excellent jelly.

CRAB APPLE, THE LANGLEY.

Gard. Chron. 1902, vol. xxxii. p. 435, fig. 147.

Raised at Langley from Crab John Downie and Apple King of the Pippins.

The fruits conical in shape, bright yellow in colour, are abundantly produced in clusters, remaining on the tree till late in the year.

The flesh crisp, juicy, and of pleasant flavour.

CRAB APPLE, VEITCHS' SCARLET.

Gard. Chron. 1904, vol. xxxvi. p. 253, fig. 108.

An ornamental variety raised at Langley from a cross between Crab Red Siberian and Apple King of the Pippins.

The fruit is highly ornamental, the skin bright red, changing to deep crimson on the sunny side; the flavour pleasant, and the variety is suitable for culinary purposes.

FRUITS

GOOSEBERRY, GOLDEN GEM.

Gard. Chron. 1897, vol. xxii. p. 113.

Raised from the two varieties Whitesmith and Antagonist, with large, long, yellow berries, of most excellent flavour.

GOOSEBERRY, LANGLEY BEAUTY.

Gard. Chron. 1896, vol. xx. p. 155, fig. 27.

A hybrid from Yellow Champagne and Railway, in which the fruits produced by the seedling exceed in size those borne by either of the parents; the berries are large, of a buff-yellow colour, semi-transparent, somewhat hairy, and of fine flavour.

GOOSEBERRY, LANGLEY GAGE.

Gard. Chron. 1896, vol. xx. p. 137 (Report of R.H.S. Fruit Committee).

Raised at Langley from the two varieties Pitmaston Green-Gage and Telegraph, the fruit a white or amber-coloured berry, smooth, of good flavour, and very sweet.

RASPBERRY-BLACKBERRY HYBRID, THE MAHDI.

Gard. Chron. 1897, vol. xxii. pp. 235, 236, fig. 71; Veitchs' Catlg. of Fruits, 1903, pp. 64, 65, fig.

A hybrid from Raspberry Belle de Fontenay, the seed parent, and the common Blackberry of English hedges.

The habit of the plant is fairly intermediate between the two, but inclines rather to that of the Blackberry. The fruits large, red-violet in colour, are of brisk and pleasant flavour, in which that of the Raspberry can be distinctly detected. The most valuable quality, the period of ripening, is after that of the Raspberry, and before the Blackberry.

RASPBERRY, NOVEMBER ABUNDANCE.

Gard. Chron. 1902, vol. xxxii. p. 375, fig. 129.

An autumn-bearer, a cross between the American variety Catawissa and the well-known Superlative.

The fruits of excellent flavour, dark red in colour, are in season during the latter part of October and the beginning of November.

RASPBERRY, QUEEN OF ENGLAND.

Syns. *Golden Queen.*

Gard. Chron. 1899, vol. xxvi. p. 63, with fig.; Veitchs' Catlg. of Fruits, 1903, p. 66, fig.

This fine Berry, first exhibited at the Hybrid Conference held at the Royal Horticultural Society's Garden, at Chiswick, on July 11th, 1899, under the name of Golden Queen, was raised at Langley from Raspberry Superlative and Rubus laciniatus. The fruit, a rich golden yellow in colour, resembles in size and shape that of Superlative—the influence of the Rubus is slight.

RASPBERRY, YELLOW SUPERLATIVE.

Veitchs' Catlg. of Fruits, 1901, p. 56, fig.

Raised by Seden from the two varieties Superlative and Yellow Antwerp, and similar to the well-known Guinea.

The fruit is large, clear yellow in colour, with an agreeable sub-acid flavour.

STRAWBERRY, LORD KITCHENER.

Gard. Chron. 1899, vol. xxvi. p. 59.

Raised from the two varieties British Queen and Waterloo, the last-named the seed parent.

A useful mid-season variety, in perfection early in July, with fruits flavoured as British Queen, in size and colour approaching Waterloo.

STRAWBERRY, PRESIDENT LOUBET.

Gard. Chron. 1903, vol. xxxiv. p. 70, fig.

Raised at Langley from a cross between the two Berries Waterloo and Lord Napier, with fruit of more than average size, dark vermilion in colour, not unlike Waterloo, and of fine flavour.

It was named in commemoration of the visit of the late President of the French Republic to this country in 1903.

STRAWBERRY, THE ALAKE.

Gard. Chron. vol. xxxvi. 1904, p. 61, fig. 26.

A variety raised at Langley from a cross between Frogmore Late Pine and Veitchs' Perfection, with fruit of very large size, variable in form, usually more or less cock's-comb-shaped in outline.

Flesh scarlet, of good flavour, produced in great quantity, the fruit in season the end of June and the first week of July.

STRAWBERRY, THE KHEDIVE.

Gard. Chron. 1902, vol. xxxii. p. 64 (Report of R.H.S. Fruit Committee); Veitchs' Catlg. of Fruits, 1903, p. 68, fig.

Raised by Seden from Lord Suffield and British Queen, the latter being the pollen parent, with fruit of medium size, conical, a dark red colour, the flavour largely that of British Queen.

STRAWBERRY, VEITCHS' PERFECTION.

Gard. Chron. 1896, vol. xx. p. 105; *id.* 1897, vol. xxii. pp. 42, 61, fig. 20.

Raised at Langley from the varieties British Queen and Waterloo, the latter the seed parent, and remarkable for an exquisite flavour, inherited from the former, but cultivation has been found most difficult.

STRAWBERRY, VEITCHS' PROLIFIC.

Gard. Chron. 1898, vol. xxiv. p. 78, fig. 20.

Raised at Langley from the two varieties Empress of India and British Queen, in shape as British Queen, of bright crimson colour, ripening to the tip—a remarkably profuse Strawberry.

CAULIFLOWER "AUTUMN GIANT"

VEGETABLES

On taking over the Chelsea Nursery of Messrs. Knight & Perry in 1853, a considerable business in vegetable and flower seeds was found to be in existence, and it at once became the persistent policy of Mr. James Veitch junior to extend that business.

With this view, stocks of seeds of the highest quality were obtained from growers in the neighbourhood of London, from Belgium, France, Germany, Italy and the United States.

The success which followed is largely due to a Scotchman, one John Davidson, who, in addition to an exceptional knowledge, was gifted with unusual business capacity. In thirty years, under his *régime*, a world-wide business was developed.

Many of the improved varieties of vegetables originated with the Veitchian house or with amateurs who from time to time placed selections at our disposal, and it is necessary to make clear that, although they appear in this section as Veitchs' Introductions, the credit of being the first to raise them is not in all cases our own, but sometimes due to those who sent us the stock. Evident as it is that the steady improvement in the varieties of vegetables rapidly causes many forms once considered first-rate to disappear from cultivation, yet, nevertheless, many originally distributed still hold a leading position in their respective groups.

The following may be quoted as examples:—Veitchs' Improved Ashleaf Kidney Potato, distributed in 1868, one of the best early potatoes in cultivation, and an excellent forcer; Veitchs' Chiswick Favourite Potato, in 1885, a very fine round variety of first-rate table quality and admirable for late use; Veitchs' Autumn Giant Cauliflower, distributed in 1870, still the best late variety and indispensable for autumn use; Veitchs' Self-Protecting Broccoli, first offered in 1876, and Veitchs' Model Broccoli in 1878, have both become standard varieties of this important vegetable; Ellam's Early Spring Cabbage sent out in 1879 is still unsurpassed as a Spring Cabbage, and Lily White Seakale remains the finest flavoured known for forcing.

That very important vegetable the Pea naturally received great attention, and as a result several marked improvements were obtained.

Amongst others in a foremost position are Chelsea Gem, offered in 1884, a dwarf-growing early variety, rarely exceeding 15 in. in height, and very prolific. The pods are large and handsome, remarkably well filled for an early sort, and contain from eight to ten good-sized peas of first-rate quality. Veitchs' Criterion Pea is a leading main crop, sent out for the first time in 1876, and Veitchs' Autocrat, distributed in 1888, is one of a rare quality for late work; owing to a strong constitution, it bears successional crops for a long period. Of the blue wrinkled marrow section, the pods are large and broad, well filled with deep grassy-green seed of a superb quality. Veitchs' Main Crop and

Veitchs' Prestige are two excellent varieties for second and general crops, the former distributed in 1892, the latter in 1903.

There is scarcely a family of culinary plants that has not at some time received attention, and been improved either by selection in the trial grounds, or by crossing distinct varieties producing new breaks.

Among Beans, Veitchs' Giant White, the largest-podded runner in cultivation, and Mammoth Scarlet, are two excellent varieties; Veitchs' Selected Red, Veitchs' Improved Black, Improved Globe, and Pragnell's Exhibition are standard Beets, and with the Brassicas, besides those previously alluded to, may be mentioned the varieties of Broccoli or Kale, such as Chelsea Exquisite Curled, Read's Improved Hearting, Veitchs' Sprouting and Veitchs' Exhibition, leading kinds of these important winter vegetables. Veitchs' Exhibition Brussel Sprouts, Veitchs' Paragon and Veitchs' Market Favourite are well tried and excellent, and Veitchs' Scarlet Model and Veitchs' Matchless Scarlet among the most widely grown of all Carrots.

Lettuce, so much in demand in all gardens, received important additions in Veitchs' Golden Queen, a desirable small early variety of the Cabbage section, and Veitchs' Perfect Gem, also a Cabbage Lettuce, valuable on account of its ability to withstand summer drought and for the beautiful crisp and tender leaves. Amongst the Cos varieties none are better than Veitchs' Superb White and Veitchs' Selected Brown, and the Self-Folding Chelsea Imperial is a great type.

Several varieties of Cucumber distributed are also now recognized as standard kinds; Veitchs' Perfection, one of the best for spring and winter work, and Allen's Favourite, Veitchs' Improved Telegraph and Tender and True are all meritorious, and more recent varieties, Veitchs' Sensation and Veitchs' Unique, promise to be indispensable.

The varieties of Melon have been numerous and excellent, and include the best raised in the first gardens of the country.

Tomatoes have benefited by the addition of such familiar kinds as Ham Green Favourite, Frogmore Selected, Hackwood Park, Golden Jubilee, and Chiswick Peach, while Veitchs' Main Crop and Selected Globe Onions are in all gardens where high-class vegetables are in demand.

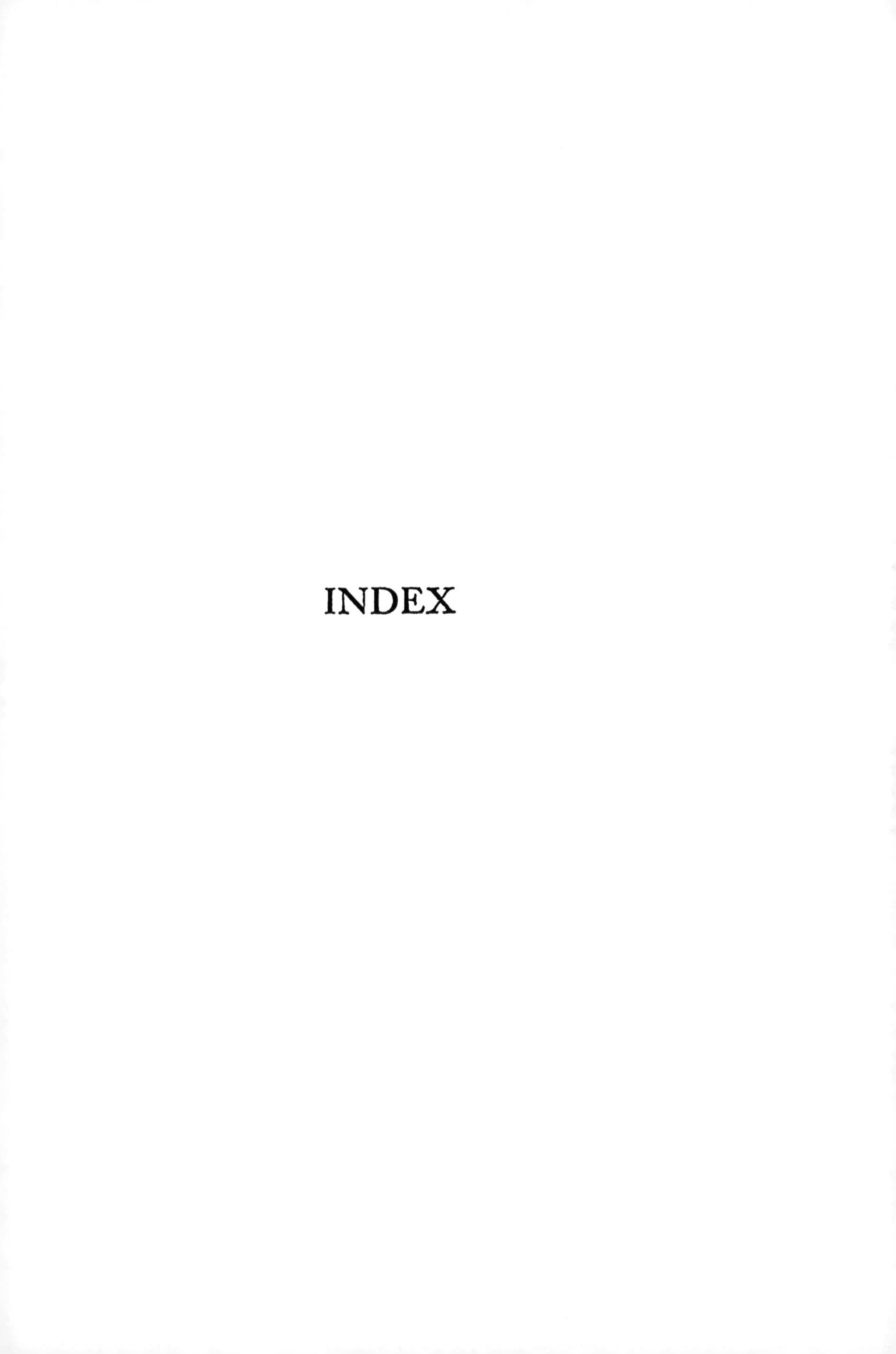

INDEX

INDEX

(The Names in *italics* are Synonyms.)

L l

INDEX

INDEX

INDEX

INDEX

INDEX

INDEX

INDEX

INDEX

INDEX

INDEX

INDEX

INDEX

INDEX

INDEX

M m

INDEX

INDEX

INDEX

INDEX

INDEX

INDEX

INDEX

INDEX

LONDON
PRINTED BY GILBERT AND RIVINGTON, LD.
ST. JOHN'S HOUSE, CLERKENWELL, E.C.

For EU product safety concerns, contact us at Calle de José Abascal, 56–1°, 28003 Madrid, Spain or eugpsr@cambridge.org.

www.ingramcontent.com/pod-product-compliance
Ingram Content Group UK Ltd.
Pitfield, Milton Keynes, MK11 3LW, UK
UKHW050728090726
473066UK00013B/1067

* 9 7 8 1 1 0 8 0 3 7 3 6 5 *